Learn By Doing: Algebra II

An Active Approach to Learning Mathematics

Ryan Hobbs

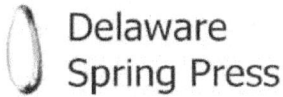

Delaware Spring Press

© 2025 by Ryan Hobbs

All rights reserved. No part of this publication may be reproduced in any form without written permission from Delaware Spring Press. www.delawarespring.org

ISBN: 978-1-7332514-5-7

This book follows the structure of Abramson, Jay. *College Algebra 2e*. OpenStax, 2021., which is available for free at https://openstax.org/details/books/college-algebra-2e.
The traditional lecture videos, linked by QR code at the end of each lesson, use the material from *College Algebra 2e*, and are available for free at https://www.youtube.com/@rphobbs2002.

The Reason for this Workbook

I have been a college math lecturer for over ten years and it didn't take me long to develop my foremost principle of education—if students aren't awake then they aren't learning. In my opinion, for most math students, the traditional lecture approach has limited value. And so, I began a journey to develop a more interactive approach to the mathematics classroom. This workbook is the result.

However, there is much more here than just a series of worksheets, as I've also strived to incorporate the other principles of math education which I've discovered over the years.

Math is a language. Good teaching must translate this language of numbers into English. Language learning also requires "comprehensible input." It is good to be challenged and stretched, but when teaching is beyond a student's level of understanding they become overwhelmed and shut down. As math progresses, it is valuable to show students how more advance concepts relate to simpler concepts which they already know and understand. And, when possible, it helps to make math visual.

Yet perhaps most importantly, good teaching comes beside students when they need help. This is why every activity has a QR code which links to a walk-through video of the active learning lesson.

In other words, good teaching is very much like good tutoring, and that is the fundamental idea which has guided the development of this course.

How to Use this Book

Each activity in this workbook has a QR code in the top righthand corner. This QR code will take you to a YouTube video where I work and explain every problem and concept in that lesson. It is my attempt to use technology to come along side of a student just as I would do in a real classroom.

So, as you work the activity, start and stop the video, as needed, for help. Or, skip through any part of the video which you don't need. The back of the workbook also contains a solution guide for checking your answers.

If after finishing an activity, you feel as if you could use some additional help, a QR code in the bottom right links to a traditional lecture for that section.

So, use the workbook as an instructional guide to a complete Algebra II course, or use it as a tutoring resource to aid in a course you are already taking.

This workbook is structured to follow the textbook College Algebra. The activity numbers correspond to the chapters and sections of that book. For more practice problems, the textbook is available for free at https://openstax.org/details/books/college-algebra-2e.

Table of Contents

Prerequisites

1.1a: Classifying Numbers and the Order of Operations 1
1.1b: Key Math Properties; Evaluating and Simplifying Expressions 7
1.2a: Exponent Rules 13
1.2b: Scientific Notation 19
1.3a: Working with Roots 25
1.3b: Rationalizing Denominators; Fractional Exponents 29
1.4a: Adding, Subtracting and Multiplying Polynomials 33
1.4b: Special Patterns when Multiplying Polynomials 37
1.5a: Factoring Polynomials 39
1.5b: Factoring the Sum or Difference of Cubes; Factoring Fractional Exponents 43
1.6a: Simplifying Rational Expressions 47
1.6b: Adding and Subtracting Rational Expressions 51

Equations and Inequalities

2.1a: Graphing by Plotting Points; x and y-intercepts 55
2.1b: The Midpoint and the Distance Formula 59
2.2a: Solving Equations 63
2.2b: Slope of a Line; Standard Form of a Line 67
2.2c: Parallel or Perpendicular Lines 71
2.3a: Word Problems Involving Lines or Distance, Rate and Time 77
2.3b: Word Problems Involving Perimeter and Area 79
2.4a: Introduction to Complex Numbers 83
2.4b: Adding/Subtracting and Multiplying Complex Numbers 87
2.4c: Dividing Complex Numbers 91
2.5a: Solving Quadratic Equations 97
2.5b: The Quadratic Formula 101
2.6a: Higher Order Roots and Equations with Rational Exponents 105
2.6b: Solving Absolute Value, Higher Order and Rational Equations 111
2.7a: Compound Inequalities 115
2.7b: Absolute Value Inequalities 121

Functions

3.1a: Introduction to Functions 127

3.1b: One-to-One Functions 133

3.2a: Domain of a Function 139

3.2b: Graphs of the Toolkit Functions; Piecewise Functions 147

3.3a: Average Rate of Change 153

3.3b: Increasing and Decreasing; Extrema of a Function 157

3.4a: Mathematical Operations on Functions; Composite Functions 161

3.4b: Domain of a Composite Function 169

3.5a: Vertical and Horizontal Transformation of a Function 175

3.5b: Combining Multiple Transformations 179

3.5c: Vertical and Horizontal Stretches/Compressions; Even/Odd Functions 183

3.6: Absolute Value Functions 191

3.7a: Inverse Functions 197

3.7b: Finding an Inverse of a Function 201

Linear Functions

4.1a: Applying Lines as Functions 205

4.1b: Understanding Lines as Transformations 209

4.1c: Investigating Parallel or Perpendicular Lines 213

4.2: Further Application with Linear Functions 215

4.3a: Introduction to Correlation and Scatterplots 217

4.3b: Introduction to Linear Regression 221

Polynomial and Rational Functions

5.1a: The Vertex Form of a Quadratic Function 223

5.1b: Intercepts of Quadratics; Max/Min of Quadratics 229

5.1c: Applications Involving Max/Min of Quadratics 233

5.2a: Introduction to Polynomial Functions 237

5.2b: End Behavior of Polynomial Functions 241

5.3a: Intercepts of Polynomial Functions; Multiplicity 247

5.3b: Graphing Polynomial Functions 253

5.4a: Dividing Polynomials Using Long Division 257

5.4b: Dividing Polynomials Using Synthetic Division 263

5.5a: The Remainder Theorem and the Factor Theorem 267

5.5b: The Rational Zero Theorem 271

5.5c: Descartes Rule of Signs... 275

5.6a: Asymptotes of Rational Functions 279

5.6b: Graphing Rational Functions... 285

5.7a: Review of Inverse Functions 291

5.7b: Restricting Functions.. 293

5.7c: Finding the Inverse of a Formula... 299

5.8: Direct and Indirect Variation... 303

Exponential and Logarithmic Functions

6.1a: Introduction to the Exponential Function.. 307

6.1b: Compound and Continuous Interest... 313

6.2a: Transformation of the Exponential Function 319

6.2b: Graphing Exponential Functions... 323

6.3: Introduction to Logarithms 329

6.4a: Graphs of Logarithmic Functions 333

6.4b: Graphing Transformations of Logarithms 339

6.5a: Properties of Logarithms.. 343

6.5b: Condensing Logarithms and the Change of Base Formula 349

6.6a: Solving Equations Involving the Exponential Function 353

6.6b: Solving Equations Involving Logarithms 357

Systems of Equations and Inequalities

7.1a: Solving Systems of Linear Equations by Graphing 361

7.1b: Solving Systems of Linear Equations by Substitution 363

7.1c: Solving Systems of Linear Equations by Elimination 367

7.2: Solving Three-Dimensional Systems of Equations 371

7.3a: Solving Systems of Non-Linear Equations 373

7.3b: Solving Non-Linear Systems of Inequalities 379

7.4a: Decomposing Fractions 383

7.4b: Decomposing Fractions which include Quadratics 387

7.5a: Introduction to Matrices 391

7.5b: Multiplying Matrices 395

7.6a: Solving Systems of Linear Equations with Augmented Matrices 399

7.6b: Solving Larger Systems of Equations with Matrices 405

7.7a: Inverse Matrices 409

7.7b: Solving Systems of Equations with Inverse Matrices 413

7.8a: Determinant of a Matrix; Cramer's Rule 417

7.8b: Properties of Determinants 423

Analytic Geometry

8.1a: Understanding the Equation for the Standard Form of an Ellipse 427

8.1b: Graphing an Ellipse 435

8.1c: Write an Equation for an Ellipse in Standard Form 439

8.2a: Understanding the Equation for the Standard Form of a Hyperbola 441

8.2b: Graphing Hyperbolas 449

8.2c: Write an Equation for a Hyperbola in Standard Form 453

8.3a: Understanding the Equation for the Standard Form of a Parabola 455

8.3b: Graphing Parabolas 463

8.3c: Write an Equation for a Parabola in Standard Form 467

Sequences, Probability and Counting Theory

9.1a: Explicit Formulas for Sequences 469

9.1b: Recursive Formulas for Sequences 473

9.2a: Introduction to Arithmetic Sequences 477

9.2b: Writing Explicit and Recursive Formulas for Arithmetic Sequences 479

9.3a: Introduction to Geometric Sequences 483

9.3b: Writing Explicit and Recursive Formulas for Geometric Sequences 485

9.4a: Summations 489

9.4b: Infinite Geometric Sequences 495

9.5: Counting Principle; Combinations and Permutations 501

9.6: The Binomial Theorem 507

9.7: Basic Probability 511

Algebra II

Active Lesson: 1.1a

In this activity, we are going to learn about different categories which a number can be defined by. As you pass through algebra and move to even higher levels, understanding these number categories is very helpful.

Let's begin learning the different classifications of numbers.

Counting Numbers- these are the first numbers you learned as a child: 1, 2, 3, 4, 5…. Counting numbers exclude zero.

Whole Numbers- these are identical to the counting numbers except that they also include zero.

Integers- these are any number without a decimal or fraction. Integers include decimals.

$$…-3, -2, -1, 0, 1, 2, 3…$$

Our next classification is rational numbers.

Rational Numbers- these are any number which can be expressed as a fraction.

$$\frac{4}{5}, -\frac{17}{20}$$

Those are obvious, but these are rational numbers too:

$$5, .02$$

Although they make not look like it, they can be expressed as fractions. Any integer can be a fraction simply by putting it over the number 1.

$$5 = \frac{5}{1}$$

And in Algebra I we learned how to make decimals into fractions.

$$.02 = \frac{2}{100}$$

a)   Express the following numbers as fractions:

121   .012   -6   .8

Our next classification is irrational numbers.

Irrational Numbers - These are numbers which cannot be expressed as a fraction. They cannot be expressed as a fraction because they contain decimals which never repeat.

$$3.1415926\ldots, -.171819\ldots$$

Rational vs. Irrational Numbers

We saw that rational numbers can be written as a fraction. In decimal form, rational numbers will always end or will have a repeated pattern.

Circle any of the following numbers which are rational. Put a square around any numbers which are irrational.

b) $4.\overline{125}$ 6.25 $2.7182818459\ldots$ $7.15423985407\ldots$

A key place where we find irrational numbers is with square roots. If a number doesn't have a perfect square-root it is an irrational number.

$\sqrt{21}$ is irrational. There are no two identical numbers which multiply to make 21. Using a calculator we find:

$$\sqrt{21} = 4.582575695\ldots$$

It makes a decimal which doesn't repeat.

c)   Circle any of the following numbers which are rational. (They will have a perfect square root.) Put a square around any numbers which are irrational.

$\sqrt{4}$

$\sqrt{48}$

$\sqrt{35}$

$\sqrt{225}$

We have one final classification (at least at this point in algebra).

Real Numbers- Any number which can be classified as a rational number or an irrational number.

The only exception to a real number is a negative under a square root.

$$\sqrt{-9}$$

This doesn't exist and isn't a real number.

If you notice, our classification system keeps building outward.

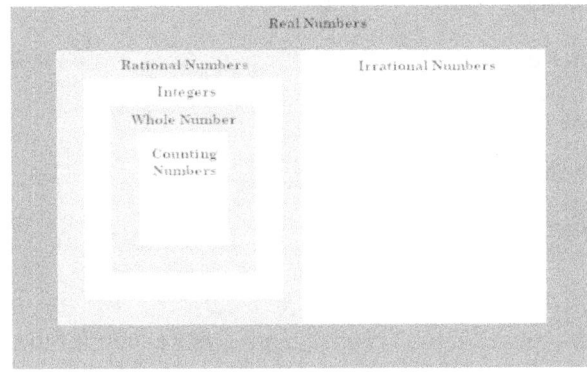

Counting numbers are also whole number, integers, rational numbers, and real numbers.

Whole numbers are also integers, rational numbers, and real numbers.

Notice that irrational numbers are only also real numbers.

For the numbers below, give the letter (or letters) which could classify the number. (Remember, it builds outward. A counting number would be everything except irrational.)

A-Counting Number   B-Whole Number   C-Integers   D-Rational Number

E-Irrational Number   F-Real Number

d) $\frac{7}{12}$ e)   5.148236427....   f) $12.\overline{3}$

g)   -11   h) $\sqrt{15}$ i) $\sqrt{9}$

j)   7   k)   0   l) $\sqrt{-16}$

Next, let's refresh your memory on the idea of the Order of Operations. Look at the problem below:

$$25 \div 5^2 + 3(10 - 8)$$

Mathematicians realized that without guidelines, two different students could tackle this problem in different ways, resulting in different answers. So, a set of instructions were developed, called the Order of Operations. It is like a recipe for how to tackle problems such as this. There is an acronym to help you remember.

Please	Parenthesis, actually any grouping symbol comes first.
Excuse	Exponents
My	Multiplication
Dear	Division
Aunt	Addition
Sally	Subtraction

In truth, multiplication and division are actually tied. So are addition and subtraction. But with anything that is tied, we move left to right across the expression, just like you were reading. Let's work the original problem now that we have the order in which we are to work.

$$25 \div 5^2 + 3(10 - 8)$$

Parenthesis

$$25 \div 5^2 + 3(2)$$

Exponents

$$25 \div 25 + 3(2)$$

Multiplication/Division

There is both. Remember 3(2) is actually multiplication. So, we'll go left to right.

$$1 + 3(2)$$

$$1 + 6$$

Addition/Subtraction

7

Try some on your own.

m) $6^2 - 3 + 2(3 + 5)$

n) $100 \div (7 + 3) - 2 \cdot 2^2 + 12$

This next set has grouping symbols inside of grouping symbols. When this happens work your way from the inside group outward.

o) $3 + 3^3 + 2[8 - 3(5 - 3)]$

p) $4[15 - 2(3 + 1)] + 6^2 \div 2$

In this problem, we see absolute value bars. We will discuss the absolute value in more depth later. For now, an absolute value works as a grouping symbol (like a parenthesis) and what is inside needs to be simplified first. Do all the work inside of an absolute value. When the work is done, an absolute value *then* turns the number inside into a positive.

q)
$$5 + |7 - 12| - 2(4 - 1)$$

Next, square roots also work like a grouping symbol. Simplify what is underneath and then take the root.

r)
$$(3 - 6)^2 + \sqrt{79 + 2} - 15$$

Finally, fraction bars are also considered a grouping symbol. Simplify what is in the numerator and simplify what is in the denominator then divide.

s)
$$\frac{24 + 3(4)}{4^2 - 4(2) - 2}$$

Algebra II

Active Learning: 1.1b

There are several essential properties of numbers which we often use in mathematics. A more detailed discussion of these properties is done in the Algebra I course. In all likelihood, you are comfortable with how to use the properties, even if you haven't memorized the names. Therefore, we will review the concepts quickly.

Add the following:

a) $5 + 9 =$ $\qquad$ $9 + 5 =$

This is called the commutative property of addition. The order which you add doesn't change the answer.

Multiply the following:

b) $3 \cdot 7 =$ $\qquad$ $7 \cdot 3 =$

This is called the commutative property of multiplication. The order which you multiply doesn't change the answer.

Add the following:

c) $2 + (5 + 8) =$ $\qquad$ $(2 + 5) + 8 =$

This is called the associative property of addition. Regardless of which two numbers "associate" first, it doesn't change the answer.

Multiply the following:

d) $2 \cdot (5 \cdot 8) =$ $\qquad$ $(2 \cdot 5) \cdot 8 =$

This is called the associative property of multiplication. Again, regardless of which two numbers "associate" first, it doesn't change the answer.

Multiply the following:

e) $5(x + 3) =$

This is called the distributive property. The five on the outside is distributed (multiplied) with each of the numbers in the parenthesis.

Add the following:

f) $9 + 0 =$

This is called the identity property of addition. By adding zero, you keep the same number (identity) as you started with.

Multiply the following:

g) $7 \cdot 1 =$

This is called the identity property of multiplication. By multiplying by 1, you keep the same number (identity) as you started with.

Add the following:

h) $8 + (-8) =$

This is called the inverse property of addition. When adding a number to its opposite, you always get zero.

Multiply the following:

i) $5 \cdot \dfrac{1}{5} =$

This is called the inverse property of multiplication. When multiplying a number by its reciprocal, you always get one.

As we saw in Algebra I, variables are values which we don't know. Since we don't know their value, we replace them with a letter. Constants are number values which we do know. In other words, they are just numbers.

In each of the examples below, indicate which values are constants and which values are variables.

j) $\quad 5x$

k) $\quad \frac{1}{4}xy$

l) $\quad m + 7$

m) $\quad \sqrt{6rs}$

Evaluating algebraic expressions means that we have a value for a variable and we substitute it in to the expression. Then, we simplify. When you substitute a value in, be sure to put it into a parenthesis. This is to make sure that you don't overlook any multiplication which had originally been in the problem.

Evaluate the following:

n) $\quad 7y - 2$ when $y = 3$

o) $\quad 3x^2$ when $x = 5$.

p) $\quad 4a - 16$ when $a = \frac{1}{2}$

q) $x^2 - xy + y^2$ when $x = 2$ and $y = 3$

r) $\dfrac{1}{2ab^2}$ when $a = 5$ and $b = 1$

s) $\sqrt{x^2 y^2}$ when $x = 4$ and $y = 5$

Using a formula follows the same concept as evaluating. Put the values for the variables into the formula and then simplify.

t) The formula for the volume of a rectangular prism is $V = w \cdot l \cdot h$. Find the volume if $w = 5\ in, l = 3\ in,$ and $h = 10\ in.$

Simplifying expressions can also consist in combining like terms. Recall that variables are like words. $3x + 5x$ each have the same "name," and so they can be combined. $3x + 5x = 8x$.

Simplify the following by combining like terms.

u) $3x + 4y + 5x - 2y$

v) $5xy - 12xy + 3x - 15y$

w) $18m - 2(5 - 3m) + 4(6 - 8n) + n$

The concept for simplifying a formula is the same as combining like terms.

Simplify the following formulas by combining like terms.

x) $\quad P = l + w + l + w$

y) $\quad Surface\ Area = 2a^2 + 2a^2 + 2a^2$

Algebra II

Active Learning: 1.2a

This section covers the rules for working with exponents. The rules are covered in-depth in Algebra I. Here, we will examine them more quickly.

The Product Rule

$$x^5 \cdot x^3$$

When you multiply exponents with the same base, you add the exponents. Here's the reason why:

$$x^5 \cdot x^3 = (x \cdot x \cdot x \cdot x \cdot x) \cdot (x \cdot x \cdot x) = x^8$$

Simplify the following:

a) $x^6 \cdot x^7$

b) $y^{15} \cdot y^{13}$

c) $z^3 \cdot z^4 \cdot z^3$

The Quotient Rule

When you divide exponents with the same base, you subtract the exponents. The reason is that we can reduce the fraction by cancelling like terms in the numerator and denominator.

$$\frac{x^5}{x^3} = \frac{x \cdot \cancel{x} \cdot \cancel{x} \cdot x \cdot x}{\cancel{x} \cdot \cancel{x} \cdot \cancel{x}}$$

$$\frac{x^5}{x^3} = x^{5-3} = x^2$$

Simplify the following:

d) $\dfrac{x^9}{x^2}$

e) $\dfrac{y^{15}}{y^{13}}$

f) $\dfrac{a^5}{a^8}$

The Power Rule of Exponents

$$(x^2)^3$$

When you have an exponent raised to an exponent, you multiply. Here's the reason why:

$$(x^2)^3 = (x^2) \cdot (x^2) \cdot (x^2) = x^6$$

Simplify the following:

g) $(x^4)^3$

h) $(y^4)^5$

i) $(z^{-2})^2$

The Zero Exponent Rule

$$x^0$$

When an exponent is raised to a power of zero, it must always equal 1. Here's why:

$$\dfrac{x^3}{x^3}$$

We learned that in a situation like this, we could subtract the exponents. So $x^{3-3} = x^0$.

However, we could also reduce this by hand.

$$\frac{x^3}{x^3} = \frac{x \cdot x \cdot x}{x \cdot x \cdot x} = 1$$

In math, we can't use two different approaches and get two different results. Therefore:

$$x^0 = 1$$

Any time, you get an exponent to a power of zero, it must equal 1.

Simplify the following:

j) y^0

k) $(xy)^0$

l) $\dfrac{a^{15}}{a^{15}}$

The Negative Rule of Exponents

$$x^{-5}$$

When a base is raised to a negative exponent, we can make the exponent positive by moving it from the numerator to the denominator.

$$x^{-5} = \frac{1}{x^5}$$

Here's why:

$$\frac{x^3}{x^5} = \frac{x \cdot x \cdot x}{x \cdot x \cdot x \cdot x \cdot x} = \frac{1}{x^2}$$

But if we had used the quotient rule of exponents, this is what we would have gotten:

$$\frac{x^3}{x^5} = x^{3-5} = x^{-2}$$

Just like before, we can't get two different answers to the same problem. Therefore:

$$\frac{1}{x^2} = x^{-2}$$

Simplify the following without negative exponents.

m) x^{-7}

n) $\dfrac{y^4}{y^{10}}$

o) $\dfrac{a^5 \cdot a^3}{a^{12}}$

The Power of a Product Rule

$$(x^3 y^2)^2$$

When a product is being raised to an exponent, we "distribute" the exponent to both individually.

$$(x^3 y^2)^2 = (x^3)^2 (y^2)^2$$

The reason is that the order in which we multiply doesn't matter.

$$(x^3 y^2)^2 = (x^3 y^2) \cdot (x^3 y^2) = x^3 \cdot x^3 \cdot y^2 \cdot y^2 = (x^3)^2 (y^2)^2$$

After you distribute the exponent, you would simplify as usual.

$$(x^3 y^2)^2 = (x^3)^2 (y^2)^2 = x^6 y^4$$

Simplify the following:

p) $(x^4 y^5)^3$

q) $(a^2 b^3)^7$

r) $(x^{-3} y^5)^2$

s) $(3y^3)^2$

The Power of a Quotient Rule

$$\left(\frac{x^4}{y^3}\right)^2$$

In much the same way as the last idea, we give the exponent to both the numerator and denominator.

$$\frac{(x^4)^2}{(y^3)^2}$$

The reasoning is the same.

$$\left(\frac{x^4}{y^3}\right)^2 = \left(\frac{x^4}{y^3}\right) \cdot \left(\frac{x^4}{y^3}\right) = \frac{x^4 \cdot x^4}{y^3 \cdot y^3}$$

After giving the exponent to both, we would then simplify as usual.

$$\left(\frac{x^4}{y^3}\right)^2 = \frac{(x^4)^2}{(y^3)^2} = \frac{x^8}{y^6}$$

Simplify the following:

t) $\left(\dfrac{x^3}{y^4}\right)^5$

u) $\left(\dfrac{2}{z^4}\right)^3$

v) $\left(\dfrac{m}{n^6}\right)^4$

Finally, use a combination of the rules to simplify the following problems. If necessary, make any negative exponents positive.

w) $\left(\dfrac{a^6}{a^4}\right)^3$

x) $\left(\dfrac{2x^2}{y^3}\right)^2$

y) $\dfrac{(x^2)^3 (x^3)^2}{(x^3)^5}$

z) $(3x^{-2}y^4)^3$

aa) $\left(\dfrac{a^{-2}b}{b^{-2}}\right)^3$

Algebra II

Active Learning: 1.2b

Write each of the following without an exponent.

a) $10^2 =$

b) $10^4 =$

Now, with the help of your work above, multiply the following:

c) $154 \times 10^2 =$

d) $154 \times 10^4 =$

Write each of the following as a decimal.

e) $10^{-2} =$

f) $10^{-4} =$

Now, with the help of your work above, multiply the following:

g) $154 \times 10^{-2} =$

h) $154 \times 10^{-4} =$

The idea we've done above leads to something called Scientific Notation. We are using the ideas of exponents to write numbers in an alternate form. The concept works because every time we multiply by a 10, our number adds a zero. (This moves the decimal place to the right.) And if we divide by 10, our number loses a zero. (This moves the decimal place to the left.)

Write the following numbers in standard form. These are large numbers. Move the decimal place to the right. Add zeros as necessary.

i) $5.647 \times 10^5 =$

j) $9.32 \times 10^7 =$

Write the following numbers in standard form. These are small numbers. Move the decimal place to the left.

k) $1.76 \times 10^{-3} =$

l) $2.951 \times 10^{-6} =$

Next, write the following numbers in Scientific Notation. These are large numbers, so the exponent will be positive. Move the decimal until there is only one leading digit. The number of places which you moved the decimal is your exponent for the Scientific Notation. Scientific notation doesn't show zeros that are at the end of a number.

$245000 = 2.45 \times 10^5$

m) $31560000 =$

n) $5106700000 =$

Do the same again. Here, however, the numbers are small, so the exponent will be negative. Move the decimal until there is only one leading digit.

$.000157 = 1.57 \times 10^{-4}$

o) $.0000215 =$

p) $.000007088 =$

Simplify the following by adding the exponents.

q) $10^3 \cdot 10^5 =$

r) $10^7 \cdot 10^2 =$

s) $10^{-2} \cdot 10^{-6} =$

t) $10^{-4} \cdot 10^9 =$

The exponent rules help us with multiplying two numbers in Scientific Notation.

$(2.8 \times 10^3)(3.9 \times 10^5)$

Multiply the decimal numbers and multiply the base-ten numbers.

$$(2.8 \cdot 3.9)(10^3 \cdot 10^5)$$

$$10.92 \times 10^8$$

But this number has more than one leading number, so it isn't completely in Scientific Notation. Let's fix that.

$$(1.092 \times 10^1) \times 10^8$$

$$1.092 \times 10^9$$

Work these problems.

u) $(1.3 \times 10^4)(6.4 \times 10^6)$

v) $(8.9 \times 10^2)(3.8 \times 10^3)$

The idea remains the same with negative exponents.

$(2.9 \times 10^{-3})(8.9 \times 10^5) = (2.9 \cdot 8.9)(10^{-3} \cdot 10^5) = 25.81 \times 10^2 = (2.581 \times 10^1)(\times 10^2)$

$$= 2.581 \times 10^3$$

Work these problems.

w) $(8.72 \times 10^7)(3.9 \times 10^{-10})$

x) $(1.5 \times 10^{-5})(3.45 \times 10^{-6})$

Simplify the following by subtracting the exponents.

y) $\dfrac{10^8}{10^5} =$

z) $\dfrac{10^{-6}}{10^4} =$

aa) $\dfrac{10^{-4}}{10^{-10}} =$

To divide numbers in Scientific Notation, we will imitate the same idea.

$$(2.5 \times 10^3) \div (3.7 \times 10^5)$$

Divide the numbers and divide the base ten.

$$\left(\dfrac{2.5}{3.7}\right)\left(\dfrac{10^3}{10^5}\right)$$

$$(.6757)(10^{-2})$$

The number isn't quite in Scientific Notation. It is actually $.006757$. To convert it fully to scientific notation, we get:

$$6.757 \times 10^{-3}$$

Work these problems. (Round to three decimal places.)

bb) $(5.9 \times 10^7) \div (2.6 \times 10^4)$

cc) $(1.5 \times 10^8) \div (8.23 \times 10^{10})$

dd) $(6.82 \times 10^{-3}) \div (3.95 \times 10^{-2})$

Algebra II

Active Lesson: 1.3a

As with all the topics in chapter one, they have been previously seen in Algebra One. For a more in-depth discussion of the reasoning behind each idea, see the activities for that course. Here, we will go through a quick review of the concepts of roots.

First, roots are a type of grouping symbol. To simplify them, you clean up anything inside. If there are grouping symbols inside of grouping symbols, you work from the inside out.

Simplify the following.

a) $\sqrt{\sqrt{81}}$

b) $\sqrt{153 + 16}$

c) $\sqrt{49} - \sqrt{25}$

Products underneath square roots can be turned into individual roots. For instance:

$\sqrt{9a^2} = \sqrt{9} \cdot \sqrt{a^2} = 3a$

Simplify the following.

d) $\sqrt{16x^2}$

e) $\sqrt{81x^4y^2}$

Not everything comes out from under the root in this problem.

f) $\sqrt{50m^2}$

g) $\sqrt{80x^3y^2}$

When multiplying roots, you multiply together anything outside the roots and multiply together anything inside the roots.

$2\sqrt{5} \cdot 3\sqrt{5} = 6\sqrt{25}$

Then you can simplify the root. (Here what comes out of the root will multiply with what his already there.)

$6\sqrt{25} = 6 \cdot 5 = 30$

Simplify the following.

h) $\sqrt{7} \cdot \sqrt{7} =$

i) $\sqrt{10} \cdot \sqrt{2} =$

j) $3\sqrt{8} \cdot 2\sqrt{2} =$

When a fraction is beneath a root, we have the option of separating it into two different roots.

$$\sqrt{\frac{16}{25}} = \frac{\sqrt{16}}{\sqrt{25}}$$

In this case, separating it will make the work easier.

$$\frac{\sqrt{16}}{\sqrt{25}} = \frac{4}{5}$$

Simplify the following by turning the fraction under the root into two separate problems.

k) $\sqrt{\dfrac{4}{121}}$

l) $\sqrt{\dfrac{50}{81}}$

m) $\sqrt{\dfrac{9x^2}{16y^4}}$

And, if we can separate roots, it is also possible to combine them.

$$\dfrac{\sqrt{8x^2}}{\sqrt{2}} = \sqrt{\dfrac{8x^2}{2}}$$

The reason we would choose this here is because we can reduce to make the problem easier.

$$\sqrt{4x^2} = 2x$$

Simplify the following by combining the two roots into one fractional root.

n) $\dfrac{\sqrt{20}}{\sqrt{45}} =$

o) $\dfrac{\sqrt{75x^5}}{\sqrt{3x}} =$

p) $\dfrac{\sqrt{21x^9y}}{\sqrt{7xy^3}} =$

Finally, adding and subtracting roots works just like adding and subtracting variables. If they are like terms they can be combined. Numbers out front are just like coefficients in front of a variable.

$$15\sqrt{2} + 3\sqrt{3} - 4\sqrt{2} + 6\sqrt{3} = 11\sqrt{2} + 9\sqrt{3}$$

Simplify the following.

q) $17\sqrt{5} + 4\sqrt{5} =$

r) $8\sqrt{7} - 4\sqrt{7} =$

s) $13\sqrt{11} + 3\sqrt{17} - 15\sqrt{11} - 5\sqrt{17} =$

t) $3\sqrt{v} - 2\sqrt{v} + 10\sqrt{v} =$

Sometimes you must first simplify the roots to see if you can add or subtract.

u) $\sqrt{32} + \sqrt{8} =$

The entire variable portion of this next problem is the same. It can be combined.

v) $10x\sqrt{x} + 2x\sqrt{x} - 3x\sqrt{x}$

w) $x\sqrt{48} - 2\sqrt{3x^2} + 5\sqrt{75x^2}$

Algebra II

Active Lesson: 1.3b

Here, we will review several additional ideas involving roots. First, we will examine rationalizing denominators. As before, a more detailed discussion can be found in the Algebra I activities.

Because it is easier to work without roots in the denominator of a fraction, it has become standard mathematical practice to make an equivalent fraction without them.

$$\frac{3}{5\sqrt{2}} \cdot \frac{\sqrt{2}}{\sqrt{2}}$$

Multiplying the numerator and denominator by $\sqrt{2}$ is the same as multiplying by 1, and therefore is legal. Then, we multiply across the top and across the bottom.

$$\frac{3}{5\sqrt{2}} \cdot \frac{\sqrt{2}}{\sqrt{2}} = \frac{3\sqrt{2}}{10}$$

Rationalize the following denominators.

a) $\dfrac{7}{3\sqrt{5}}$

b) $\dfrac{12}{6\sqrt{3}}$

Here, simplify first, and then you can rationalize.

c) $\dfrac{5}{\sqrt{8}}$

The next set of problems will have a binomial in the denominator.

$$\frac{5}{2 - \sqrt{3}}$$

The trick is to multiply the numerator and the denominator by the conjugate of the denominator. The conjugate is the same numbers but with the opposite sign.

$$\frac{5}{2-\sqrt{3}} \cdot \frac{2+\sqrt{3}}{2+\sqrt{3}}$$

Multiplying then involves distributing (FOIL).

$$\frac{10+5\sqrt{3}}{4-3} = \frac{10+5\sqrt{3}}{1} = 10+5\sqrt{3}$$

The conjugate is the correct choice because it will always cause the roots to drop out of the denominator.

Rationalize the following denominators using the conjugate.

d) $\dfrac{2}{(3-4\sqrt{2})}$

e) $\dfrac{\sqrt{2}}{(5+2\sqrt{3})}$

f) $\dfrac{\sqrt{x}}{7+2\sqrt{15}}$

So far, we have limited our roots to square roots. Square roots ask what two of the same number multiply together to get the value inside the root. However, there can be any kind of root. A cube root requires three of the same number to get the value inside.

$$\sqrt[3]{27} = \sqrt[3]{3 \cdot 3 \cdot 3} = 3$$

Simplify the following roots.

g) $\sqrt[5]{32} =$

h) $\sqrt[3]{x^3} =$

i) $\sqrt[3]{8x^3} =$

All of the properties of square roots extend to any type of root. Separate the following into a root in the numerator and the denominator.

j) $\sqrt[3]{\dfrac{8x^6}{27}} =$

Roots can be written as fractional exponents.

$$\sqrt[3]{x} = x^{\frac{1}{3}} \qquad \sqrt{y} = y^{\frac{1}{2}}$$

Where the "index" of the root becomes the denominator of the fraction.

$$\sqrt[3]{x^2} = x^{\frac{2}{3}} \qquad \left(\sqrt[3]{x}\right)^2 = x^{\frac{2}{3}}$$

If there is a power involved, it becomes the numerator of the fractional exponent. Notice that the exponent can be inside the root or outside the root and the fraction is still the same.

Write the following with a fractional exponent.

k) $\sqrt[3]{y^5}$

l) $\left(\sqrt[5]{a}\right)^3$

m) $\dfrac{1}{\sqrt{x^3}}$

Simplify the following fractional exponents. It is easiest to take the root first and then work with the exponent.

$$27^{\frac{2}{3}} = \left(\sqrt[3]{27}\right)^2$$

n) $81^{\frac{3}{4}} =$

The value of making roots into fractional exponents is that all the properties of exponents then apply.

$$\left(3x^{\frac{1}{3}}\right)\left(2x^{\frac{1}{2}}\right)$$

When we multiply like bases, we add the exponents. It is the same here.

$$\left(3x^{\frac{1}{3}}\right)\left(2x^{\frac{1}{2}}\right) = 6x^{\frac{1}{3}+\frac{1}{2}} = 6x^{\frac{2}{6}+\frac{3}{6}} = 6x^{\frac{5}{6}}$$

Try these problems on your own.

o) $\left(4x^{\frac{3}{4}}\right)\left(5x^{\frac{1}{8}}\right)$

p) $\left(8y^{\frac{5}{4}}\right)\left(5y^{\frac{1}{3}}\right)$

Algebra II

Active Learning: 1.4a

Here, we will review concepts related to polynomials. Polynomials are an expression filled with terms involving addition, subtraction, multiplication, and exponents. The exponents are never negative and we don't divide by a variable.

$$a_n x^n + a_{n-1} x^{n-1} + \cdots + a_1 x^1$$

The degree of a polynomial is the highest exponent among all the terms. (It technically involves adding up the degrees of each term, but in practice, this doesn't occur very frequently.)

Find the degree of the following polynomials. (It is simply the highest exponent.)

a) $9x^5 - 3x^3 + 4x - 3$

Degree:

b) $14x^2 + 5x^7 - 3x + 15$ (Don't let the order fool you. It is still the highest exponent.)

Degree:

Adding polynomials simply involves combining the like terms. Add the following polynomials.

c) $(5x^3 - 2x^2 + 12x - 15) + (8x^3 + 18x^2 - 2x + 5)$

d) $(3x^4 - 2x^2 + 8x - 21) + (5x^3 - 7x^2 + 2x - 7)$

To subtract polynomials, the subtraction sign works as a negative which gets distributed to everything which follows it.

$(4x^2 + 5x - 2) - (x^2 - 3x + 5) = 4x^2 + 5x - 2 - x^2 + 3x - 5 = 3x^2 + 8x - 7$

Subtract the following polynomials.

e) $(12x^3 + 2x^2 + 9x + 4) - (10x^3 - 4x^2 + 5x - 2)$

f) $(x^3 + 9x^2 - 6x + 15) - (5x^2 - 8x + 9)$

To multiply polynomials, we distribute.

$3x(4x^2 + 2x + 5) = 3x \cdot 4x^2 + 3x \cdot 2x + 3x \cdot 5 = 12x^3 + 6x^2 + 15x$

Multiply the following.

g) $6x(3x^2 - 4x + 8)$

h) $5x^2(6x^3 + 3x - 10)$

If the polynomial you are multiplying by has more than one term, you distribute to each term.

$(4x + 2)(5x^2 + 3x + 3) = 4x \cdot 5x^2 + 4x \cdot 3x + 4x \cdot 3 + 2 \cdot 5x^2 + 2 \cdot 3x + 2 \cdot 3$

$20x^3 + 12x^2 + 12x + 10x^2 + 6x + 6$

Then, combine like terms.

$20x^3 + 22x^2 + 18x + 6$

i) $(7x + 4)(3x^2 + 6x + 10) =$

j) $(3x-5)(2x^2-6x-7)$

k) $(x^2+3)(x^2+8x-4)$

Algebra II

Active Learning: 1.4b

When multiplying some binomials, special patterns can emerge. (For a full discussion, see the topic in the Algebra I course.)

An example of such a special pattern is when a binomial is multiplied by itself.

$$(x-3)^2$$

It isn't necessary to memorize the pattern. Foiling it out will get you the desired result. However, the pattern is helpful.

$$(x-3)(x-3) = x^2 - 3x - 3x + 9$$

When you multiply a binomial by itself, you always get two identical middle terms.

$$x^2 - 3x - 3x + 9 = x^2 - 6x + 9$$

Recognizing that can save you time.

$$(x+6)^2 = x^2 + 12x + 36$$

I knew that I would get two $6x$ terms, so I added them to create the $12x$.

Multiply the following. I recommend you try the shortcut, but it is okay to FOIL if necessary.

a) $(x+9)^2$

b) $(y-5)^2$

c) $(3x-6)^2$

d) $(a-2b)^2$

The next idea actually follows something which we previously saw with roots. Multiplying a binomial by its conjugate, the same numbers with a different sign, will cause the middle term to drop out. It isn't necessary to memorize this either, but it is quite helpful.

$$(x - 5)(x + 5) = x^2 - 5x + 5x - 25 = x^2 - 25$$

Multiply the following.

e) $(x + 7)(x - 7)$

f) $(y - 3)(y + 3)$

g) $(3x - 7)(3x + 7)$

h) $(4a - 5b)(4a + 5b)$

Finally, we can often be asked to multiply binomials with multiple variables. (Which I've already done to you in several problems.) Nothing changes. Simply distribute through and combine like terms.

Multiply the following.

i) $(3x + 5)(4x - 2y + 6)$

Algebra II

Active Learning: 1.5a

Here, we will have a quick review of factoring. The most basic type of factoring is a Greatest Common Factor. We pull out any shared factors from each of the terms in a polynomial.

$3y^3 + 6y^2 = 3y^2(y + 2)$

Try some on your own. Factor.

a) $14x^2y^2 - 28x^2y + 21xy$

b) $12a^3 + 8a^2 + 4a$

Factoring Trinomials becomes trickier. We will start with those trinomials where the leading term has a 1. These are the easiest type.

$$x^2 - 2x - 15$$

Factoring a trinomial with a leading 1 is a game. You ask what two numbers multiply to get the last term (-15) and add to get the middle coefficient (-2). Here, the two numbers are -5 and +3.

$$x^2 - 2x - 15 = (x - 5)(x + 3)$$

Try some on your own.

c) $x^2 + 12x + 20 =$

d) $x^2 - 12x + 35 =$

Trinomials with a leading coefficient other than 1 are best factored by the A/C method.

$$2x^2 + 1x - 10$$

First, multiply the leading coefficient (A) with the last coefficient (C). In this problem, we have

$2 \cdot -10 = -20$.

Next, we play a similar game to the basic trinomials. What two numbers multiply to get -20 and add to get the middle term (1). Those two numbers are 5 and -4. Break the middle term into those two numbers. It isn't cheating because they will add up to get what was previously there.

$$2x^2 + 5x - 4x - 10$$

Now, pull the GCF of the first two terms and then the GCF of the last two terms.

$$\underline{2x^2 + 5x} - \underline{4x - 10}$$

$$2x(x+5) - 2(x+5)$$

You know you did it right if you have a matching binomial. That binomial is now a GCF and you can pull it out.

$$(x+5)(2x-2)$$

Factor these by the A/C method.

e) $\quad 6x^2 - 11x + 4$

f) $\quad 8x^2 + 10x - 3$

Next, we look at a special type of trinomial called a perfect square trinomial.

$$y^2 + 6y + 9$$

What makes them special is that they have square roots at both the front and the back terms. The middle term needs to be twice the two roots multiplied together. Here the front root is y and the back root is 3. Twice those roots would be $2 \cdot 3 \cdot y = 6y$. We have a perfect square trinomial.

When we do, the factor just involves the roots:

$$(y+3)(y+3)$$

The two factors are exactly the same with a perfect square binomial. And, the sign will always be the same as the trinomials middle term. If our original problem had been:

$$y^2 - 6y + 9$$

Our factors would have become:

$$(y-3)(y-3)$$

Factor the following.

g) $x^2 + 10x + 25$

h) $x^2 - 22x + 121$

For this review, the final type is called a difference of squares.

$$x^2 - 9$$

Here, we have a square root on both the front and rear terms, however there is no middle term. The factor looks like this:

$$(x+3)(x-3)$$

If you FOIL this out, the opposite signs cause the middle term to drop out. Be careful, to have a difference of squares, it must be subtraction. If there is an addition sign between them, it is prime and can't be factored.

Factor the following.

i) $x^2 - 121$

j) $9x^2 - 4$

Algebra II

Active Learning: 1.5b

The next type of factor problem which we would like to review involves a sum or difference of cubes.

$$x^3 - 27$$

To factor a binomial which has cube roots on the first term and the last term, you need to know two things:

1) The pattern $(a\ b)(a^2\ ab\ b^2)$.
2) And the acronym: S.O.A.P.

Let's walk through a problem.

$$x^3 - 27$$

1) Does it have a cube root on the front? Yes. x.
2) Does it have a cube root on the back? Yes. 3.

Note: These can have a positive or a negative in-between!

Now, we will fit this to our pattern $(a\ b)(a^2\ ab\ b^2)$. a equals the cube root of the first term. b equals the cube root of the last term. So:

$$a = x$$
$$b = 3$$

Then we put those into our pattern:

$a = x$

$b = 3$

$a^2 = x^2$

$ab = 3x$

$b^2 = 9$

$$(x\ 3)(x^2\ 3x\ 9)$$

43

But we also need signs between the terms. That's where S.O.A.P. comes in. The acronym stands for:

S = Same

O = Opposite

A = Always

P = Positive

So, first we put in the same sign from the original problem. $x^3 - 27$ The sign was negative, so our first sign is negative.

$$(x - 3)(x^2 \; 3x \; 9)$$

Opposite means the next sign is always the opposite of the first.

$$(x - 3)(x^2 + 3x \; 9)$$

And Always Positive means that the last sign is always a positive.

$$(x - 3)(x^2 + 3x + 9)$$

Try one.

a) $x^3 - 8$

Try another. Notice that the sign in-between is a positive this time.

b) $y^3 + 64$

Nothing changes if both the terms are variables:

$$x^3 - y^3$$

$a = x$

$b = y$

$a^2 = x^2$

$ab = xy$

$b^2 = y^2$

c) Finish this problem.

Finally, we will look over a more difficult factoring problem which you may not be familiar with. The problem has fractional exponents.

$$x^{-\frac{1}{4}} + x^{\frac{3}{4}}$$

The key is to always take the smallest exponent as the GCF. The $-\frac{1}{4}$ is the smallest exponent. When you take out GCF, you are really dividing. And dividing exponents with a similar base gives:

$$\frac{x^{\frac{3}{4}}}{x^{-\frac{1}{4}}} = x^{\frac{3}{4} - -\frac{1}{4}} = x^{\frac{3}{4} + \frac{1}{4}} = x^1$$

And so:

$$x^{-\frac{1}{4}}(1 + x)$$

Factor these:

d) $x^{-\frac{1}{2}} + x^{\frac{1}{2}}$

e) $x^{-\frac{2}{3}} + x^{\frac{1}{3}}$

The concept can extend to pulling out a GCF that is a binomial.

$$3(x-2)^{-\frac{1}{3}} + 5x(x-2)^{\frac{2}{3}}$$

Pull out the lowest exponent.

$$(x-2)^{-\frac{1}{3}}(3(1) + 5x(x-2))$$

Taking the negative exponent made the second binomial into $(x-2)^1$.

$$(x-2)^{-\frac{1}{3}}(3 + 5x(x-2))$$

There is some simplifying we can do now.

$$(x-2)^{-\frac{1}{3}}(3 + 5x^2 - 10x)$$

$$(x-2)^{-\frac{1}{3}}(5x^2 - 10x + 3)$$

Factor these:

f) $2x(x+5)^{-\frac{2}{3}} + 7(x+5)^{\frac{1}{3}}$

g) $2(x-9)^{-\frac{1}{5}} + 3x(x-9)^{\frac{4}{5}}$

Algebra II

Active Learning: 1.6a

Chapter One ends with a review of some difficult topics. It might well be worth your time to examine the complete lessons from Algebra I. Here, again, will be a shorter review. We'll begin with rational expressions.

Rational expressions are essentially fractions made from polynomials, one in the numerator and one in the denominator.

$$\frac{3x-21}{x^2-49}$$

We want to reduce the rational expressions in the same way in which we would reduce any fraction. When we reduce fractions, what we are really doing is cancelling factors.

$$\frac{15}{25} = \frac{3 \cdot \cancel{5}}{5 \cdot \cancel{5}}$$

$$\frac{3}{5}$$

Polynomials can be factored too.

$$\frac{3x-21}{x^2-49} = \frac{3(x-7)}{(x-7)(x+7)}$$

And matching factors can be cancelled.

$$\frac{3(\cancel{x-7})}{(\cancel{x-7})(x+7)} = \frac{3}{(x+7)}$$

Simplify the following rational expressions by factoring and then reducing.

a) $\dfrac{5z+25}{z^2+3z-10}$

b) $\dfrac{2x^2+3x-9}{x^3-9x}$

Multiplying rational expressions follows all the same principles as multiplying any fractions. Multiply the top times the top and the bottom times the bottom. Normally, you probably multiply and then reduce afterward. Here it is easier to reduce first. Any factor in the numerator can cancel with any factor in the denominator.

$$\frac{6}{5} \cdot \frac{10}{8} = \frac{\cancel{2} \cdot 3}{\cancel{5}} \cdot \frac{\cancel{2} \cdot \cancel{5}}{\cancel{2} \cdot \cancel{2} \cdot 2} = \frac{3}{2}$$

The concept is the same with rational expressions. First, factor. Then, cancel.

$$\frac{4x-8}{x+6} \cdot \frac{x^2+6x}{x^2-4} = \frac{4(\cancel{x-2})}{(\cancel{x+6})} \cdot \frac{x(\cancel{x+6})}{(\cancel{x-2})(x+2)}$$

$$\frac{4x}{x+2}$$

Simplify the following.

c) $\quad \dfrac{2x-4}{6x} \cdot \dfrac{4x^2}{8x-16}$

d) $\quad \dfrac{4x-6}{2x+10} \cdot \dfrac{x^2+3x-10}{2x^2-x-6}$

Dividing rational expressions also follows the pattern of dividing fractions.

$$\frac{5}{7} \div \frac{15}{28}$$

The trick is to Keep, Change, Flip. Keep the first fraction as it is. Change the sign from division to multiplication, and flip the last fraction upside down.

$$\frac{5}{7} \cdot \frac{28}{15}$$

Then it becomes the same as a multiplication problem. Let's do the same with division of rational expressions.

$$\frac{4x+12}{6x-18} \div \frac{3x+9}{5x-15}$$

We Keep, Change, Flip:

$$\frac{4x+12}{6x-18} \div \frac{3x+9}{5x-15} = \frac{4x+12}{6x-18} \cdot \frac{5x-15}{3x+9}$$

Then we factor, reduce, and multiply across.

$$\frac{4(\cancel{x+3})}{6(\cancel{x-3})} \cdot \frac{5(\cancel{x-3})}{3(\cancel{x+3})}$$

$$\frac{20}{18} = \frac{10}{9}$$

Try one. Simplify.

e) $\quad \dfrac{14x-28}{4x+4} \div \dfrac{x^2+2x-8}{2x+2}$

Algebra II

Active Learning: 1.6b

As with multiplying and dividing, adding and subtracting rational expressions also follows the identical principles as working with simple fractions.

$$\frac{1}{4}+\frac{1}{6}$$

To add these fractions, you must have a common denominator. In all likelihood, you do this in your head. But really, we are working with factors.

$$\frac{1}{2\cdot 2}+\frac{1}{2\cdot 3}$$

We need each denominator to have the same factors. So, the left fraction needs a 3. If you add it to the bottom, you must add it to the top. The right fraction needs another 2. If you add it to the bottom, you must add it to the top.

$$\frac{1\cdot 3}{2\cdot 2\cdot 3}+\frac{1\cdot 2}{2\cdot 3\cdot 2}$$

$$\frac{3}{2\cdot 2\cdot 3}+\frac{2}{2\cdot 3\cdot 2}=\frac{5}{12}$$

Follow the same pattern to add rational expressions.

$$\frac{x}{x^2-9}+\frac{2}{3x+9}$$

Factor first.

$$\frac{x}{(x-3)(x+3)}+\frac{2}{3(x+3)}$$

Making the denominators match. (Being sure to multiply the numerators also.)

$$\frac{x\cdot 3}{(x-3)(x+3)3}+\frac{2(x-3)}{3(x+3)(x-3)}$$

It is easier to leave the denominator factored, but we clean up the numerators.

$$\frac{3x}{3(x-3)(x+3)}+\frac{2x-6}{3(x+3)(x-3)}$$

And because the denominators are a match, we can add the tops.

$$\frac{5x-6}{3(x-3)(x+3)}$$

Add the following rational expression.

a) $\dfrac{x}{x^2 + 7x + 12} + \dfrac{3}{2x + 8}$

Subtraction follows the same idea. Just be sure that the subtraction sign is distributed to all of the terms which follow it.

$$\dfrac{5}{(x-2)} - \dfrac{3}{(x+3)}$$

$$\dfrac{5(x+3)}{(x-2)(x+3)} - \dfrac{3(x-2)}{(x-2)(x+3)}$$

$$\dfrac{5x + 15}{(x-2)(x+3)} - \dfrac{3x - 6}{(x-2)(x+3)}$$

$$\dfrac{5x + 15 - 3x + 6}{(x-2)(x+3)}$$

$$\dfrac{2x + 21}{(x-2)(x+3)}$$

Subtract the following.

b) $\dfrac{7}{x+5} - \dfrac{4}{x-2}$

Try one more. Sometimes when you get to the end of the problem, your final rational expression can be reduced. That will occur here.

c) $\dfrac{x}{x^2 - 16} - \dfrac{1}{2x - 8}$

Finally, we have something called a complex rational expression.

$$\dfrac{1 - \dfrac{1}{x}}{\dfrac{x}{y} + \dfrac{y}{x}}$$

This is a combination of several types of problems which we already know how to work. First, we work the subtraction problem on the top.

$$\dfrac{1}{1} - \dfrac{1}{x} = \dfrac{x}{x} - \dfrac{1}{x} = \dfrac{x-1}{x}$$

Then we work the addition problem on the bottom.

$$\dfrac{x}{y} + \dfrac{y}{x} = \dfrac{x \cdot x}{y \cdot x} + \dfrac{y \cdot y}{x \cdot y} = \dfrac{x^2 + y^2}{xy}$$

Now we have: $\dfrac{\frac{x-1}{x}}{\frac{x^2+y^2}{xy}}$. And this is one fraction being divided by another fraction.

$$\dfrac{x-1}{x} \div \dfrac{x^2 + y^2}{xy}$$

So, we Keep, Change, Flip.

$$\dfrac{x-1}{x} \cdot \dfrac{xy}{x^2 + y^2} = \dfrac{y(x-1)}{x^2 + y^2}$$

These problems look terrible. However, we already know how to do all the individual parts. It is really nothing more than three problems in one.

Simplify the following complex rational expression.

d) $\dfrac{\dfrac{1}{x}+\dfrac{1}{y}}{x-\dfrac{y^2}{x}}$

Algebra II

Active Learning: 2.1a

Here, we review the basics of graphing with the Cartesian Coordinate System. The concept is based on creating a grid system like city blocks. The primary horizontal line is called the x-axis. The primary vertical line is called the y-axis. We call the center (where the two axes intersect) the origin. This system creates four boxes called quadrants, which are numbered in a counter-clockwise fashion.

To plot a point, we have two instructions. The first tells us how far to go in the x direction. Positive numbers are to the right. Negative numbers are to the left. The second number tells us how far to go in the y direction. Positive is up. Negative is down.

$(2, -3)$

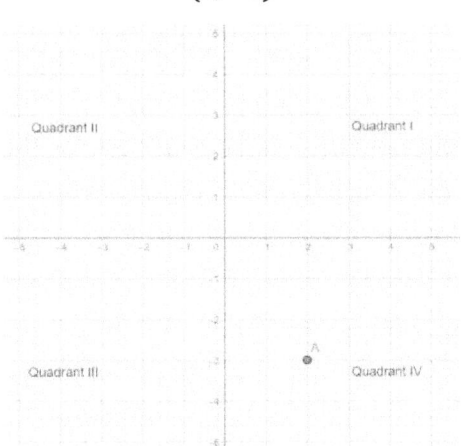

Graph the following points *and* state which quadrant they would be in:

a) $(-2, 3)$ b) $(1, 3)$ c) $(-1, -3)$

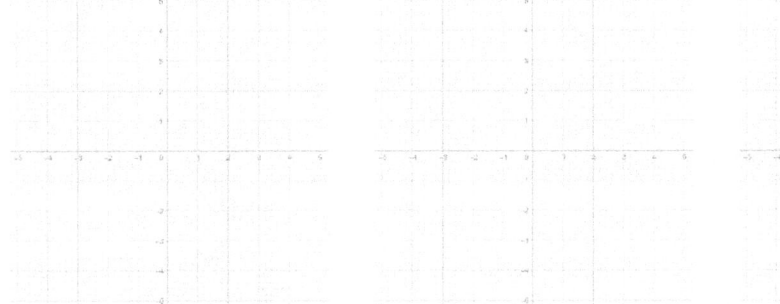

Any equation with two variables can be graphed by plotting its points. The (x, y) coordinates are a set of ordered pairs. In other words, the x and y values are a team. If we evaluate an equation at any value of x, we can find the partner value of y. Repeating this for several different values of x will get us enough points to graph.

$$y = 4x - 3$$

This equation will make a line. For a line, three sets of ordered pairs is considered sufficient to graph. Evaluate the line at the three points I've put in the table. (I've done the first one for you.)

d)
$$y = 4(0) - 3$$
$$y = -3$$

x	y
0	-3
-1	
2	

Plot your ordered pairs and connect them to make the line. Your line should extend past the final points and go off forever in both directions.

e)

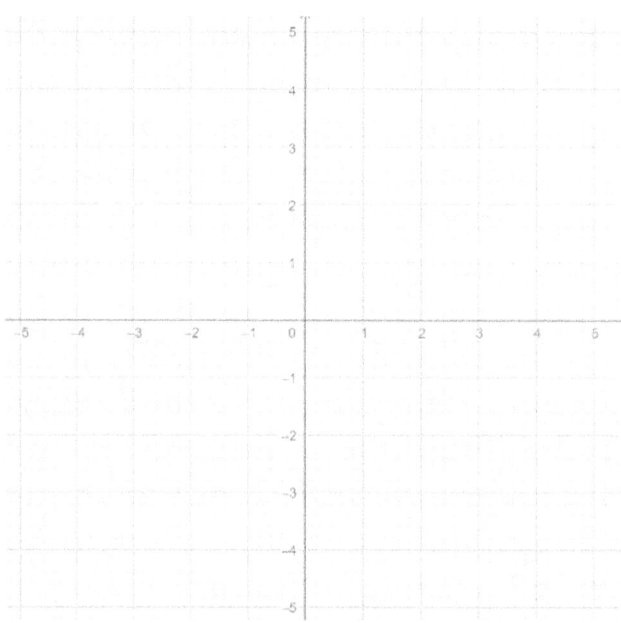

When plotting points, any values of x will work, but it is best to choose those values which make the math the easiest.

Try another on your own.

f) Graph by plotting points.

$$y = 3x - 4$$

x	y

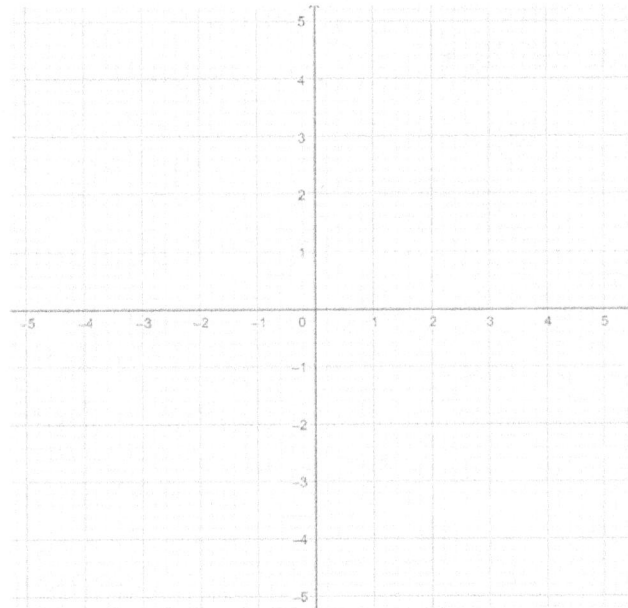

Where a line hits the x-axis is called the x-intercept. Where a line hits the y-axis is called the y-intercept. This concept will prove to be very important. Here is the graph of the line $y = 2x + 2$.

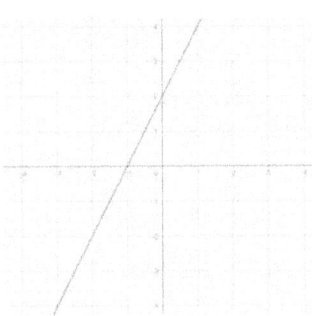

g) Read the graph to give the value of the intercepts.

x-intercept: (___, 0)

y-intercept: (0, ___)

If you aren't given the graph, you can find the values of the intercepts by evaluating. To find an y-intercept, evaluate the equation when $x = 0$. To find an x-intercept, evaluate the equation when $y = 0$.

h) Find the intercepts mathematically by evaluating. Complete the table.

$$y = 2x + 2$$

x	y
0	
	0

Intercepts are points. Be sure to give them as such.

i) Find the x and y intercepts for the following line.

$$y = \frac{1}{2}x + 3$$

x-intercept: (___, 0)

y-intercept: (0, ___)

Algebra II

Active Learning: 2.1b

The Midpoint Formula

On the number line below, mark the numbers 2 and 8.

Visually, find the midpoint (middle) between the two numbers.

a) Midpoint:

In the space below, take the average of 2 and 8.

b) Average:

It turns out that a midpoint is nothing more than an average. To better understand how to find a midpoint between two points on a Cartesian Plane, complete the following steps:

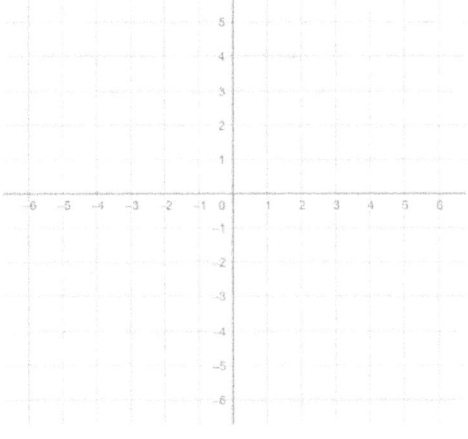

1. On the Graph to the left, mark the points: $(1, 1)$ and $(5, 3)$.
2. Draw a line between the points and mark where you believe the midpoint to be.
3. Add the point $(5, 1)$.
4. Draw a dashed line from the point $(1, 1)$ to the point $(5, 1)$.
5. Draw a dashed line from the point $(5, 3)$ to the point $(5, 1)$. You have now created a right triangle.
6. The bottom of your triangle runs parallel to the x-axis. Find the midpoint of this dashed line. (Remember, it is an average of the ends.)

c)   Midpoint:

7. The right side of your triangle runs parallel to the y-axis. Find the midpoint of this dashed line.

d)   Midpoint:

8. Make the two midpoints into an ordered pair and plot it on the graph.
(Midpoint along x, Midpoint along y)

This is the midpoint of the line segment. It is simply the midpoint (average) of the x coordinates and the midpoint (average) of the y coordinates.

Find the midpoint of the following ordered pairs. (Hint: Nothing changes when you have negative values. Continue finding the average. The math will work out the negatives for you.)

e)   (5, 6) and (11, 12)

f)   (−1, 3) and (5, −7)

The Distance Formula

Next, we want to be able to find the length of a line segment. As we will see, it is not a new formula but rather something you have known for a long time. Complete the following steps:

1. On the graph, mark the points (1, 1) and (6, 3).
2. Draw a line between the points.
3. Add the point (6, 1).
4. As we did above, add dashed lines to this point in order to create a right triangle.

g)   5. We have created a right triangle. What is the name of the side opposite the 90° angle in a right triangle?

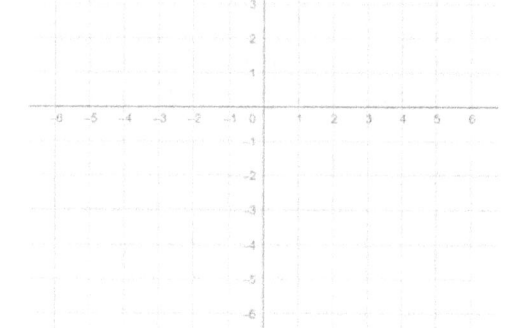

h)   6. We will call the bottom of the triangle side A. How long is side A?

i)   7. We will call the right side of the triangle side B. How long is side B?

j)   8. We have the lengths of the two sides. Find the length of side C, use the Pythagorean Theorem to find the length of the missing side.

This length is identical to the distance between the two points. So, the distance formula is really just the Pythagorean Theorem.

Here is the Pythagorean Theorem.

$$a^2 + b^2 = c^2$$

k) Leaving all the variables, solve the Pythagorean Theorem for c. (Instead of c^2.)

To find the length of side a, you subtract the two x values.

$$Side\ A = x_2 - x_1$$

To find the length of side b, you subtract the two y values.

$$Side\ B = y_2 - y_1$$

Putting those lengths into the Pythagorean Theorem creates the distance formula:

$$distance = \sqrt{(x_2 - x_1)^2 + (y_2 - y_1)^2}$$

Again, the distance formula is really just the Pythagorean Theorem. Use the distance formula to find the distance between the following ordered pairs.

l) $(4, 8)$ and $(7, 12)$

m) $(-2, 3)$ and $(4, -1)$

Algebra II

Active Learning: 2.2a

In this section, we will review how to solve equations involving one variable. When I teach this idea, I explain that there are always three steps:

1) Clean Up Both Rooms- Do any distributing and combining like terms on both sides of the equation.
2) Get the Variable Back Together- If there are variables on both sides of the equation, move them all to one side.
3) Get the Variable Alone- Get down to a factor puzzle and solve.

Here is an example.

$$8(x + 3) + 4 = 2(x - 2) + 2$$

1) Clean Up Both Rooms.

$$8x + 24 + 4 = 2x - 4 + 2$$
$$8x + 28 = 2x - 2$$

2) Get the Variable Back Together.

$$8x - 2x + 28 = -2$$
$$6x + 28 = -2$$

3) Get the Variable Alone. First, get it down to a factor puzzle.

$$6x = -2 - 28$$
$$6x = -30$$

Then solve.

$$x = -5$$

Solve the following equations.

a) $2(2x - 3) + 7 = 2(x + 5) - 1$

b) $7(2y - 1) + 2 = 5(y - 7) + 5$

Rational equations involve fractions.

$$1 - \frac{3}{x} = \frac{4}{x^2}$$

If each of the fractions has the same denominator then the denominator has nothing to do with solving.

$$\frac{1}{1} - \frac{3}{x} = \frac{4}{x^2}$$

$$\frac{1 \cdot x^2}{1 \cdot x^2} - \frac{3 \cdot x}{x \cdot x} = \frac{4}{x^2}$$

$$\frac{x^2}{x^2} - \frac{3x}{x^2} = \frac{4}{x^2}$$

Now that the denominators match, all we need to do is solve:

$$x^2 - 3x = 4$$

$$x^2 - 3x - 4 = 0$$

$$(x - 4)(x + 3) = 0$$

$$x = 4 \text{ or } x = -3$$

Solve this problem on your own.

c) $$\frac{1}{x} + \frac{2}{x^2} = \frac{x + 3}{2x^2}$$

The same concept can apply to more difficult problems.

$$\frac{x+5}{x^2+2x-15} = \frac{7}{x+5} - \frac{5}{x-3}$$

Here, factor any denominators which can be factored.

$$\frac{x+5}{(x+5)(x-3)} = \frac{7}{x+5} - \frac{5}{x-3}$$

Now, we make the denominators match. If we do something to the bottom, we must do the same to the top.

$$\frac{x+5}{(x+5)(x-3)} = \frac{7(x-3)}{(x+5)(x-3)} - \frac{5(x+5)}{(x-3)(x+5)}$$

Since the denominators match, we only need to solve this equation.

$$x + 5 = 7(x-3) - 5(x+5)$$

d) Finish the problem in the space below.

Finally, it is possible that you can get solutions to a rational equation that are not allowed. If a solution causes the denominator of any fraction in the problem to become zero that solution is discarded. It would create a situation which is undefined. Work this problem. If there are no permitted solutions indicate that the problem has no solution.

e)
$$\frac{x-17}{x^2+x-6} = \frac{4}{x+3} - \frac{2}{x-2}$$

Algebra II

Active Learning: 2.2b

Next, we want to review the concept of a line. The equation for a line can have many forms, each with their own uses. However, the most basic form is called the Standard form.

$$Ax + By = C$$

Here's an example:

$$3x + 6y = 9$$

Any two points on a line will always have a consistent slope. A slope is the change in the y direction over the change in the x direction. (For y, up is positive, down is negative. For x, right is positive, left is negative.) The formula looks like this:

$$m = \frac{y_2 - y_1}{x_2 - x_1}$$

a) Identify the two points in the following graph:

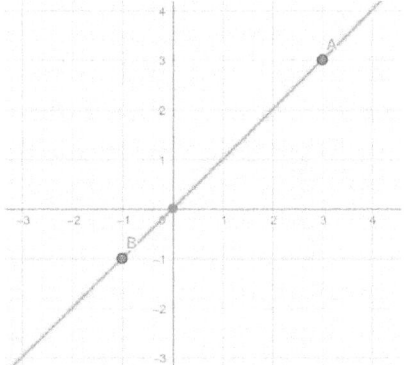

A:

B:

b) Use the two points to find the slope.

$$m =$$

c) Find the slope between the following two points: $(3, -2)$ and $(1, 4)$.

A second form of the equation for a line comes from rearranging the slope formula.

$$m = \frac{y_2 - y_1}{x_2 - x_1}$$

Dividing the denominator up, gives us:

$$m(x_2 - x_1) = y_2 - y_1$$

And if we don't know anything about point #2, we could write this as:

$$y - y_1 = m(x - x_1)$$

This is called the point-slope equation for a line. We use it whenever we know a slope and a single point.

d) If the slope of a line $m = \frac{1}{2}$ and the point $(4, -2)$ is on the line, give the equation for the line. (When using the point-slope equation, we typically end by cleaning up. Distribute m and get y alone.) Work the problem in the space below.

If you didn't have the slope, but did have two points, you could still use the point-slope formula. First, find the slope. Then, use that slope and either of the two points in the equation.

e) Given the points $(6, -4)$ and $(3, -1)$, find the equation for the line using the point-slope formula.

You may be asked to put an equation in Standard Form.

$$y - 2 = \frac{3}{4}x + 6$$

Standard form doesn't involve fractions. So, to clear the 4 from the denominator, multiply every term in the problem by 4.

$$4 \cdot y - 4 \cdot 2 = 4 \cdot \frac{3}{4}x + 4 \cdot 6$$

$$4y - 8 = 3x + 24$$

Finally, to arrange it in standard form we get the x and y terms on one side:

$$4y - 3x = 32$$

Convert the following line to Standard Form.

f) $\quad 2y + 3 = \dfrac{2}{5}x - 4$

Finally, vertical and horizontal lines have special slopes.

g) 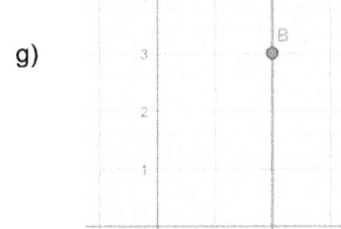 Vertical lines have an undefined slope. Put the two points from the graph into the slope formula.

h) Why do vertical lines have an undefined slope?

i) 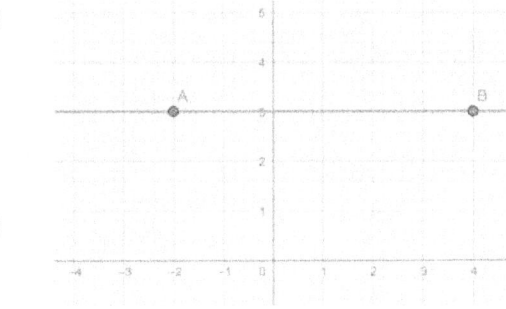 Horizontal lines have a zero slope. Put the two points from the graph into the slope formula.

j) Why will horizontal lines always get a slope of zero?

Algebra II

Active Learning: 2.2c

Find the equation for the two lines on the graph below, using the slope and a point. Put the equations in slope-intercept form.

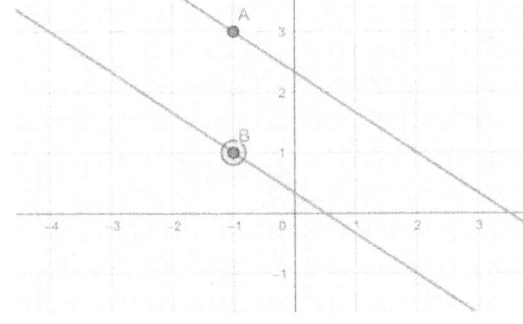

a)

Equation One

b)

Equation Two

c) These two lines are parallel. Looking at your equations, what feature causes the two lines to be parallel? Explain why.

Find the equation for the two lines on the graph below, using the slope and a point. Put the equations in slope-intercept form.

d) Equation One

e) Equation Two

f) These two lines are perpendicular. Looking at your equations what feature causes the two lines to be perpendicular? Explain why.

We want to graph a line parallel to the one graphed. Point A is the *y*-intercept of the line we want to graph.

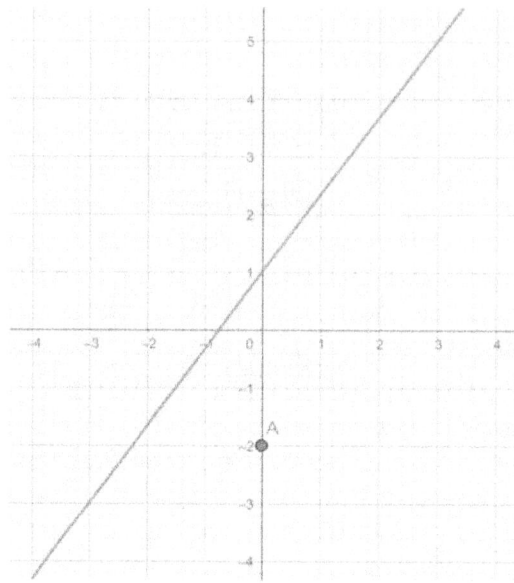

g) Find the slope of the graphed line.

h) If our line is parallel to the one already graphed, what do we know about the slope of our line?

i) What is the slope of the parallel line we want to graph?

j) Using that slope and point A, graph the parallel line on the graph above.

k) Let's do the same thing without the picture. Graph a line parallel to $y = -\frac{2}{3}x + 3$ with a *y*-intercept of $(0, -1)$. What is the slope of the line we want to graph?

l) Graph it below.

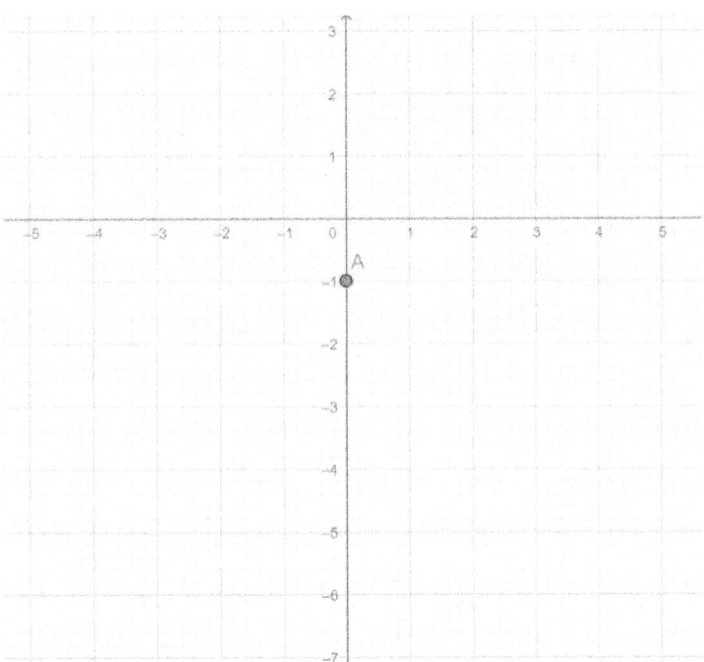

Let's do one more. This time, we aren't going to graph it; we are just going to find the equation for our new line using the point-slope formula.

Find the equation for a line parallel to the line $y = \frac{1}{2}x + 4$ and going through point $(2, 1)$.

m) What is the slope of our line?

n) Use the point-slope formula with our slope and the point $(2, 1)$. (Rearrange it to put it into slope-intercept form.)

Next, we want to graph a line perpendicular to the one graphed. Point A is the y-intercept of the line we want to graph.

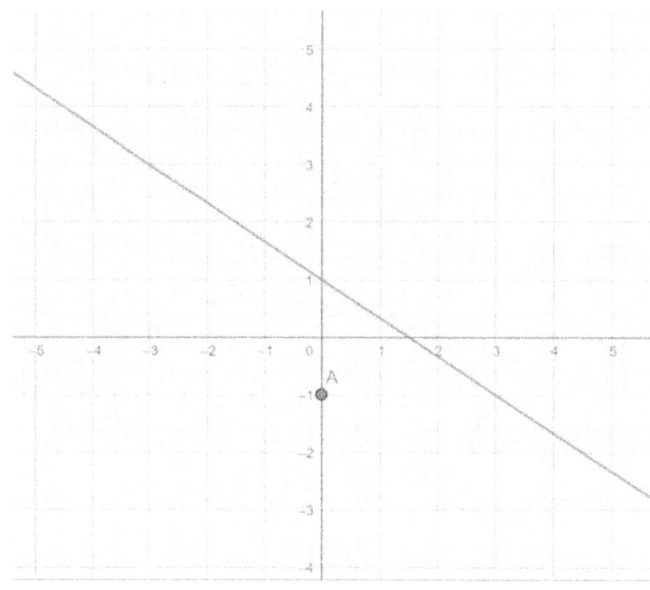

o) Find the slope of the graphed line.

p) If our line is perpendicular to the one already graphed, what do we know about the slope of our line?

q) What is the slope of the perpendicular line we want to graph?

r) Using that slope and point A, graph the perpendicular line on the graph above.

Again, let's do the same thing without the picture. Graph a line perpendicular to $y = -3x + 3$ with a y-
s) intercept of $(0, -1)$. What is the slope of the line we want to graph? (Remember -3 is the same as $-\frac{3}{1}$.)

t) Graph it below.

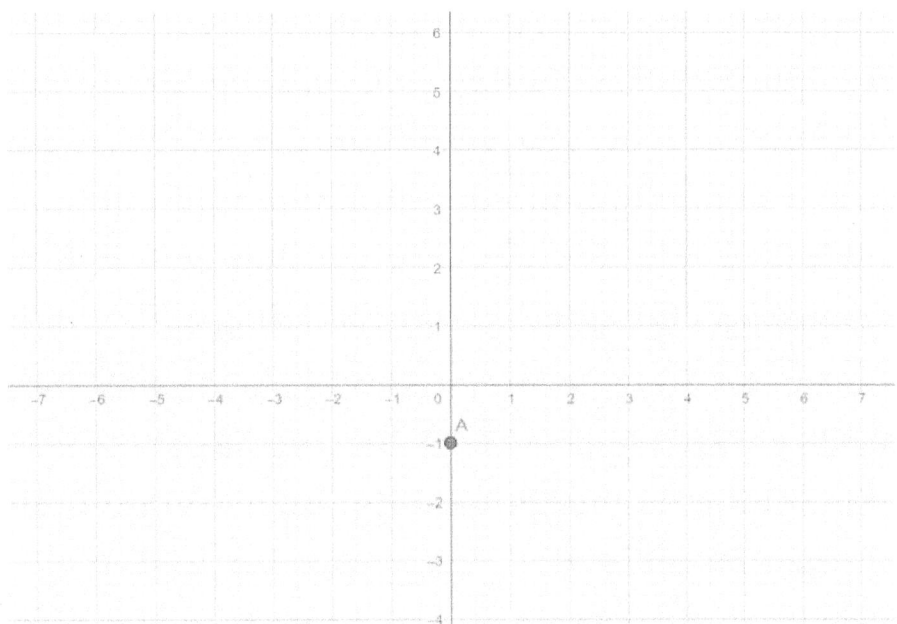

Like we did before, let's find a perpendicular line without graphing. Find the equation for a line perpendicular to the line $y = \frac{1}{2}x + 4$ and going through point $(2, 1)$.

u) What is the slope of our line?

v) Use the point-slope formula with our slope and the point $(2, 1)$. (Rearrange it to put it into slope-intercept form.)

Algebra II

Active Learning: 2.3a

Perhaps the main goal of mathematics is to explain the real-world using numbers. When math offers an explanation, it is called a mathematical model. The most basic mathematical model is a line.

$$y = mx + b$$

Here's an example.

A tow truck company charges $100 for a service call plus $.15 per mile.

$$Price = .15(miles) + 100$$

Changing fees are slopes. Flat fees are y-intercepts.

Create an equation for each of the following situations.

a) Customers of the Alpha Electric Company pay $20 plus .15 per kilowatt hour. Write an equation for the total cost of electricity with Alpha.

b) At Beta Electric, customers are charged .25 per kilowatt hour and have a flat fee of $10. Write an equation for the total cost of electricity with Beta.

c) If you expect to use 90 kilowatt hours, which company is the better deal.

d) At what amount of kilowatt hours will the cost for the two companies be the same?

Linear applications can take many forms. This can include distance problems.

$$Distance = rate \cdot time$$

As we saw in Algebra I, we handle problems of this type with a chart. Here's an example.

The drive to grandma's house took two hours. Bad traffic on the way home caused the drive to take 2.5 hours. The average rate was 10 miles per hour less. Find the rate to grandma's house and the rate home.

	Rate	Time	Distance
To Grandma's	r	2	d
Home	r-10	2.5	d

e) Write two equations from the chart.

f) The distance back and forth to grandma's house is the same. Set the two distances equations equal to one another and solve for r.

g) Try one on your own.

A trip to the beach took 5 hours. It took 5.5 hours to get home. The average rate on the way home was 5 miles per hour less. Find the rate to the beach and the rate home.

	Rate	Time	Distance
To The Beach			
Home			

Algebra II

Active Learning: 2.3b

This next section follows all the same principles as the previous section, this time with rectangles. First, let's look at the perimeter of a rectangle. Remember, perimeter is like adding up fencing around the outside of the shape. Since the lengths are the same, and the widths are the same, the perimeter could be summed up as:

$$P = 2L + 2W$$

Now, suppose we knew the perimeter and either the length or the width. For example:

The perimeter of a rectangle is 120 meters. The width is 20 meters. Find the length.

$$120 = 2L + 2(20)$$

a) Finish solving this for length.

Try these on your own.

b) The perimeter of a rectangle is 180 feet. The width is 65 feet. Find the length.

c) The perimeter of a rectangle is 240 yards. The length is 85 yards. Find the width.

Next, we don't know either the length or the width. We will need to build one from the other.

The perimeter of a rectangle is 200 meters. The length is 5 meters less than the width. Find the width.

Here's what I call the dictionary:

Math	English
x-5	Length
x	Width

This is what the equation would look like:

$$200 = 2(x - 5) + 2x$$

Notice that you would have a lot of clean up to do to solve for x.

$$200 = 2x - 10 + 2x$$

$$200 = 4x - 10$$

$$210 = 4x$$

$$52.5 = x$$

And, you must give both the length and the width. So, using the dictionary:

Math	English
(52.5-5)=47.5	Length
52.5	Width

Work these on your own.

d) The perimeter of a rectangle is 250 meters. The length is 15 meters less than the width. Find the width.

e) The length of a rectangle is five more than twice the width. The perimeter is 70 feet. Find the width.

Here, we will solve area problems involving rectangles. The area of a rectangle is the length times the width.

$$A = L \cdot W$$

The ideas are the same.

If the area of a rectangle is 100 $meters^2$, and the length is 20 meters. Find the width.

f) Set up an equation and solve.

Work this problem on your own.

g) If the area of a rectangle is 250 cm^2, and the width is 25 cm. Find the width.

Finally, let's look at a problem involving the volume of a rectangular prism. (In other words, a box.)

$$V = lwh$$

If the volume of a rectangular prism is 1200 in^3, the height is twice the length and the width is 10 inches, find the width.

Here is our dictionary.

Math	English
L	length
10	Width
2L	Height

h) Plug these facts and the volume of 1200 into the formula and solve for the length.

Algebra II

Active Learning: 2.4a

There is a strange problem that occurs in mathematics. You cannot have a negative under a square root.

$\sqrt{-9}$ $\qquad$ $\sqrt{-4}$ $\qquad$ $\sqrt{-16}$

a) Why is this a problem?

For this reason, negatives under square roots aren't real because this doesn't seem to occur in the "real world." But it creates a problem for mathematicians because in math language negatives under square roots come up rather easily. So, a simple solution was created. The solution was to call $\sqrt{-1} = i$.

Mathematicians can then finish the square root and leave the $\sqrt{-1}$ behind.

$\sqrt{-9} = \sqrt{-1 \cdot 3 \cdot 3} = 3\sqrt{-1}$ $\qquad$ $\sqrt{-4} = \sqrt{-1 \cdot 2 \cdot 2} = 2\sqrt{-1}$

And in order to make working with such numbers easier, they replace the $\sqrt{-1}$ with i.

$3\sqrt{-1} = 3i$ $\qquad$ $2\sqrt{-1} = 2i$

Simplify each of the following using i. (Be careful, on the last two, not everything can come out from under the root.)

b) $\sqrt{-121}$ $\qquad$ c) $\sqrt{-36}$ $\qquad$ d) $\sqrt{-8}$ $\qquad$ e) $\sqrt{-24}$

Numbers involving i are called imaginary numbers and mathematicians work with them frequently. And when real numbers are combined with imaginary numbers, we have something called a complex number. Complex numbers look like $a + bi$. Where the a refers to the real portion of the number and the b is the imaginary portion.

In the following complex numbers, identify the real portion and the imaginary:

f) $-5 + 17i$ $\qquad$ g) $14 - 7i$ $\qquad$ h) $9i$

i) Growing up, you frequently graphed sets of ordered pairs, like (2, 4), on a coordinate plane. Where there was the *x*-axis and the *y*-axis. Plot the point on the graph below.

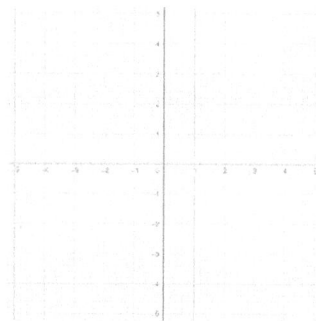

But mathematicians created a complex plane. The difference is that the *x*-axis is now the real portion of a complex number, the *a*. And the *y*-axis is now the imaginary portion of the complex number, the *b*. Here are two examples.

$2 + 3i$

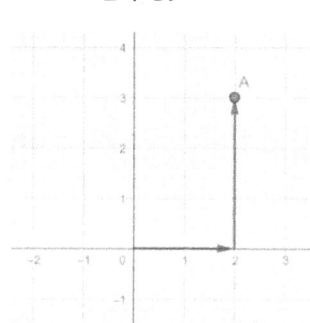

$3 - 4i$

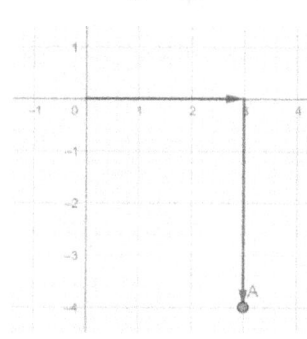

Plot the following numbers on the complex plane:

j) $5 + 2i$

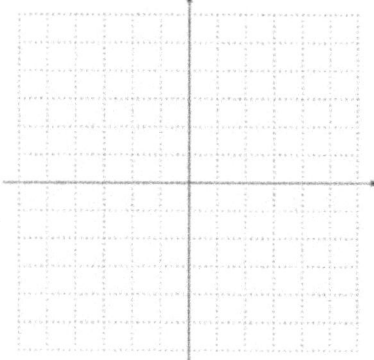

k) $-4 - 2i$

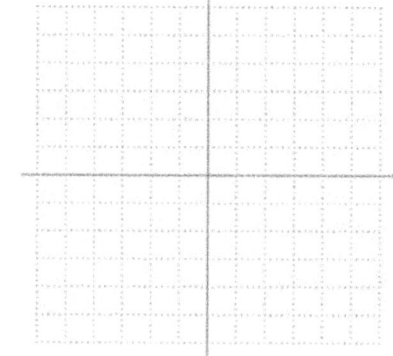

Finally, we need to look at adding and subtracting complex numbers. This is actually pretty easy. The real portions are added/subtracted together and the imaginary portions are added/subtracted together. For instance:

$$(2 - 6i) + (3 + 5i)$$
$$(2 + 3) + (-6 + 5)i$$
$$5 - 1i$$

Subtraction works the same, but the negative sign has to be distributed just like in basic algebra.

$$(4 + 6i) - (2 - 7i)$$
$$(4 + 6i) + (-2 + 7i)$$
$$(4 - 2) + (6 + 7)i$$
$$2 + 13i$$

Add or subtract the following complete numbers:

l) $(9 + 5i) + (3 + 7i)$

m) $(10 - 6i) + (-2 - 3i)$

n) $(16 + 7i) - (15 + 19i)$

o) $(-7 - 2i) - (-1 + 7i)$

p) $(20 + 50i) - (24i)$

Algebra II

Active Learning: 2.4b

In our last lesson we were introduced to the idea of complex numbers: $a + bi$. Then we learned that we could add and subtract complex numbers by adding/subtracting their real parts and adding/subtracting their imaginary parts. Work the following:

a) $(2 - 22i) + (-11 + 13i)$

b) $(16 - 3i) - (-5 - 21i)$

Multiplication of complex numbers is a bit harder but not much. Let's first refresh the key idea that you will need. Multiply the following:

c) $3(2x + 5)$

d) $-5(12y - 6)$

e) What is the name of the property which you used to simplify these expressions?

Multiplying a complex number by a real number

To multiply a complex number by a real number, just multiply the real number to each part of the complex number.

$$2(3 + 5i) = 2 \cdot 3 + 2 \cdot 5i = 6 + 10i$$

Try these:

f) $4(-3 + 20i)$

g) $5(4 - 6i)$

h) $-2(9 - 6i)$

Multiplying a complex number by a complex number

In order to multiply two complex numbers, the idea is based on a concept you already know. Multiply the following:

i) $(x-3)(x+2)$

j) $(y-7)(y-6)$

k) What is the name of this concept?

l) Multiplying two complex numbers follows the same pattern. But where $x \cdot x = x^2$, when working with i we get something different: $i \cdot i = i^2 = -1$. In the space below, show why $i^2 = -1$. (Hint: start with $\sqrt{-1} \cdot \sqrt{-1}$)

Here's an example of how multiplying two complex numbers would work.

Multiply $(3+2i)(4-5i)$

$$(3+2i)(4-5i) = 3 \cdot 4 + 3 \cdot (-5)i + 2i \cdot 4 + 2i \cdot (-5i)$$

$$12 - 15i + 8i - 10i^2$$

But $-10i^2$ turns into $+10$ because $i^2 = -1$ so $-10(-1) = 10$.

$$12 - 15i + 8i + 10$$

$$22 - 7i$$

Multiply the following:

m) $(2+3i)(-5+4i)$

n) $(4 + 2i)(6 - 5i)$

o) $(10 - 3i)(10 + 3i)$

Powers of i

In the last section we saw that $i^2 = -1$. Here we are going to discover an interesting thing which happens with powers of i.

p) Show why $i^3 = -i$. Start with $\sqrt{-1} \cdot \sqrt{-1} \cdot \sqrt{-1}$.

q) Show why $i^4 = 1$. Start with $\sqrt{-1} \cdot \sqrt{-1} \cdot \sqrt{-1} \cdot \sqrt{-1}$.

Notice that $i^5 = i$ because $i^5 = i^4 \cdot i^1$ and $(i^4) \cdot i^1 = (1) \cdot i^1 = i$.

Powers of i are cyclical, meaning they go in a circle, because every $i^4 = 1$. And so, we get the following:

$i^5 = i^4 \cdot i^1 = i$

$i^6 = i^4 \cdot i^2 = -1$

$i^7 = i^4 \cdot i^3 = -i$

$i^8 = i^4 \cdot i^4 = 1$

Therefore, if you have a power of i just divide the power by 4. The remainder is the only thing which matters. For instance:

i^{35} divide 35 by 4 we get 8 R3. Only the remainder matters $i^3 = -i$.

$i^{35} = (i^4)^8 + i^3 = (1)^8 + i^3 = 1 + i^3 = -i$

i^{25} divide 25 by 4 we get 6 R1. Only the remainder matters $i^1 = i$.

Try these:

r) i^{13}

s) i^{18}

t) i^{47}

u) i^{32}

Algebra II

Active Learning: 2.4c

So far, we've learned how to add, subtract, and multiply complex numbers. Simplify the following:

a) $(40 + 12i) - (21 - 8i)$

b) $(3 + 4i)(-5 - 6i)$

The final mathematical operation which we need for complex numbers is division.

<u>Dividing complex numbers by a real number</u>

Dividing complex numbers by a real number is much like this.

$$\frac{12x - 10}{2}$$

We divide the two into both terms of the numerator and get:

$$\frac{12x - 10}{2} = 6x - 5$$

Divide the following:

c) $\dfrac{15x - 25}{5}$

d) $\dfrac{4x + 24}{3}$

Dividing a complex number by a real number is exactly the same.

$$\frac{-6 + 20i}{2} = -3 + 10i$$

Try these:

e) $\dfrac{21 + 49i}{7}$

f) $\dfrac{-12 - 16i}{-4}$

(On this next problem, the best you can do is fractions.)

g) $\dfrac{5 + 23i}{3}$

Complex Conjugates

In order to divide a complex number by another complex number, we need to refresh our memory on the idea of a conjugate. We worked with conjugates when working with roots. For instance, the conjugate of $2 + \sqrt{5}$ has the same terms just with the opposite sign. So, $2 - \sqrt{5}$. And the conjugate of $9 - \sqrt{3}$ is $9 + \sqrt{3}$. Give the conjugate for the following:

h) $121 - \sqrt{47}$

i) $19 + \sqrt{13}$

The reason we work with conjugates is that they have a very nice property. Multiply the following:

j) $(x+3)(x-3)$

k) $(2y+4)(2y-4)$

l) $(3x-5)(3x+5)$

Since conjugates have opposite signs, when you FOIL them, the middle term always drops out and you get the square of the first term minus the square of the last term. Try this one:

m) $(2+\sqrt{5})(2-\sqrt{5})$

Complex conjugates are exactly the same. For instance, the complex conjugate of $(3+4i)$ is $(3-4i)$. Give the complex conjugate of the following:

n) $(21+11i)$

o) $(-3-26i)$

p) $(10+i)$

q) $(-2i)$

Multiply each of the following complex numbers by their complex conjugate.

r) $(3 + 4i)$

s) $(2 - 5i)$

t) $(7 + 3i)$

If you remembered that $i^2 = -1$ then you found that multiplying complex conjugates results in the creation of a real number.

Dividing Complex Numbers

Dividing complex numbers is exactly like rationalizing the denominator of a fraction. We rationalize this denominator by multiplying the top and bottom by the conjugate of $2 - \sqrt{3}$.

$$\frac{4}{2 - \sqrt{3}} \cdot \frac{2 + \sqrt{3}}{2 + \sqrt{3}} = \frac{4(2 + \sqrt{3})}{4 - 3} = \frac{8 + 4\sqrt{3}}{1}$$

Work this one. Rationalize the denominator.

u) $\dfrac{5}{3 + \sqrt{2}}$

To divide two complex numbers, multiply the numerator and the denominator by the complex conjugate of the denominator. For instance:

$$\frac{3 + 5i}{2 - i}$$

$$\frac{3 + 5i}{2 - i} \cdot \frac{2 + i}{2 + i} = \frac{6 + 3i + 10i + i^2}{4 - i^2} = \frac{6 + 13i - 1}{4 - (-1)}$$

$$\frac{5 + 13i}{5}$$

This looks very complicated, but you know all of the individual ideas. The top is just FOIL. The bottom is multiplying complex conjugates. Try this one, you should get $\frac{26-7i}{25}$ or $\frac{26}{25} - \frac{7}{25}i$

v) $\dfrac{5 + 2i}{4 + 3i}$

Work one more on your own.

w) $\dfrac{3 - 2i}{5 + 4i}$

Algebra II

Active Learning: 2.5a

We are beginning a section that works with quadratics. Quadratics are polynomials that have a degree of two. (This means the highest exponent in the quadratic is a 2.) For instance, $x^2 - 3x + 15$ or $x^2 - 2$. Quadratics always take the shape of a parabola, which gives them lots of applications in mathematics and physics.

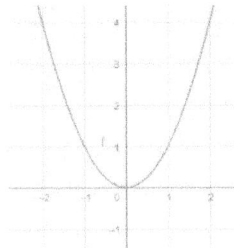

There is a lot more detail we will discuss later about quadratics, but first we are going to work on solving quadratic equations. A quadratic equation has an equal sign involve. Such as:

$x^2 + x = 6$

If it isn't already done, arrange the equation so zero is on one side:

$x^2 + x - 6 = 0$

Now, we want to solve for x.

Solving Quadratic Equations by Graphing

When we set the quadratic equal to zero, it means the y-value (the output) is zero. When the y-value is zero, we have the x-intercepts.

a) Using Geogebra, graph $x^2 + x - 6$. There are two x-intercepts, write them in the space below.

b) Using Geogebra, graph $x^2 + 8x + 15$. Give the x-intercepts.

c) The values of x at the x-intercepts are the solutions to the quadratic equations. In the space below, put the values of x from the x-intercepts into $x^2 + 8x + 15 = 0$ and show that they do make $0 = 0$.

Solving Quadratic Equations by Factoring

Graphing is a great way to solve quadratic equations, but we will often want to use a pure algebra approach. One of the easiest ways is to check and see if the quadratic equation can be factored.

If we factor $x^2 + x - 6 = 0$ we get $(x + 3)(x - 2) = 0$.

And if two factors are being multiplied together and they equal zero, either of the factors might be the reason. (This is called the Zero Product Property.) So, we set each factor equal to zero and solve them.

$(x + 3) = 0$ and $(x - 2) = 0$

$x = -3$ and $x = 2$

Solve the following by factoring:

d) $x^2 - 5x - 6 = 0$

e) $x^2 - 4x - 21 = 0$

The idea is the same for quadratics that have something other than a 1 before the x^2, but the factoring is a bit harder. Remember, the AC method is easiest for factoring in these cases.

Solve the following by factoring:

f) $4x^2 + 15x + 9 = 0$

g) $12x^2 + 11x + 2 = 0$

Sometimes you will be asked to solve equations which are higher than quadratics and they don't appear to be factorable. In these cases, check to see if it has a GCF.

$$x^3 + 2x^2 - 15x = 0$$
$$x(x^2 + 2x - 15) = 0$$
$$x(x + 5)(x - 3) = 0$$

The zero-product property would also apply to the x in front. So, we have:

$$x = 0 \text{ and } (x + 5) = 0 \text{ and } (x - 3) = 0$$
$$x = 0, -5, \text{ and } 3$$

Work this problem by factoring out a GCF and then using the zero-product property.

h) $3x^3 + 6x^2 - 24x = 0$

Solving Quadratic Equations by the Square Root Property

Give the answer to each of the following. (And be careful, each one has two answers.)

i) $\sqrt{9}$ j) $\sqrt{121}$ k) $\sqrt{81}$

If you have a quadratic equation with **only** a squared term on one side, you can solve it by taking the square root of both sides. But, again, be careful, a square root always has two answers.

Try these:

l) $x^2 = 16$

m) $x^2 = 8$

(Here, isolate the x^2 and then take the square root of both sides.)

n) $2x^2 + 1 = 51$

(On this problem, isolate the $(x-4)^2$ and then take the square root of both sides. You will then send the 4 to the other side.)

o) $3(x-4)^2 = 15$

Solving a Quadratic Equation by Completing the Square

Our final approach is completing the square. Here's a refresher on the idea.

$x^2 + 4x + 1 = 0$

First, organize the x terms and move the constant to the other side.

$x^2 + 4x = -1$

Now, we want to add a number that is half the 4 and then square it. Half of 4 is 2. $2^2 = 4$. In order to keep the equation balanced we add the 4 to both sides.

$x^2 + 4x + 4 = -1 + 4$

$x^2 + 4x + 4 = 3$

This has created a perfect square trinomial on the left. Factor the perfect square trinomial.

$(x + 2)^2 = 3$

p) Use the Square Root Property to find the answers.

Solve by completing the square.

q) $x^2 - 6x - 13 = 0$

Algebra II

Active Learning: 2.5b

As we continue with quadratic equations, we are going to learn one more approach to solving them—using the quadratic formula. It is an approach that can solve any quadratic equation and it is created by taking a generic quadratic equation and completing the square. How it is created is a bit complicated, but here is what it leaves us with:

$$x = \frac{-b \pm \sqrt{b^2 - 4ac}}{2a}$$

With this formula, all we need to do is plug in the appropriate $a, b,$ and c and we can solve for x. The $a, b,$ and c are taken from a quadratic equation in standard form:

$$ax^2 + bx + c = 0$$

The a is the number in front of the x^2. The b is the number in front of the x. And, the c is the number by itself.

Here's an example:

$x^2 + 3x + 1 = 0$

$$x = \frac{-3 \pm \sqrt{3^2 - 4(1)(1)}}{2(1)}$$

$$x = \frac{-3 \pm \sqrt{5}}{2}$$

Try this one:

a) $x^2 + 7x + 2 = 0$

Try another. This one involves imaginary numbers.

b) $x^2 + x + 3 = 0$

Solve the following using the quadratic formula.

c) $9x^2 + 3x + 2 = 0$

The Discriminant

As you worked on the previous problems, you may have noticed that the portion of the quadratic equation under the square root decides if there are complex numbers involved in the solution. The portion under the square root is called the discriminant. It predicts what types of solutions we will get. Here is the equation for the discriminant: $b^2 - 4ac$.

Calculate the value of the discriminant in each of the following:

d) $x^2 + 4x + 4 = 0$

This means you will get only one value of x from the quadratic equation.

e) $8x^2 + 14x + 3 = 0$

You would then get a perfect square that you would add and subtract to the rest of the quadratic equation to create two rational solutions.

f) $3x^2 - 5x - 3 = 0$

This value doesn't have a perfect square, so you would have square roots involved in the two solutions.

g) $3x^2 - 10x + 15 = 0$

This value is negative. So, a negative under the square root will get you two complex solutions.

Here is a summary of what happens depending on your discriminant.

$b^2 - 4ac = 0$	The quadratic formula will give one rational solution.
$b^2 - 4ac > 0$, perfect square	The quadratic formula will give two rational solutions.
$b^2 - 4ac > 0$ non perfect square	The quadratic formula will give two answers involving square roots.
$b^2 - 4ac < 0$	The quadratic formula will give two complex solutions.

Use the discriminant to tell the type of solutions you would get to the following quadratic equations.

h) $3x^2 - 5x + 1$

i) $2x^2 + 5x + 6$

j) $x^2 + 4x + 4$

Algebra II

Active Lesson: 2.6a

Two square roots multiply together to make a complete number. For example:

$$\sqrt{x} \cdot \sqrt{x} = x$$

Now, follow the rules of exponents to simplify the following expression. (Remember, you add the exponents here.)

a) $x^{\frac{1}{2}} \cdot x^{\frac{1}{2}}$

In a similar way, it takes three cube roots to make a complete number. For example:

$$\sqrt[3]{y} \cdot \sqrt[3]{y} \cdot \sqrt[3]{y} = y$$

Again, follow the rules of exponents to simplify the following expression.

b) $y^{\frac{1}{3}} \cdot y^{\frac{1}{3}} \cdot y^{\frac{1}{3}}$

Mathematically, roots and fractional exponents must be the same thing. Write each of the following as a fractional exponent.

c) $\sqrt{z} =$

d) $\sqrt[3]{r} =$

e) $\sqrt[4]{x} =$

We can then extend the idea, combining roots and regular exponents:

$$(\sqrt{x})^3 = x^{\frac{3}{2}}$$

Or

$$\sqrt{x^3} = x^{\frac{3}{2}}$$

So, we can simplify the following. (Taking the root first will make the work easier.)

$$27^{\frac{2}{3}} = (\sqrt[3]{27})^2 = 3^2 = 9$$

Simplify these:

f) $8^{\frac{4}{3}}$

g) $16^{\frac{5}{4}}$

Fractions in the exponents are called Rational Exponents and they aid us in working with roots by allowing the use of all the rules of exponents.

Knowing that we multiply a power to a power can allow us to solve a problem like this:

$$x^{\frac{3}{2}} = 27$$

$$\left(x^{\frac{3}{2}}\right)^{\frac{2}{3}} = (27)^{\frac{2}{3}}$$

$$x = 9$$

Solve the following:

h) $x^{\frac{5}{3}} = 32$

i) $y^{\frac{3}{2}} = 125$

To extend the idea to equations, we always begin by isolating the difficult portion. In the problem below, first add the 3 to the other side and then divide by 5 in order to isolate the $x^{\frac{2}{3}}$. Once it is isolated, use the previous idea.

j) $5x^{\frac{2}{3}} - 3 = 42$

Try another.

k) $3x^{\frac{1}{4}} - 12 = -3$

Here, the difficult portion is already isolated. So, immediately raise both sides to the $\frac{3}{2}$ power.

l) $(x - 2)^{\frac{2}{3}} = 25$

In this next problem, we will need to factor out a GCF.

$x^{\frac{3}{4}} = x^{\frac{1}{4}}$

We need a zero on one side.

$x^{\frac{3}{4}} - x^{\frac{1}{4}} = 0$

Take a GCF. With fractional exponents, we need to take the lowest fraction. As we saw in an earlier section, pulling out a GCF is division. $\frac{x^{\frac{3}{4}}}{x^{\frac{1}{4}}} = x^{\frac{3}{4} - \frac{1}{4}}$

$x^{\frac{1}{4}}\left(x^{\frac{2}{4}} - 1\right) = 0$

The zero-product property allows us to set each equal to zero.

$x^{\frac{1}{4}} = 0$ or $x^{\frac{2}{4}} - 1 = 0$

m) Finish solving the two equations to find our two solutions.

This problem will have one more step.

$$2x^{\frac{3}{4}} = x^{\frac{1}{2}}$$

$$2x^{\frac{3}{4}} - x^{\frac{1}{2}} = 0$$

Before pulling out a GCF, the fractional exponents must have the same denominator.

$$2x^{\frac{3}{4}} - x^{\frac{2}{4}} = 0$$

n) Pull out the lowest fraction as the GCF and finish the problem. (You will get a fraction for an answer.)

Next, look at the following equation:

$$3x^4 = 27x^2$$

Here, we can solve this equation with a GCF and with the zero-product property.

$$3x^4 - 27x^2 = 0$$

o) The GCF is $3x^2$. Pull it out from the left side and then solve the two equations using the zero-product property.

The next idea is solving equations involving roots. This topic was handled in Algebra I. See that activity for a more in-depth discussion.

$$\sqrt{x+1} + 1 = x$$

Like with many equations, the first step is to isolate the difficult portion. Here, the difficult portion is the root.

$$\sqrt{x+1} = x - 1$$

Now that it is isolated, we get rid of a root by squaring. If we do it to one side, we must do it to the other.

$$\left(\sqrt{x+1}\right)^2 = (x-1)^2$$

$$x + 1 = x^2 - 2x + 1$$

p) Next, make one side zero and use the zero-product property to solve. (You will need to factor.)

q) When we square a root to solve an equation, it creates the possibility that an answer is not actually true. Test the two answers you got for the previous problem. One of them will not actually work and is discarded.

Finally, we want to work a very difficult problem involving two roots.

$$\sqrt{x-1} + \sqrt{x+1} = 2$$

We can't isolate a single square root, but isolating one is still the best we can do.

$$\sqrt{x-1} = 2 - \sqrt{x+1}$$

Now, we square both sides.

$$\left(\sqrt{x-1}\right)^2 = \left(2 - \sqrt{x+1}\right)^2$$

The left is easy, but the right we would need to FOIL out $(2 - \sqrt{x+1})(2 - \sqrt{x+1})$. We get:

$$x - 1 = 4 - 4\sqrt{x+1} + x + 1$$

If we simplify, we get:

$$-6 = -4\sqrt{x+1}$$

We will use our squaring trick again.

$$(-6)^2 = \left(-4\sqrt{x+1}\right)^2$$

$$36 = 16(x+1)$$

r) Finish solving for x. Because we squared roots, we will need to check to see if our answer is really a solution to the equation.

Try one on your own.

s) $\sqrt{x-2} + \sqrt{x-1} = 1$

Algebra II

Active Lesson: 2.6b

When combining absolute values and algebra, we have something like this:

$$|x| = 3$$

We know that absolute values make the number inside a positive. So, there are two possible values for x which could be 3.

$$x = 3 \text{ or } x = -3$$

If there is an equation inside, we need to work it twice because what is inside could be a positive or it could be a negative.

$$|2x - 8| = 4$$

$$2x - 8 = 4 \text{ or } 2x - 8 = -4$$

a) Finish solving this equation for the two possible values of x.

Work this problem.

b) $|4x + 2| = 10$

In a more complex problem, make sure you get the absolute value alone before working the problem twice.

$$2|x + 4| - 6 = 12$$

Imagine this is the problem

$$2X - 6 = 12$$

The 6 would be added to both sides. Then, we would divide both sides by 2.

c) Follow that idea to get the absolute value alone. At that point, work the problem once for the positive and once for the negative.

$$2|x+4| - 6 = 12$$

There are two unusual problems when working with absolute values. Here is one:

$$|x - 2| = -3$$

The left side of this equation must be a positive number, but the left is a negative. This could never happen, so the answer is "no solution." Here is the other:

$$|x - 6| = 0$$

Working this problem twice would generate this:

$$x - 6 = 0 \text{ or } x - 6 = -0$$

There is no such thing as -0, so in this case we would only get one answer. Work these problems.

d) $|2x - 4| + 6 = 6$ \hspace{2cm} e) $4|x - 8| + 12 = 8$

Some equations don't appear as if they can be solved.

$$2x^4 + x^2 - 6 = 0$$

However, a clever trick of substitution can sometimes help. Suppose we make $u = x^2$. Our equation then becomes:

$$2u^2 + u - 6 = 0$$

This can be factored and solved.

$$(2u - 3)(u + 2) = 0$$

$$2u - 3 = 0 \text{ or } u + 2 = 0$$

$$u = \frac{3}{2} \text{ or } u = -2$$

But we can't stop here, we need to undo our substitution.

$$x^2 = \frac{3}{2} \text{ or } x^2 = -2$$

And solving for x, we get:

$$x = \pm\sqrt{\frac{3}{2}} \text{ or } x = \pm\sqrt{-2}$$

Some simplifying gets us:

$$x = \pm\frac{\sqrt{6}}{2} \text{ or } x = \pm\sqrt{2}i$$

Try one. Solve using substitution.

f) $x^4 + x^2 - 6 = 0$

Substitution can also be used on this problem.

$$(x+1)^2 + 2(x+1) - 8 = 0$$

We can make the substitution $u = (x+1)$, and the problem becomes:

$$u^2 + 2u - 8 = 0$$

g) Finish the problem. Remember to "undo" your substitution.

In Algebra I, we learned to solve rational equations like this:

$$\frac{x+5}{x^2+2x-15} = \frac{7}{x+5} - \frac{5}{x-3}$$

The approach I teach is to get a common denominator and then the denominators don't matter.

$$\frac{x+5}{(x+5)(x-3)} = \frac{7}{x+5} - \frac{5}{x-3}$$

$$\frac{x+5}{(x+5)(x-3)} = \frac{7(x-3)}{(x+5)(x-3)} - \frac{5(x+5)}{(x-3)(x+5)}$$

Now, we can ignore the denominators and solve this equation:

$$x+5 = 7(x-3) - 5(x+5)$$

h) Finish solving for x. Remember, any values of x which would cause a denominator to become zero, at any point in the problem, aren't allowed. (That may or may not occur here.)

i) In this next problem, the concept is identical. The only difference is that after dropping the denominator, you will be left with a quadratic. Solve the quadratic by factoring and the zero-product property.

$$\frac{2x}{x-1} + \frac{2}{x+1} = \frac{-4}{x^2-1}$$

Algebra II

Active Learning: 2.7a

In Algebra, inequalities create a range of answers.

$$x > 2$$

This is all numbers greater than 2.

$$x \leq -5$$

This is all numbers less than and including -5.

$$-5 < x \leq 2$$

This is all numbers between -5 and 2.

Throughout this course, solutions to inequalities will be shown in interval notation. We saw interval notation in Algebra I. Here's a quick review.

$$x > 2$$
$$(2, \infty)$$

Interval notation always goes from left to right up the number line. If we don't want to include a number we use a parenthesis instead of a bracket. The infinity symbol means it goes on forever. Infinity always gets a parenthesis because we can never get there.

$$x \leq -5$$
$$(-\infty, 5]$$

Here we are starting at negative infinity and because we want to include the 5, it gets a bracket.

$$-5 < x \leq 2$$
$$(-5, 2]$$

Next, we want all the numbers from -5 to 2. We don't want to include -5, so it gets a parenthesis. We do want to include 2, so it gets a bracket.

Give the interval notation for the following:

a) $x < 3$ b) $x \geq 12$ c) $-2 \leq x < 7$ d) $-9 < x < 11$

There is only one difference between solving equations involving inequalities and those with equal signs. Inequality signs change direction whenever you multiply or divide by a negative number.

Solve the following. Give your answers in interval notation.

e) $3x - 2 > -4(x + 4)$

f) $8(x - 3) \leq 10x + 2$

On this problem, clear the fractions by multiplying each term by the LCD.

g) $\frac{3}{4}x - 6 \leq \frac{2}{3}x + 2$

Finally, let's look at a concept known as compound inequalities.

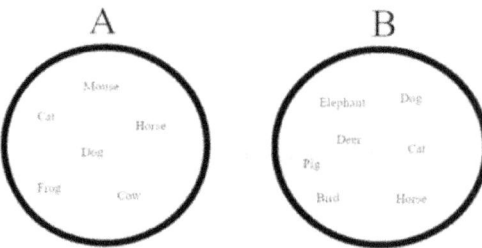

h) List the names of all the animals which are in circles A *and* B. (At the same time.)

i) Add those names which are in both to the center of the intersecting circles.

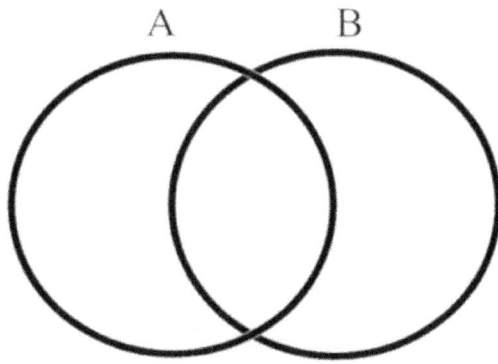

In both logic and mathematics, items that are in both circles (sets) are said to be in the intersection. The symbol for intersection is ∩.

$$Intersection\ of\ A\ and\ B = A \cap B$$

Some compound inequalities are *and* inequalities.

$$x > -5 \text{ and } x \leq 2$$

So, we need where they overlap. If you can't see this in your mind, use a number line.

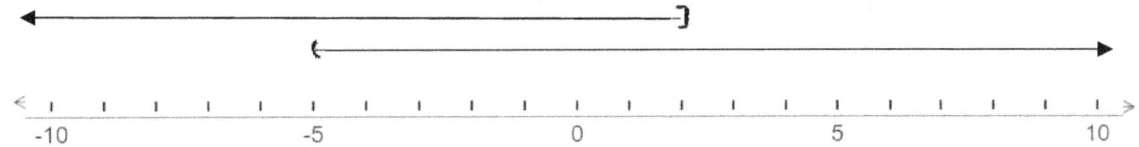

Our interval notation become: $(-5, 2]$. This is where the two lines overlap.

j) List the names of the animals which are in circle A or circle B. (Don't list repeats twice.)

In logic and mathematics, items that are in either one circle or the other circle are called the union. The symbol for union is ∪.

$$Union\ of\ A\ and\ B = A \cup B$$

Some compound inequalities are *or* inequalities.

$$x < -3 \text{ or } x > 7$$

So, we don't need them to overlap. We are happy with either. On a number line it would look like this.

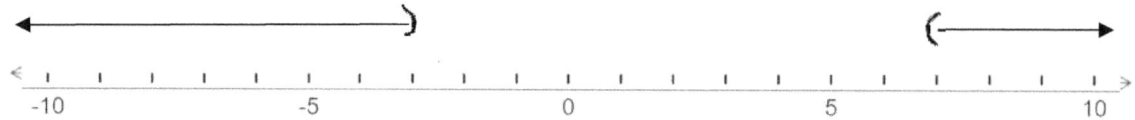

Written in interval notation, we would have:

$$(-\infty, -3) \cup (7, \infty)$$

Try the following. Write the compound inequalities in interval notation.

k) $x > -1$ and $x < 6$ l) $x \leq 2$ or $x \geq 10$

The next two are oddities. One is "all real numbers," which in interval notation is $(-\infty, \infty)$. The other is "no solution," which can be written as $\emptyset$. Determine which is which.

m) $x < -1$ and $x > 2$ n) $x < 4$ or $x > -5$

Solve the following compound inequalities and give their solutions in interval notation. (Work the problems individually. For "and" problems, find their overlap. For "or" problems, write their union.)

o) $8x - 2 > 3x + 13$ and $3(2x + 1) < 7(x - 3)$

p) $-6x + 15 \geq 39$ or $-2(x + 6) - 4 > -3x + 8$

Finally, you will often see a two-sided inequality like this:

$$20 + 2x > 3x + 5 > x + 19$$

These are *and* compound inequalities. These can be worked in both directions at the same time. However, I recommend you simply turn it into two separate inequalities with the *and* in-between.

$$20 + 2x > 3x + 5 \text{ and } 3x + 5 > x + 19$$

q) Notice that the center portion of the equation has been repeated; once in each direction. Solve the two separate inequalities and give the final interval notation. (Since it is *and* you need the overlap.)

Work this problem on your own.

r) $5x - 20 \leq 4x - 6 < 8x - 38$

Algebra II

Active Learning: 2.7b

Solve the following compound inequalities and write their solution regions in interval notation.

a) $-2x + 4 \leq -8$ or $3x + 5 \leq 8$

b) $3x - 5 < 19$ and $-2x + 3 < 23$

c) $2x - 4 < 8$ and $2x - 4 > -8$

d) $4x + 2 > 6$ or $4x + 2 < -6$

Solve the following absolute value equations.

e) $|2x - 4| = 8$

f) $|4x + 2| = 6$

Now we want to solve the following absolute value inequality: $|2x - 4| < 8$

g) Start by ignoring the inequality sign and solve it as if it were an equal sign. However, we already did that in problem e. Using a vertical line, mark your two answers from problem e on the number line below.

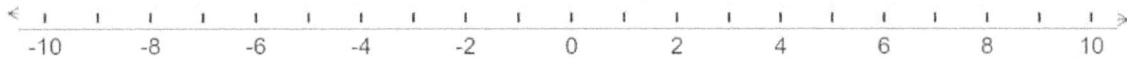

Notice that those values break the number line into three regions. Next, we want to test to see if numbers in each of those regions make the inequality true. Put the following values of x into $|2x - 4| < 8$. Write true or false depending on whether or not it makes the inequality give a true statement. These test points will be true for the entire region which they are in. Add true or false to the number line above.

h) $\quad x = -3$

i) $\quad x = 0$

j) $\quad x = 8$

k) Using interval notation, write the region which made the inequality true. (Hint: it should be the same as one of the first four problems you worked.)

Next, we want to solve: $|4x + 2| > 6$

l) Again, start by ignoring the inequality sign and solve it as if it were an equal sign. However, we already did that in problem f. Add your two answers from problem f to the number line below.

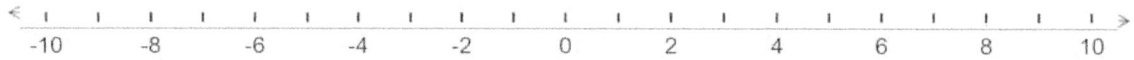

Once again, notice that those values break the number line into three regions. Next, we want to test to see if numbers in each of those regions make the inequality true. Put the following values of x into $|4x + 2| > 6$. Write true or false depending on whether or not it makes the inequality give a true statement and add the true or false to the number line above.

m) $\quad x = -4$

n) $\quad x = 0$

o) $\quad x = 2$

p) Using interval notation, write the region which made the inequality true. (Hint: it should be the same as one of the first four problems you worked.)

Absolute value inequalities with a < or ≤ are the same as "and" compound inequalities. Absolute value inequalities with a > or ≥ are the same as "or" compound inequalities.

Solve the following: (Your answers should be given in interval notation.)

q) $|3x - 3| \leq 12$

r) $|4x - 2| > 10$

There are special cases where absolute value inequalities are not problems to work, but rather logic puzzles to be solved. If you have an absolute value on one side and a zero or negative on the other, **don't work the problem like normal.** Reasoning will solve the problem. We will look at each of the cases.

$|x + 3| > -4$

s) We know that an absolute value is always positive. So, if the left side is always positive, which of the following must be true?

   a) A positive is never greater than a negative.
   b) A positive is sometimes greater than a negative.
   c) A positive is always greater than a negative.

t) Based on the choice you picked, what would be the interval notation?

$|3x - 2| < -9$

u) We know that an absolute value is always positive. So, if the left side is always positive, which of the following must be true?

   a) A positive is never less than a negative.
   b) A positive is sometimes less than a negative.
   c) A positive is always less than a negative.

v) Based on the choice you picked, what would be the interval notation? (Hint: this one really isn't an interval but rather an answer.)

$|5x - 10| \leq 0$

w) This absolute value can never be a negative, but it can be zero. So, the solution is only the number which would make it zero. What is that number?

$|2x - 14| > 0$

x) This absolute value is greater than zero everywhere except at one number. It excludes the number which makes the equation equal to zero. What number makes this equal to zero?

y) The interval notation would be all numbers except that single number. Write the interval notation. (Hint: This will require a union and parenthesis.)

Solve the following absolute value inequalities and give their solution sets in interval notation, if possible. (Notice that they are the special cases, so they are logic puzzles, not algebra problems.)

z) $|2x + 12| < -3$ aa) $|5x - 25| > 0$

bb) $|3x + 21| \leq 0$ cc) $|2x - 11| > -6$

These special cases only occur when there is nothing left but the absolute value on one side. On this problem, you first move the 10 to the other side, which "fixes" the problem and prevents the special case from occurring. Solve this inequality and give your answer in interval notation. (It will involve fractions.)

dd) $|2x - 3| - 10 > -2$

If you recall, an absolute value is asking the question, "How far away is this number from zero?"

$$|3| = 3$$

$$|-3| = 3$$

Both 3 and -3 are three away from zero.

When it becomes an algebra problem, we are asking, "What numbers are this many steps from zero?"

$$|x| = 3$$

There are two, $x = 3$ or $x = -3$.

But we can change the center. It doesn't have to be zero. This changes the question to, "What numbers are four steps from the center? "

$$|x - center| = 4$$

For instance:

$$|x - 2| = 4$$

This is asking what two numbers are four steps away from 2. Therefore, $x = 6$ or $x = -2$.

Finally, we could use a $\leq$ to change the question to, "What numbers are within four steps from the number 2?"

$$|x - 2| \leq 4$$

The answer would be the interval notation $[-2, 6]$.

Write an equation which answers the following:

ee) Use an absolute value equation to describe all x values that are within a distance of 5 from the number 3.

ff) Use an absolute value equation to describe all x values that are within a distance of 10 from the number 2.

Algebra II

Active Learning: 3.1a

Solve each of the following equations for the variable y.

a) $4x + 2y = 8$

b) $x^2 + y^2 = 16$

For the following questions, look at the graph and fill in the chart below.

c)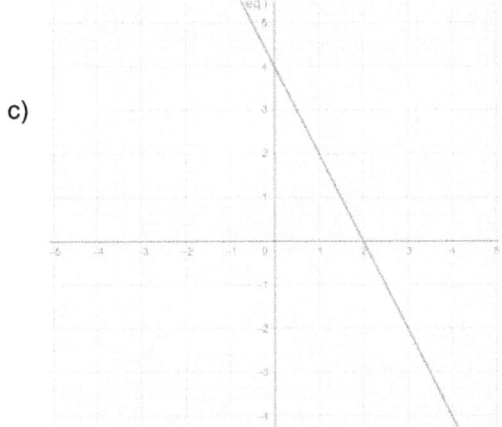

x	y
-1	
0	
1	
2	
3	

d) Fill in the circles below to show the relationship between the x's and the y's. (I've done the x circle for you. At the end of the arrow, write down the y which goes with that x.)

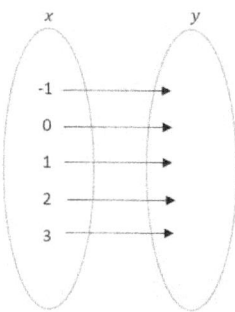

Fill in the chart below. (The y values won't always be whole number. That's okay. Just estimate the best you can. If an x has two partner values, put them both in the box.)

e)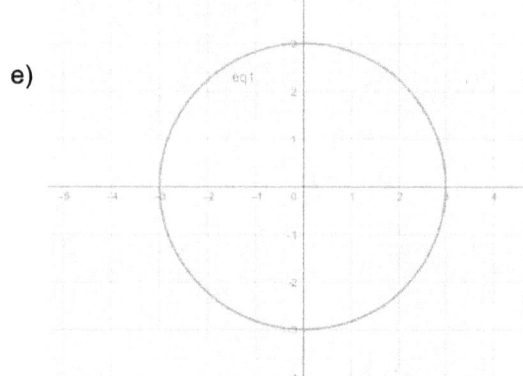

x	y
-2	
-1	
0	
1	
2	

f) Now show the relationship between the x's and y's for the circle.

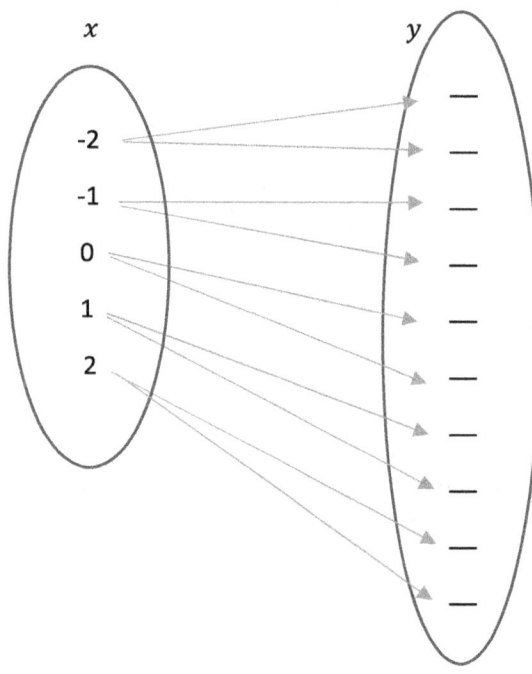

g) What is different about the relationship between the x's and y's in the first graph (the line) and in the second graph (the circle)?

h) A line is called a function (y is a function of x). A circle is not a function (y is not a function of x). The difference is in the relationship between the x's and y's. What does it take to have a function?

A function is a special kind of relationship. (In math, we just call it a relation.) First, the relation is set up like a machine, with the equation solved for y:

$$y = 3x + 2$$

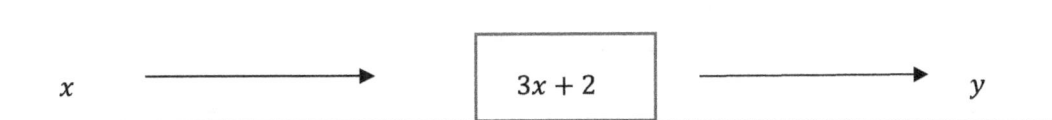

Next, x goes into the machine. Then, it gets calculated. Here, it is multiplied by 3 and then has 2 added. Finally, the transformed number comes out the other side of the machine as the value of y.

You have a function if only one value always comes out the other side of the machine. It wouldn't be a "functional" machine if you put something in one side and miraculously two things came out the other side.

A line is a function. For any value of x which you put in the machine, only one value of y will come out. However, a circle is not a function. When it is set up like a machine, you get:

$$y = \pm\sqrt{16 - x^2}$$

If you put an x into the machine, you get two values out for y. That is not a "functional" machine, and therefore, not a function.

If you have a function then we say that the "output is a function of the input." So, in the case of a line, y is a function of x.

The idea can be extended to any context. This is the price of used cars at a dealer:

Car	Price
2011 Lexus	11,999
2018 Chevy	14,999
2015 Toyota	13,999
2009 Ford	7,999

i) Is price a function of car? (In other words, if you start with the car, do you get out more than one price. If so, it is not a function. If you only get one price, it is a function. For instance, does any car on the list ever have two prices at the same time?)

Here is a menu at a local food stand:

Item	Cost
Hot Dog	2.99
Hamburger	3.99
Grilled Cheese	2.99
French Fries	1.99

j) Is cost a function of item? (In other words, if you start with the item, do you get out more than one cost. If so, it is not a function. If you only get one cost, it is a function. For instance, does any item on the list ever have two costs at the same time?)

k) Is item a function of cost? (In other words, if you start with the cost, do you get out more than one Item. If so, it is not a function. If not, it is a function. For instance, does any cost ever have two items at the same time?)

When something is a function, mathematicians wanted it to be clear and so they gave functions a special notation. This is the equation for a line:

$$y = 3x + 2$$

Lines are functions and so it gets the function notation:

$$f(x) = 3x + 2$$

Students can have a lot of trouble with this idea. But, $f(x)$ is still y. So, both of these indicate an ordered pair from the line:

$$(x, y)$$
$$(x, f(x))$$

Here is the reason that mathematicians thought this confusing notation was worth it:

$$f(2)$$

This tells us that we are set up like a machine. And that the number 2 is going in for x.

$$f(2) = 3(2) + 2$$

And here, we see that 2 went in for x and 8 came out the other side.

$$f(2) = 8$$

But 8 is still y, just like it always was. So, the ordered pair for this line is:

$$(2, 8)$$

Answer the following for this function:

$$f(x) = 5x - 2$$

l) $f(3) =$

m) $f(5) =$

n) $f(-2) =$

All of your answers were still values of y.

Anytime an x goes into the machine and only one value of y comes out, it is a function. This can happen for more complicated machines. Answer the following for this function:

$$f(x) = x^2 - 2x + 3$$

o) $f(1) =$

p) $f(5) =$

q) $f(-1) =$

A function doesn't have to be a story about x and y. It can be a story about anything. So, the notation can change:

$$h(p)$$

This would mean that h is a function of p. So, we could have:

$$h(p) = 3p + 2$$

```
p  ———————→  [ 3p + 2 ]  ———————→  h
```

When we plug a value into the machine, we are evaluating the function. Evaluate the following function:

$$h(p) = 2p^2 - 4$$

r) $h(1) =$

s) $h(-3) =$

Finally, although it may not be obvious why this would be helpful, we can put variables inside of a function machine.

$$f(x) = x^2 - 5x$$

$$f(a) = (a)^2 - 5(a) = a^2 - 5a$$

And we can even put whole expressions. (This is an idea which we will need more of in Calculus.)

$$f(a - h) = (a - h)^2 - 5(a - h) = (a - h)(a - h) - 5a + 5h = a^2 - 2ah + h^2 - 5a + 5h$$

Notice, that I had to F.O.I.L. out $(a - h)^2$.

Try a problem on your own.

$$f(x) = x^2 + 3x + 1$$

t) $f(a) =$

u) $f(a - h) =$

Algebra II

Active Learning: 3.1b

In our last activity, we learned that a function is a special type of relation(ship). Two variables are set-up like a machine, with an x variable being what goes into the machine, and the y variable being what comes out of the machine. If an x goes into the machine and only one value comes out the other side for y then we have a "functional" machine and therefore a function.

It turns out that having the graph of a relation makes it very easy to determine whether or not we have a function. Here is the graph of a circle. I've added a vertical line.

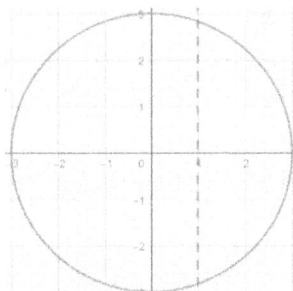

We saw previously that circles are not functions. The vertical line that I've added proves that. Vertical lines are x-lines. (The equation of the line I've added is $x = 1$.) When x is equal to 1, we hit the circle twice. That means when x is equal to 1 there are two values of y. Functions don't have two values of y for one value of x, so it isn't a function.

Look at the following graphs. Imagine drawing a vertical line. If at any point, a vertical line would hit the graph in two spots it isn't a function. If a vertical line would never hit in two spots, it is a function. Below each graph, write "function" or "not a function," depending on this vertical line test.

a)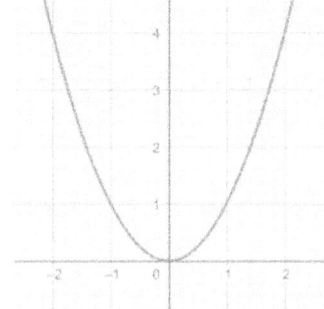

b)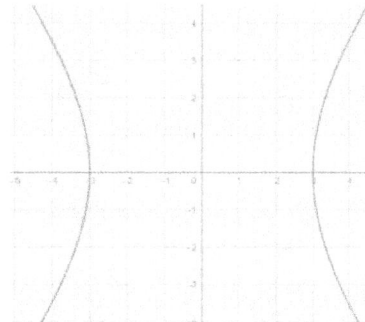

c) 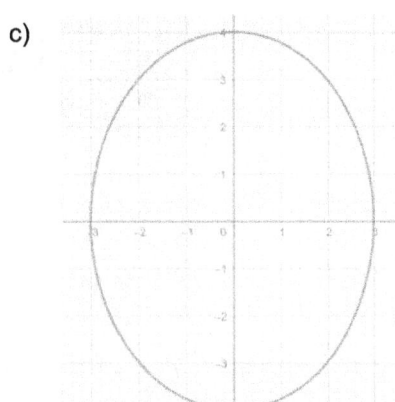   d)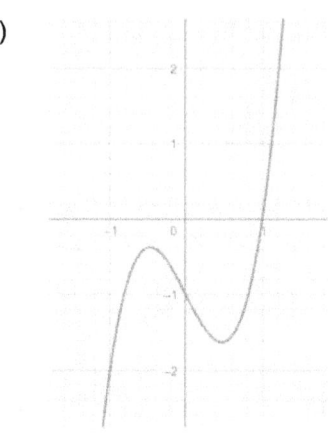

Next, we want to discuss a second kind of test called the horizontal line test. I think it would be best to ignore this test at this point, however the textbook brings it up here, so we will too. As we've explained a function, we've said when an x goes into the machine only one y comes out. Look at this function:

x	y
2	12
3	17
4	21
5	12

e) What is happening with the output when x equals 2 and when x equals 5?

The table still represents a function; each x has only one y. But mathematicians also give a name to a function where there is a unique x which leads to a unique y. They call such a function a one-to-one function.

x	y
2	12
3	17
4	21
5	12

This table is a function, but it is **not** a one-to-one function.

Look at this table:

x	y
-5	8
-2	12
0	7
3	11

This table is a function. No x leads to more than one y. But it is also a one-to-one function. No two x's ever give the same value of y. It is easy to find a one-to-one function if you have the graph. Just as a vertical line would tell us if any value of x ever had two values of y, a horizontal line would tell us if any value of y ever has two values of x.

A one-to-one function will also pass a horizontal line test. The function below is one-to-one.

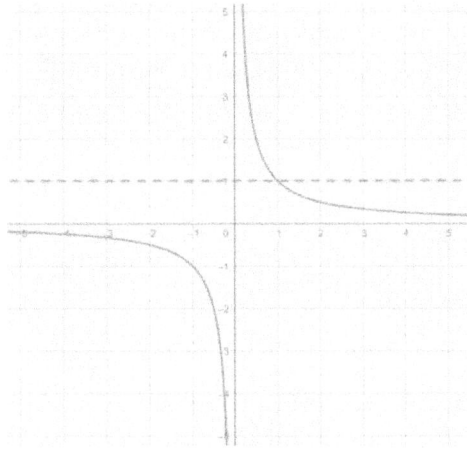

No matter where we drew a horizontal line, we will never touch the graph in two places at once. All of the following graphs are functions. Circle them if they would also pass the horizontal line test, making them one-to-one functions:

f)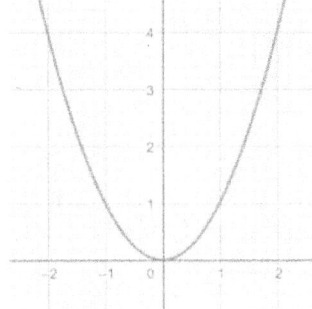

g)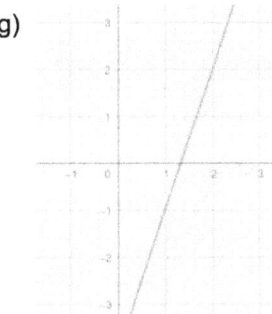

h)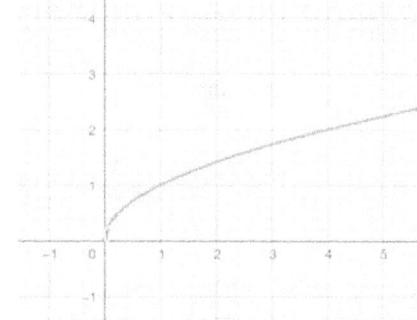

i)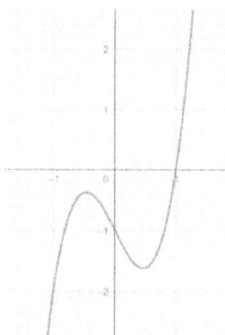

Finally, we want to read function values from a graph. It is pretty straightforward, provided you remember the notation. Let's look at the graph of a square root function. The notation $f(4)$ means that 4 was put into the machine as x. Then, we read the graph like normal.

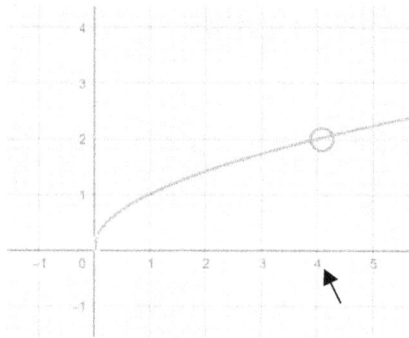

When we put in 4, we get out a y of 2. So, we have:

$$f(4) = 2$$

Evaluate the same function at 1.

j) $f(1) =$

Now, evaluate this quadratic function as indicated:

k) 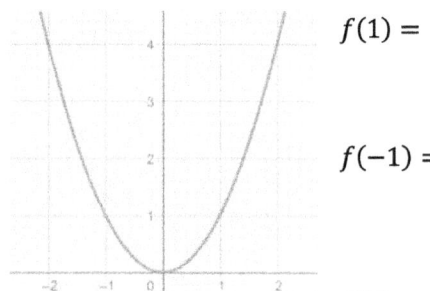 $f(1) =$

l) $f(-1) =$

m) $f(2) =$

Our last type of problem asks you to go backward. You will know the y value and you must give the x value which was put into the machine to make it.

n)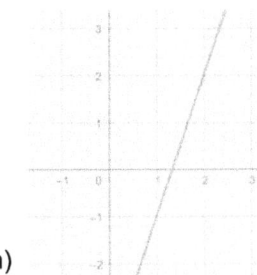

Use the graph to the left to solve the following: (I've done the first one for you.)

Solve $f(x) = -1$

The answer would be $f(1)$. 1 is the input which makes an output of -1.

Solve $f(x) = 2$

On this last problem, they have asked you to solve for the input. Be careful, there are two values of x for each problem.

o)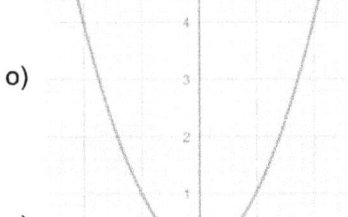

Solve $f(x) = 1$

p)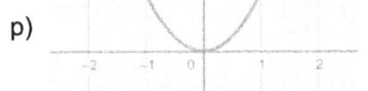

Solve $f(x) = 4$

Algebra II

Active Learning: 3.2a

Two of the following expressions have a mathematical "problem." Circle the two with a problem.

a) $\dfrac{175}{0}$

b) $\sqrt[3]{-8}$

c) $\sqrt[2]{-6}$

d) $\dfrac{1}{\sqrt{9}}$

Look at the following functions. Two of the functions could "break" if they took on x values that would create a problem similar to what we saw in the previous section. Circle the two which could "break."

e) $h(x) = \sqrt[2]{x}$

f) $g(x) = |x|$

g) $f(x) = \dfrac{1}{x}$

As we've discussed, a function is a machine. We put the values of x into the machine and get the values of y out the other side:

$$f(x) = 3x + 2$$

x $\longrightarrow$ $\boxed{3x+2}$ $\longrightarrow$ y

This machine is a line. Any value of x can go into it without a problem. But that isn't the case for every machine.

Here is the function $\frac{1}{x}$:

$$f(x) = \frac{1}{x}$$

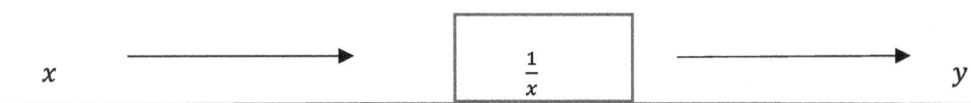

Not every value of x can go into the machine. If $x = 0$ then the machine would break. It would make the function undefined, which isn't allowed.

The domain of a function are the acceptable values which are allowed to be put into the machine. And, we write the domain in interval notation. Interval notation is a topic we look at in Algebra I, but here is a quick review.

-We list the number from left to right, going up the number line.

-The negative end of the number line is negative infinity: $-\infty$.

-The positive end of the number line is positive infinity: ∞.

-If we want to include a number on one end of the interval notation, we use a].

-If we don't want to include a number on one end of the interval notation, we use a).

-Positive and negative infinity always get a) because we can never reach them.

Match the following with their proper interval notation:

h)   All Real Numbers $(5, \infty)$

i) $x > 5$ $[5, \infty)$

j) $x \geq 5$ $(-\infty, 5]$

k) $x < 5$ $(-\infty, \infty)$

l) $x \leq 5$ $(-\infty, 5)$

In the set of functions, I'm going to give you the domain. Look at the function and explain why the domain has been limited.

m) $f(x) = \sqrt[2]{x-5}$ Domain: $[5, \infty)$

n) $f(x) = \frac{1}{x+3}$ Domain: $(-\infty, -3) \cap (-3, \infty)$

At this point in the course, we will only look at two ways a function "machine" can break: square roots and denominators. Unfortunately, we work each type of problem in an opposite fashion.

Denominators

To find the domain of a function with a denominator, we find the values that are **not** allowed.

$$f(x) = \frac{3x}{2x-4}$$

Set the denominator equal to zero in order to find the values that will break the machine.

$$2x - 4 = 0$$
$$2x = 4$$
$$x = 2$$

The value of 2 is not allowed into the machine or the machine will break. So, the domain is all the numbers but 2.

$$(-\infty, 2) \cup (2, \infty)$$

Notice that I didn't do anything with the $3x$ in the numerator. It is part of the machine, but it is not a part which could ever cause the function to break.

Try these. Write the domain for the following functions:

o) $f(x) = \frac{5x}{x-8}$

Domain:

p) $f(x) = \dfrac{x+9}{3x-12}$

Domain:

q) $f(x) = \dfrac{x}{4-2x}$

Domain:

Square-roots

To find the domain of a function with a square-root, we find the values which **are** acceptable.

$$f(x) = 3x + \sqrt{4-2x}$$

The $3x$ in the equation will not break the function. (Any value of x is acceptable there.) So, we focus on the square root. The value under a square root can't be a negative number. Therefore, we make an inequality:

$$4 - 2x \geq 0$$

This says the portion under the square root can't be a negative number. Now we just solve for x. (Be careful, this follows normal inequality rules. At one point, we will divide by a negative so the inequality sign will flip.)

$$-2x \geq -4$$

$$x \leq 2$$

In interval notation, the domain of the function is $(-\infty, 2]$.

Try these. Write the domain for the following functions:

r) $f(x) = \sqrt{3x+9}$

Domain:

s) $f(x) = 4x + \sqrt{6 - 2x}$

Domain:

t) $f(x) = x^2 - \sqrt{4x - 10}$ (You will get a fraction here and that is okay.)

Domain:

Both Denominators and Square-Roots

The most difficult problems in this section will have multiple issues which could break the function.

$$f(x) = \frac{\sqrt{x - 2}}{x - 3}$$

First, find the domain for the square root:

$x - 2 \geq 0$

$x \geq 2$

Domain: $[2, \infty)$

Second, find the domain for the denominator:

$x - 3 = 0$

$x = 3$

Domain: $(-\infty, 3) \cup (3, \infty)$

Now, we need the domain for the entire function. It will be the values of x which won't break either part of the machine. Here, I've marked the two domains on the same number line.

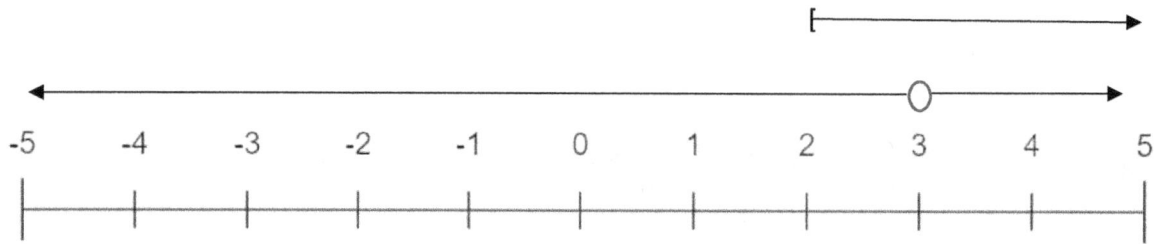

The domain for the entire function includes those number which are acceptable to both machines. So, this is where the two domains overlap:

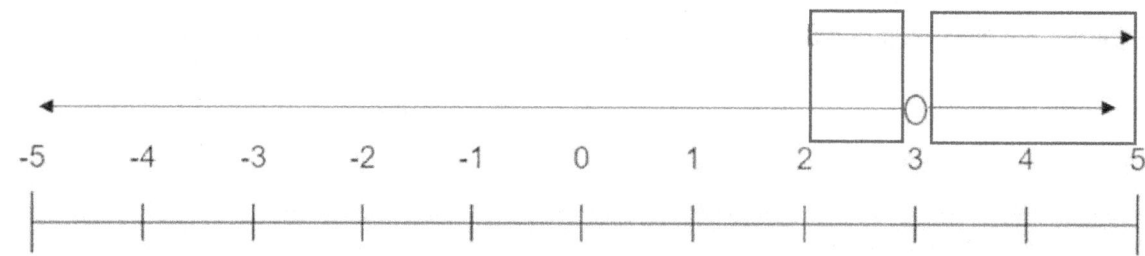

Domain: $[2, 3) \cup (3, \infty)$

The two domains start to overlap at the number 2. They continue to overlap everywhere to the right, except at 3.

Try one on your own.

u) Find the domain of the following function. (If you need it, I've included a number line to help.)

$$f(x) = \frac{\sqrt{2x - 6}}{x - 4}$$

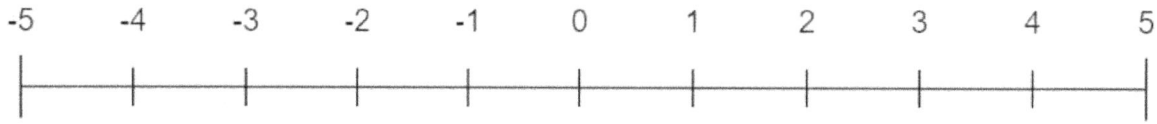

Domain:

If the domain are the values which go into the function machine, the range are the values which come out. Finding these values from a graph is much easier than doing it mathematically.

Domain

To find the domain, imagine a vertical line running from left to right across the x-axis.

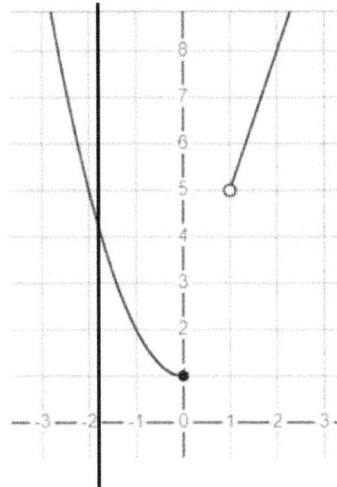

- Although we can't see it, my vertical line would hit the graph starting at negative infinity.
- As I move across the graph, the vertical line would stop at 0. The dot at 0 is filled-in, so I would use a bracket.
- Then, my vertical line would begin hitting again at 1. This is an open circle, so I would use a parenthesis.
- Again, although the graph doesn't show it, my vertical line would continue hitting the graph until positive infinity.

Domain: $(-\infty, 0] \cup (1, \infty)$

Range

For the range, we make a horizontal line. Now, we move from down to up.

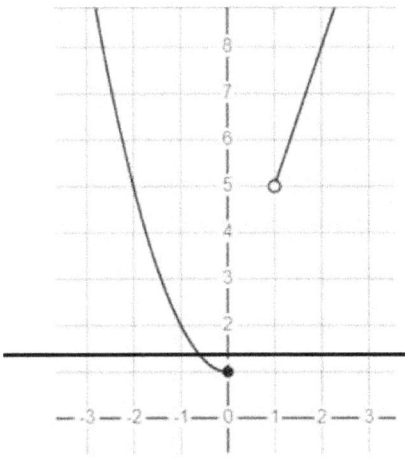

- As we move up the y-axis, we would first hit our graph at 1. It is a filled in circle and so we would give it a [
- As we continue moving up, we would hit *something* the rest of the way. (There is an open circle at 5, but we hit on the other side.)
- Students struggle with the range. We are no longer looking at the x-values. We are moving up the y-axis and are concerned about the y-values.

Range: $[1, \infty)$

Try these on your own.

Find the domain and range of the following graphs. (Be sure to give them in interval notation.)

v) 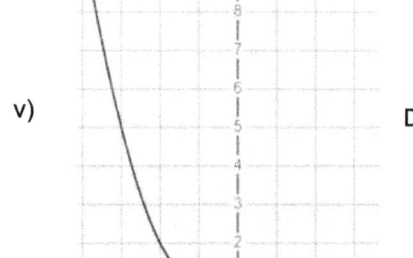   Domain:

w)   Range:

x) 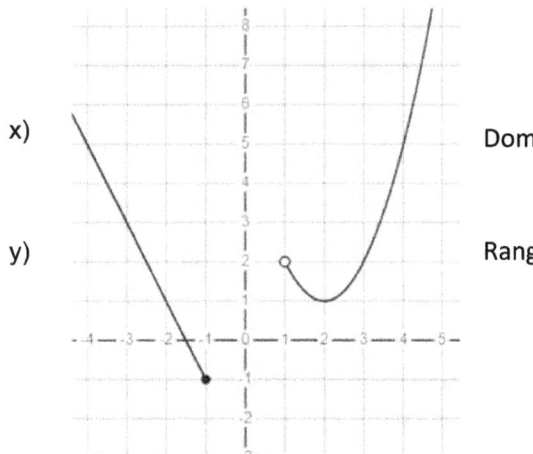   Domain:

y)   Range:

Algebra II

Active Learning: 3.2b

As we move forward, we will repeatedly work with a number of key functions. And, the basic shapes of these functions will be important to have memorized. Below, I will give you these key functions. If you are unsure of their shape, use Geogebra or Desmos to help you graph them. Again, you will be responsible to know these shapes.

Identity Function

a) $f(x) = x$

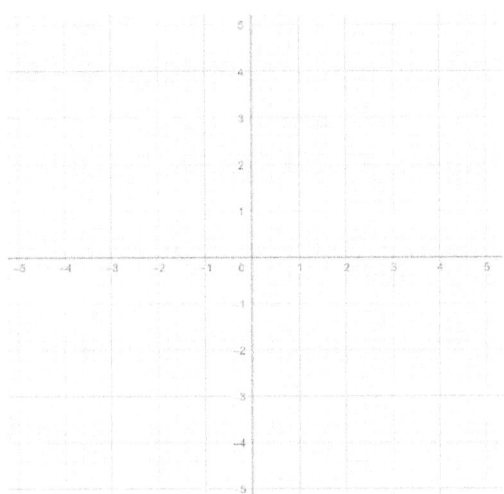

Absolute Value Function

b) $f(x) = |x|$

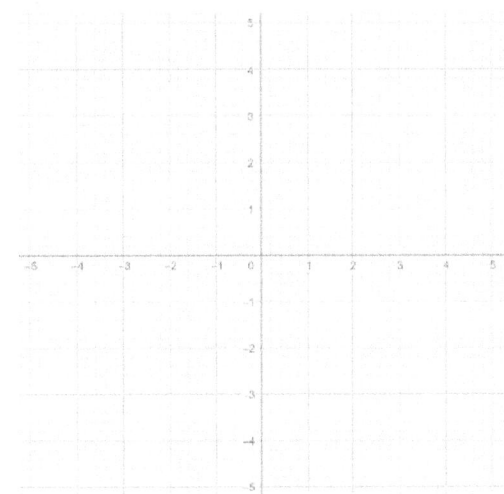

Quadratic Function

c) $f(x) = x^2$

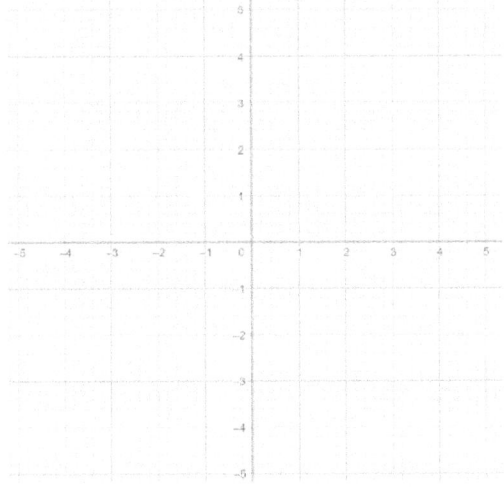

Cubic Function

d) $f(x) = x^3$

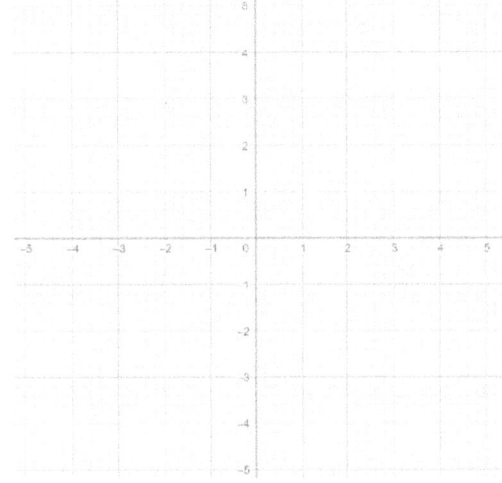

Reciprocal Function

e) $f(x) = \frac{1}{x}$

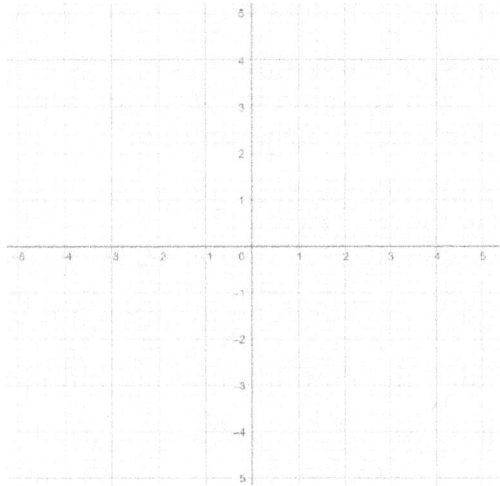

Square Root Function

f) $f(x) = \sqrt{x}$

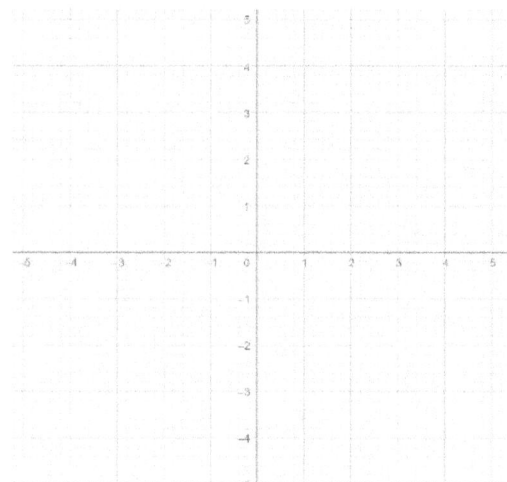

Reciprocal Squared Function

g) $f(x) = \frac{1}{x^2}$

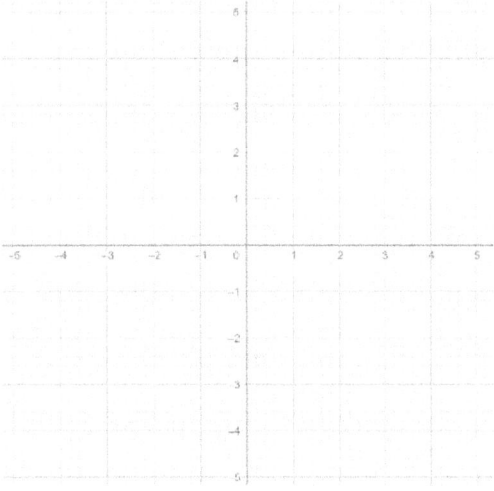

Cube Root Function

h) $f(x) = \sqrt[3]{x}$

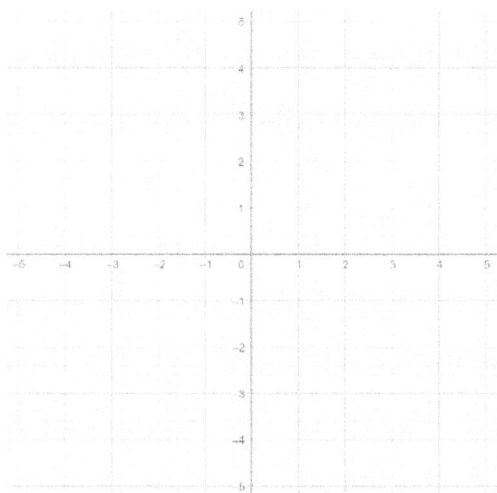

Piecewise Functions

Sometimes functions are made up of "pieces" of other functions and they are merged together over different domains. Here's an example:

$$f(x) = \begin{cases} -2x + 1 & x < -2 \\ \frac{1}{2}x + 1 & 0 \le x \end{cases}$$

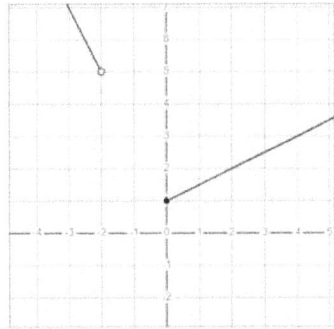

This piecewise function is made up of two different lines. One line is true when values of x are less than -2. The other line is true when value of x are greater than or equal to 0.

To graph a piecewise function, the key is to mark the boundaries of the domains. I've put the boundaries in a rectangle. We want to mark what happens when $x = -1$. And, we want to mark what happens when $x = 0$.

$$f(x) = \begin{cases} -3x - 5 & \boxed{x < -1} \\ \frac{3}{2}x + 1 & \boxed{0 \le x} \end{cases}$$

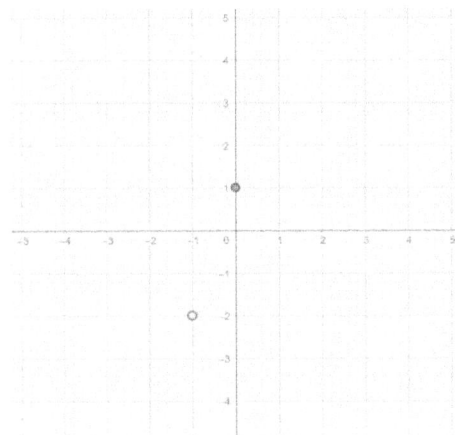

To find what happens when $x = -1$, we put -1 into the top equation. We get: $-3(-1) - 5 = -2$. On the graph, I marked the point $(-1, -2)$. Notice that I've put an open circle, for our inequality is $x < -1$ and it doesn't include an equal sign. This is our top boundary.

To find what happens when $x = 0$, we put 0 into the bottom equation. We get: $\frac{3}{2}(0) + 1 = 1$. On the graph, I marked $(0, 1)$. Here I've put a closed circle, for our inequality is $0 \le x$, and it does include an equal sign. This is our bottom boundary.

$$f(x) = \begin{cases} \boxed{-3x - 5 \quad x < -1} \\ \frac{3}{2}x + 1 \quad 0 \leq x \end{cases}$$

The top piece of the function is happening to the left of the boundary point $(-1, -2)$, since we want x's less than -1. So, I've graphed the line $y = -3x - 5$. Again, only the portion going to the left.

$$f(x) = \begin{cases} -3x - 5 \quad x < -1 \\ \boxed{\frac{3}{2}x + 1 \quad 0 \leq x} \end{cases}$$

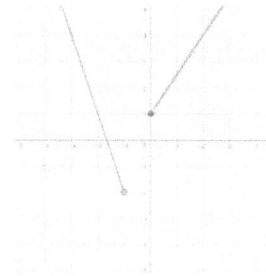

The bottom piece of the function is happening to the right of the boundary point $(0, 1)$, since we want x's greater than or equal to 0. So, I've graphed $y = \frac{3}{2}x + 1$. Again, only the portion going to the right.

Try one on your own. Graph the following piecewise function. I'll help with the boundary points.

$$f(x) = \begin{cases} -2x - 1 \quad x < -2 \\ \frac{1}{2}x + 1 \quad 0 \leq x \end{cases}$$

i)

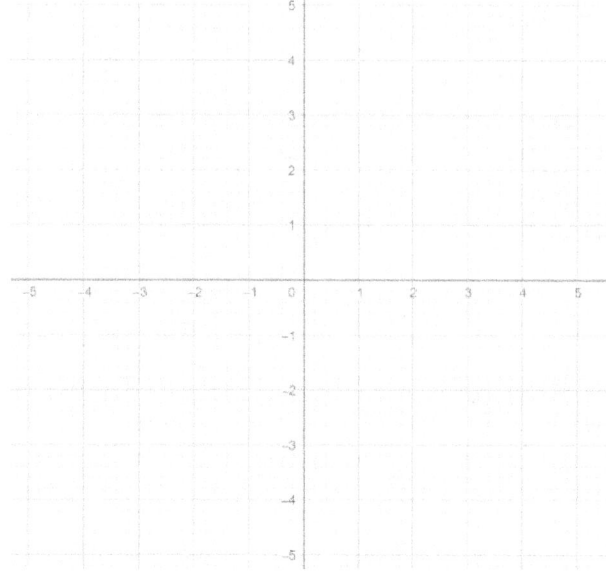

Your first boundary will be an open circle at $(-2, 3)$.

Your second boundary will be a closed circle at $(0, 1)$.

Piecewise functions aren't limited to two "pieces." This function has three.

$$f(x) = \begin{cases} -x+2 & x \leq -2 \\ 3 & -2 < x < 1 \\ x-1 & 1 < x \end{cases}$$

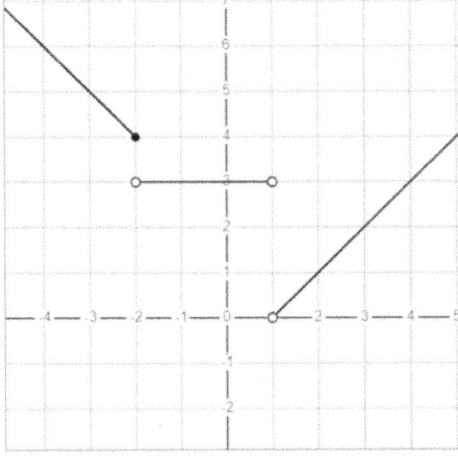

There would be three boundaries. Putting -2 into the top gets: $-(-2)+2 = 4$. And makes a boundary point $(-2, 4)$.

The middle has two boundaries, but since it is a constant function, we get the points: $(-2, 3)$ and $(1,3)$. These are both open circles because there are no equal signs.

Putting 1 into the bottom gets: $1 - 1 = 0$. And makes a boundary point $(1, 0)$. It is an open circle because it doesn't include an equal sign.

The top function is then graphed to the left. The middle function is graphed between -2 and 1. And, the bottom function is graphed to the right.

Graph the following piecewise function:

$$f(x) = \begin{cases} 2x+5 & x \leq -2 \\ 2 & -2 < x < 2 \\ 2x-4 & 2 \leq x \end{cases}$$

j)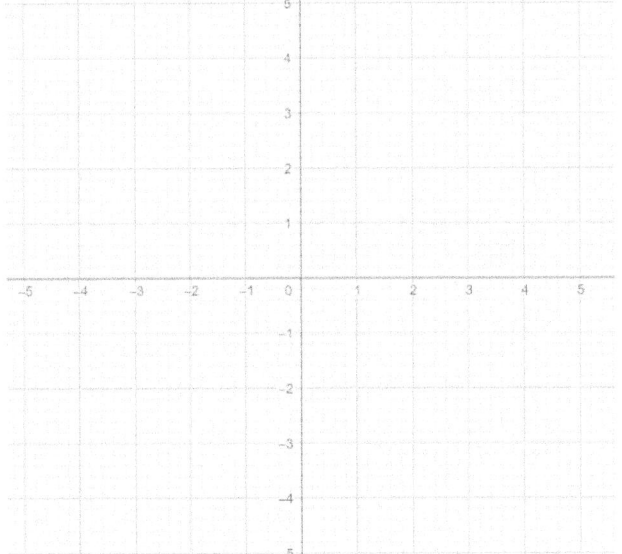

Finally, suppose we were asked to evaluate a piecewise function.

$$f(x) = \begin{cases} -x & x \leq -2 \\ 5 & -2 < x < 2 \\ 3x - 2 & 2 \leq x \end{cases}$$

Evaluate the function at the following values:

$f(-4) =$ We look to see which "piece" of the domain this would fit into. Because our x value is -4, it would fit into the top.

So, we will evaluate using the top equation: $-(-4) = 4$
$f(-4) = 4$

Work the following.

k) $f(5) =$ Here, we would use the bottom equation.

l) $f(0) =$ Here, the middle equation.

m) $f(10) =$

n) $f(-3) =$

o) $f(-2) =$ This one is tricky. We would want the top equation.

p) $f(1) =$

q) $f(2) =$ This one is also tricky. Be careful.

Algebra II

Active Learning: 3.3a

Below is the slope-intercept form of the equation for a line.

$$y = mx + b$$

We want to find the equation of the line between the two points below.

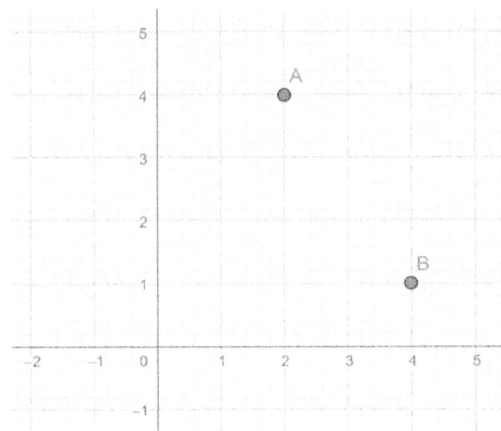

a) What are the (x, y) coordinates of point A?

b) What are the (x, y) coordinates of point B?

c) What is the slope between the two points?

d) Now that you know the slope of the line, you can use the slope intercept equation. Plug in either point A or point B, and find the missing y-intercept. Find the value of the b from the slope intercept equation.

e) What is the final equation for the line?

The graph below represents the height of a ball after it has been thrown from the top of a 5 km cliff. We want to know how, on average, the ball is changing between points A and B.

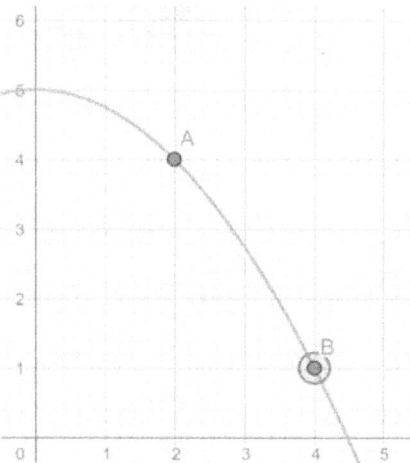

f) What are the coordinates of point A?

g) What are the coordinates of point B?

Do your best to add a straight line to the graph between points A and B.

h) Although the graph was not of a line, we are interested in knowing the average change (called the average rate of change) between the two points. To find it, we simply need to find the slope of a line between the points. (Notice these are the same two points from the earlier problem.) Use the slope formula and the coordinates of A and B to find the average rate of change.

You will be asked to find the average rate of change with only the equation for the graph. A problem would look like this.

Find the average rate of change for the function $f(x) = x^2 - 3$ on the interval [1,4].

The interval here is a domain, both numbers are x values. To find the average rate of change, we need the points at the end of the interval. This is easy. If we have the x values, we can simply plug them into the function to find the y-coordinate which is its partner.

Find the y-values at the endpoints.

i) $f(1) =$

Ordered Pair: (1,   )

j) $f(4) =$

Ordered Pair: (4,   )

k) Now that you have the ordered pairs, find the average rate of change. (Use your two points and the slope formula.)

Try another on your own.

l) Find the average rate of change of the function $f(x) = x - \sqrt{x}$ on the interval [1, 9].

In this next problem, one of the endpoints is unknown. That is okay. Simply evaluate the function with the variable and continue using the slope formula.

Find the average rate of change of the function $f(x) = x^2 + x$ on the interval $[2, a]$.

The points are:

$(2, 2^2 + 2) = (2, 6)$

And:

$(a, a^2 + a)$

Use the slope formula to find the average rate of change.

$$\frac{(a^2 + a) - 6}{a - 2} = \frac{a^2 + a - 6}{a - 2}$$

m) This can be factored and simplified. Do so in the space below:

Try one on your own.

n) Find the average rate of change of the function $f(x) = x^2$ on the interval $[3, a]$.

Algebra II

Active Learning: 3.3b

We read a graph just like we would read a book, moving from left to right. Imagine a car is driving on the graph from left to right and answer the following questions.

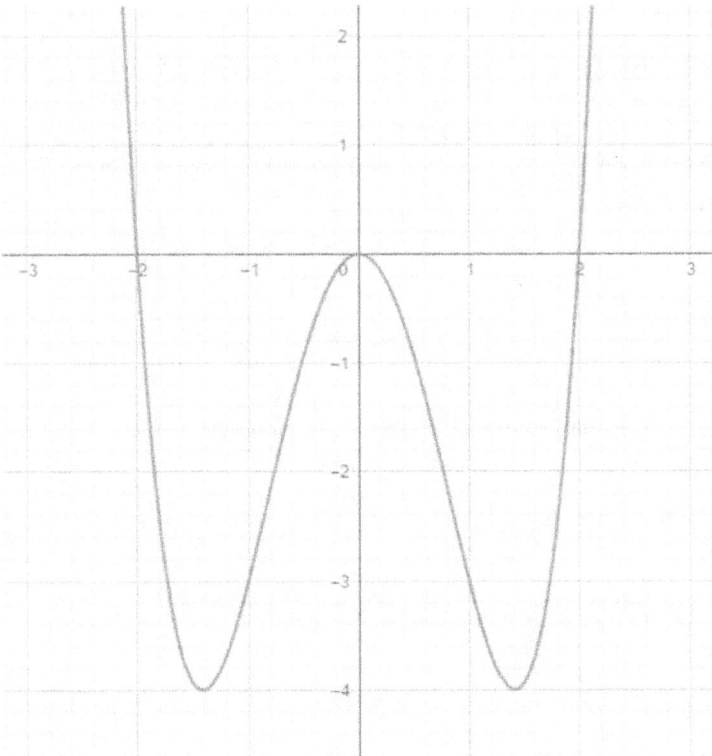

a) What are the coordinates for any points that are at the top of a hill?

b) What are the coordinates for any point that is at the bottom of a valley? (The x coordinates aren't perfectly clear. Just make your best guess.)

c) I want you to state the intervals on which you would be driving up hill. This is where the function is increasing. It works just like when we did domains. It is only the x-values which matter. Put your answers in interval notation.

d) In interval notation, write the intervals on which you would be driving downhill. (Decreasing.)

Extrema is the name which mathematicians give to the high and low points on a graph. Absolute extrema are the absolute highest (called absolute maxima) or lowest points (called absolute minima) on a graph. Relative extrema are any high (called local maxima) or low points (called local minima) on the graph, not necessarily the absolute highest or lowest.

e) Look at the point you indicated for the top of the hill. How should it be classified: absolute maxima, absolute minima, local maxima, or local minima?

f) Look at the points you indicated for the bottom of the hills. How should they be classified: absolute maxima, absolute minima, local maxima, or local minima? (It is okay that there are two points at the same place. They both tie for the classification.)

Try another.

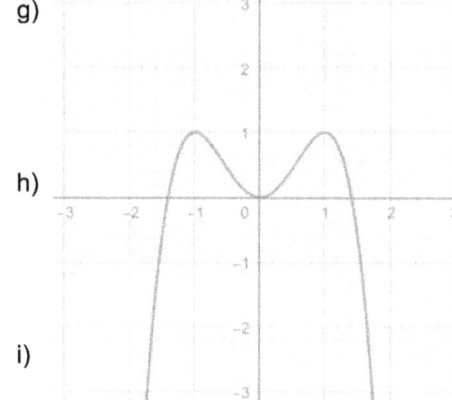

g) Give any intervals on which the graph is increasing.

h) Give any intervals on which the graph is decreasing.

i) If there are any absolute maxima, give their coordinates. If there are none, say none.

j) If there are any absolute minima, give their coordinates. If there are none, say none.

k) If there are any local maxima, give their coordinates. If there are none, say none.

l) If there are any local minima, give their coordinates. If there are none, say none.

Work one final problem.

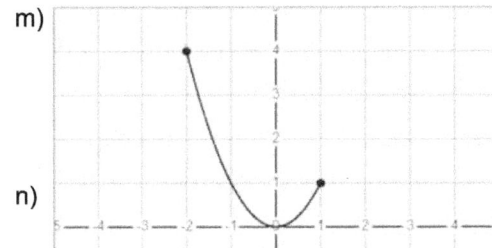

m) Give any intervals on which the graph is increasing.

n) Give any intervals on which the graph is decreasing.

o) If there are any absolute maxima, give their coordinates. If there are none, say none.

p) If there are any absolute minima, give their coordinates. If there are none, say none.

q) If there are any local maxima, give their coordinates. If there are none, say none.

r) If there are any local minima, give their coordinates. If there are none, say none.

Algebra II

Active Learning: 3.4a

Suppose I gave you these two functions.

$$f(x) = x$$
$$g(x) = 5$$

You can add these two functions, and it really is as easy as it seems. The only confusing part is the notation. Here are the two functions added together.

$$(f + g)(x) = x + 5$$

Now, find the following. Don't overthink it.

a) $(f - g)(x) =$

b) $(f \cdot g)(x) =$

c) $\left(\dfrac{f}{g}\right)(x) =$

Try another. Just be sure to simplify anything which can be simplified.

$$f(x) = x^2$$
$$g(x) = x$$

Find:

d) $(f + g)(x) =$

e) $(f - g)(x) =$

f) $(f \cdot g)(x) =$

g) $\left(\dfrac{f}{g}\right)(x) =$

Here is a problem where the functions are a bit more difficult:
$$f(x) = x^2 - 4$$
$$g(x) = x - 2$$

Find:

h) $(f + g)(x) =$

i) $(f - g)(x) =$

j) $(f \cdot g)(x) =$

On this next problem, be sure to factor and simplify.

k) $\left(\dfrac{f}{g}\right)(x) =$

Look at the following two functions and give me their domains:

l) $f(x) = \sqrt{x + 2}$

Domain:

m) $g(x) = x$

Domain:

n) Now, give me $(f \cdot g)(x) =$

Since these two functions have now been combined, the domain of the new function $(f \cdot g)(x)$ is where their individual domains overlap. (If you combine the two machines, the new machine has the flaws of both of the individual machines.) Draw the domains of $f(x)$ and $g(x)$ on the number line below.

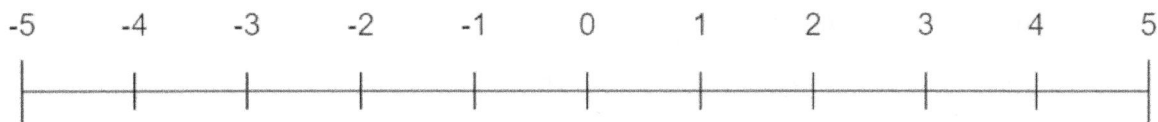

o) What is the domain of $(f \cdot g)(x)$?

p) Finally, find $\left(\frac{f}{g}\right)(x) =$

q) Since, $g(x)$ was put in the denominator, we have created one more number which must be excluded from the domain. (Remember, we can't make a denominator equal to zero.) What is that number?

To find the domain of $\left(\frac{f}{g}\right)(x)$, we must find the intersection of the domains of $f(x), g(x)$, and remove the number which makes the denominator zero. Use the number line below to help you, if necessary.

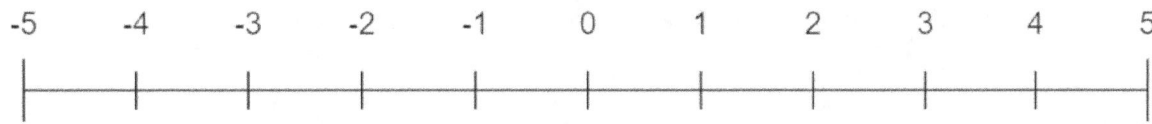

r) Give me the domain of $\left(\frac{f}{g}\right)(x)$ in interval notation.

Composite Functions

Next, evaluate this function.

s)
$$f(5) = 3x - 12$$

Now, put the answer that you just got in as x on this function.

t)
$$g(answer\ from\ before) = 5x + 21$$

When the output of one function rolls directly into (as the input) of another function, we've created a composite function. If a function is like a machine, a composite function is like an assembly line.

As we saw above, the answer from the first function becomes the value we input into the second function. There are two versions of the notation. I will use this one:

$$g(f(x))$$

With this notation, the function on the inside is first in the assembly line. So, $f(x)$ is first. We could flip that:

$$f(g(x))$$

Here, $g(x)$ is first.

The other version of the notation, which I stay away from but some textbooks use, is this:

$$(g \circ f)(x)$$

With this notation, the second function, here $f(x)$, is first in the assembly line.

Evaluate the following composite functions. I'll show you the first one.

$$f(x) = x - 4$$
$$g(x) = 3x$$

$g(f(1)) =$

$$f(1) = 1 - 4 = -3$$
$$g(-3) = 3(-3) = -9$$
$$g(f(1)) = -9$$

u) $g(f(4)) =$

v) $g(f(-2)) =$

On this next problem, the assembly line now starts with g.

w) $f(g(2)) =$

x) $f(g(-3)) =$

Next, we will use graphs to evaluate composite functions. The graphs of $f(x)$ and $g(x)$ are below:

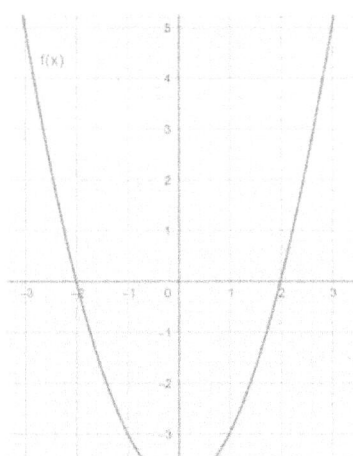 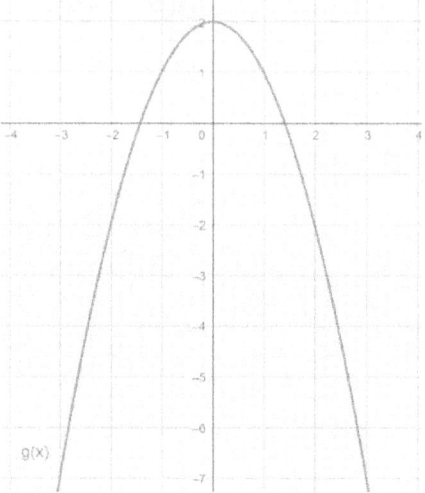

Let's evaluate $g(f(1))$.

The f function is on the inside, so it is first. So, we will evaluate $f(1)$.

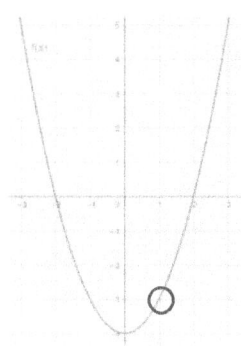

When x equals 1, we get a y equal to -3.

$$f(1) = -3$$

Now, we evaluate $g(-3)$.

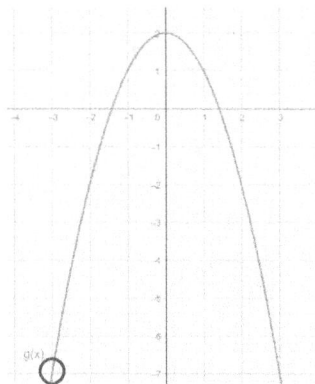

When x equals -3, we get a y value of -7.

$$g(-3) = -7$$

And so:

$$g(f(1)) = -7$$

So, the y value from the first graph becomes the x value on the second graph.

Try these on your own.

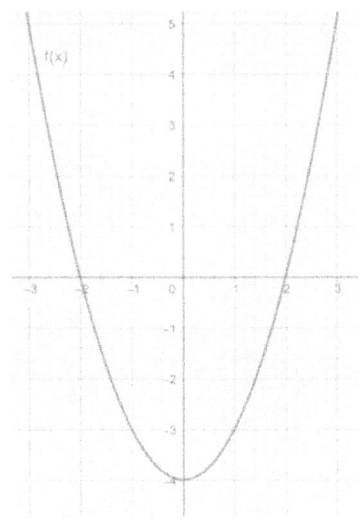 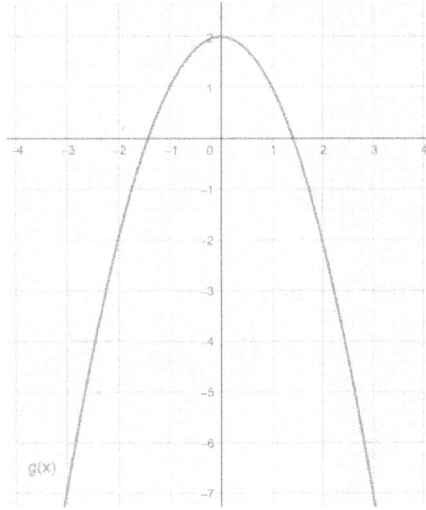

y) $g(f(2)) =$

z) $g(f(-1)) =$

Now, you will begin with the g graph.

aa) $f(g(0)) =$

bb) $f(g(-1)) =$

Finally, we could follow the same idea with a table.

x	f(x)	g(x)
1	3	2
2	1	4
3	2	9
4	8	1

$g(f(1)) =$

We put 1 into the f function and get $f(1) = 3$.

x	f(x)	g(x)
1	③	2
2	1	4
3	2	9
4	8	1

Then we put 3 into the g function and get $g(3) = 9$.

x	f(x)	g(x)
1	3	2
2	1	4
3	2	⑨
4	8	1

Try these:

x	f(x)	g(x)
1	3	2
2	1	4
3	2	9
4	8	1

cc) $g(f(2)) =$

dd) $g(f(3)) =$

Now, you will start with the g function.

ee) $f(g(1)) =$

ff) $f(g(4)) =$

Algebra II

Active Learning: 3.4b

Let's review the idea of a composite function.

Evaluate this function.

a)
$$f(3) = 2x + 4$$

Now, put the answer that you just got in as x on this function.

b)
$$g(answer\ from\ before) = 4x - 6$$

We've just evaluated $g(f(3))$.

c)
$$g(f(3)) =$$

Let's do it again. This time with harder functions. Evaluate this function.

d)
$$f(4) = x^2 + 11$$

Now, put the answer that you just got in as x on this function.

e)
$$g(answer\ from\ before) = \sqrt{x - 2}$$

We've just evaluated $g(f(4))$.

f)
$$g(f(4)) =$$

One more. Suppose I gave you this function, but I didn't give you anything to evaluate.

$$f(x) = x^2$$

We want to put what comes out of that function into this one.

$$g(x) = x + 3$$

So:

$$f(x) = \boxed{x^2}$$

Goes into $g(x)$.

$$g(x^2) = (x^2) + 3 = x^2 + 3$$

Therefore, $g(f(x)) = x^2 + 3$.

Try these on your own.

$$f(x) = x + 1$$
$$g(x) = 2x + 4$$

Find:

g) $\quad g(f(x)) = $ \hfill (Be careful, you have to distribute the 2.)

h) $\quad f(g(x)) = $ \hfill (Now, go the other direction, putting g into f.)

Let's take a look at what would happen with the domain when you build these composite functions.

Give the domain of the following function:

i) $f(x) = x - 2$

Domain:

If $g(x) = \frac{1}{x}$ then the composite function $g(f(x))$ would be:

$$g(f(x)) = \frac{1}{(x-2)}$$

j) What numbers would need to be exclude from this new combined machine?

A composite function is an assembly line. And, on this assembly line, we can't break the first machine *and* we can't break the new combined machine. So, the domain is the intersection of the first machine and the new combined machine. (Remember, domains are written in interval notation.)

k) The domain for the composite function $g(f(x))$ is:

Try another on your own.

$$f(x) = \frac{x-5}{x-3}$$

$$g(x) = \frac{1}{x}$$

Find the domain for the composite function $g(f(x))$. Remember, you need the overlap of the domains for $f(x)$ and the new combined machine $g(f(x))$. You don't need to include the domain for $g(x)$. (Hint: after you put f into g, you will then need to Keep, Change, Flip.)

l) The domain for the composite function $g(f(x))$ is:

Try one involving a square root.

$$f(x) = x^2$$

$$g(x) = \sqrt{x+3}$$

This time, find the domain of $f(g(x))$. You will need the overlap of the domains for g and for the combined machine.

m) What is the domain of $f(g(x))$?

Lastly, we will "de-compose" a composite function. In other words, we will take them apart. For instance, separate $h(x)$ in to $g(x)$ and $f(x)$, where $g(f(x))$.

$$h(x) = \sqrt{x^2 + 1}$$

There are multiple correct answers to a problem like this. However, simply select a complicated portion of the function involving x, and we will pull it out.

$$h(x) = \sqrt{\boxed{x^2 + 1}}$$

We'll make this $f(x)$.

$$f(x) = x^2 + 1$$

Then, replace what we removed with x.

$$g(x) = \sqrt{x}$$

If you aren't sure that you got it right, simply rebuild the composite function and see if you get back to where you started. Try these:

n) Separate $h(x)$ in to $g(x)$ and $f(x)$, where $g(f(x))$.

$$h(x) = \frac{1}{\sqrt{x}}$$

o) Separate $h(x)$ in to $g(x)$ and $f(x)$, where $g(f(x))$.

$$h(x) = 3 + \sqrt{x}$$

p) Separate $h(x)$ in to $g(x)$ and $f(x)$, where $g(f(x))$. (This one is a bit harder. The value of $f(x)$ you pull out will be replaced by x in two places.)

$$h(x) = \frac{x+1}{2 - \sqrt{x+1}}$$

Algebra II

Active Lesson: 3.5a

Here are six of our toolkit functions which we will be using for this lab.

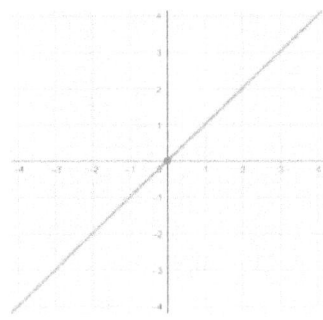

Identity Function: $y = x$

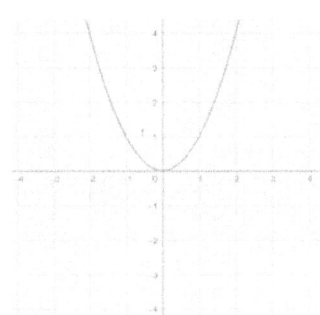

Square Function: $y = x^2$

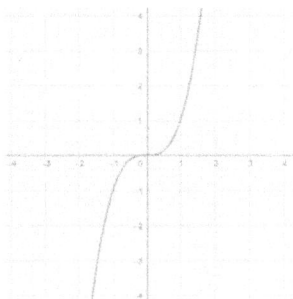

Cube Function: $y = x^3$

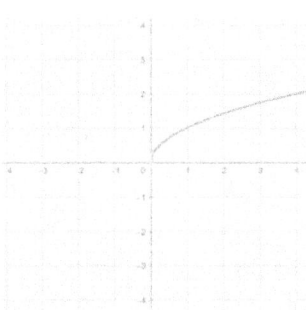

Square Root Function:

$y = \sqrt{x}$

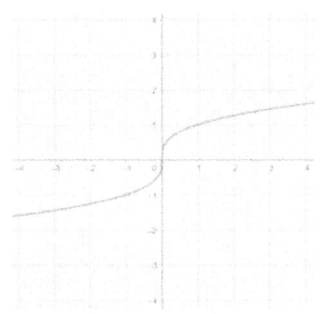

Cube Root Function:

$y = \sqrt[3]{x}$

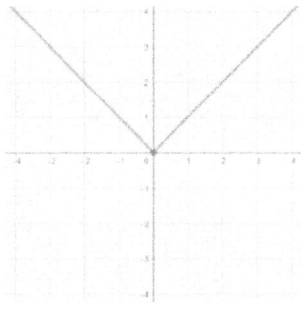

Absolute Value Function:

$y = |x|$

There is really only one basic version of each of these functions, and the basic version is modified (transformed) in various ways. In this lab we will be discovering how these graphs are transformed.

Vertical Shifts

Using Geogebra, graph the cube function. Now, modify the graph and sketch it as indicated below.

a) $f(x) = x^3 + 3$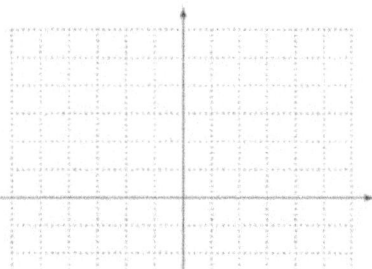

b) $g(x) = x^3 - 2$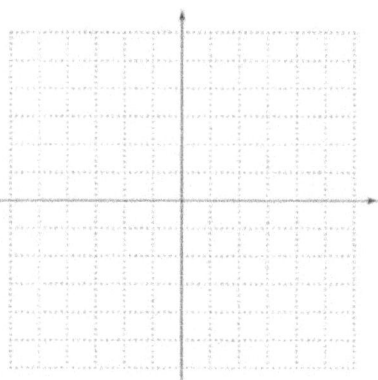

Now, graph the absolute value function. Then, modify the graph and sketch it as indicated below.

c) $h(x) = |x| - 4$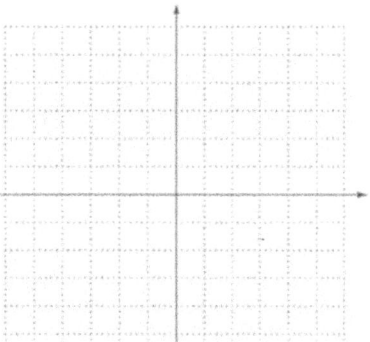

d) $r(x) = |x| + 1$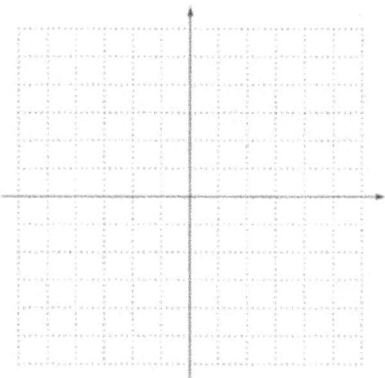

e) Explain how vertical shifts work.

Horizontal Shifts

Using Geogebra, graph the square root function. Now, modify the graph and sketch it as indicated below.

f) $f(x) = \sqrt{x-2}$

g) $g(x) = \sqrt{x+4}$

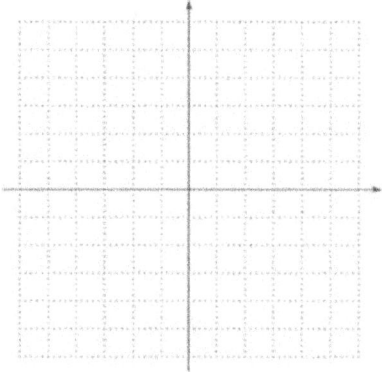

Next, graph the quadratic function. Then, modify the graph and sketch it as indicated below.

h) $h(x) = (x-2)^2$

i) $r(x) = (x+3)^2$

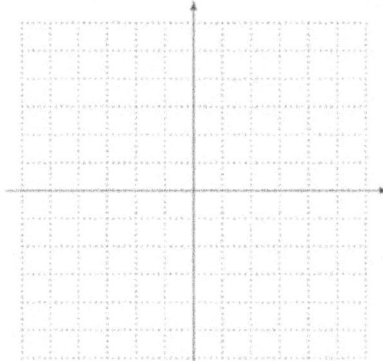
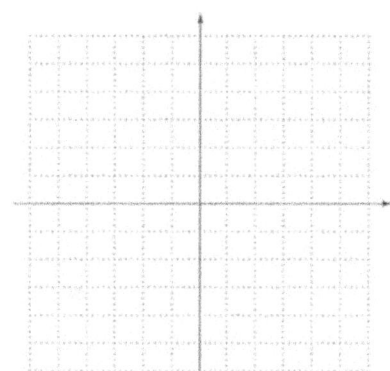

You move a graph horizontally if you add or subtract to the x value. Answer the following:

j) • Adding to x moves the graph to the _____.

k) • Subtracting from x moves the graph to the _____.

Give the equation for the following transformations. I've shown the first one:

l) Shift the absolute value function down 3.
$$f(x) = |x| - 3$$

m) Shift the quadratic function 2 to the right.

n) Shift the cubic function to the left 5.

o) Shift the square root function up 10.

The textbook will sometimes shift a generic function. You must then explain how it was shifted. Work the following, I've shown the first two problems.

$f(x - 12)$
The function has been shifted 12 units to the right.

$f(x) - 7$
The function has been shifted down 7.

p) $f(x + 5)$

q) $f(x) + 5$

This problem has two shifts.

r) $f(x - 3) + 2$

Algebra II

Active Learning: 3.5b

Reflections

Using Geogebra, graph the square function. Now, modify the graph and sketch it as indicated below.

a) $f(x) = -x^2$

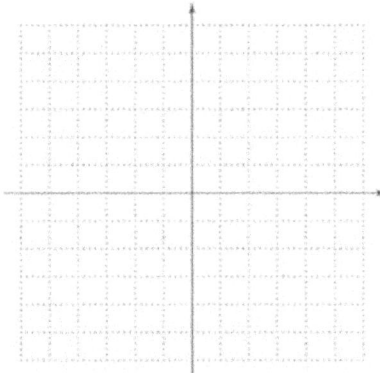

Next, graph the square root function. Then, modify the graph and sketch it as indicated.

b) $f(x) = -\sqrt{x}$

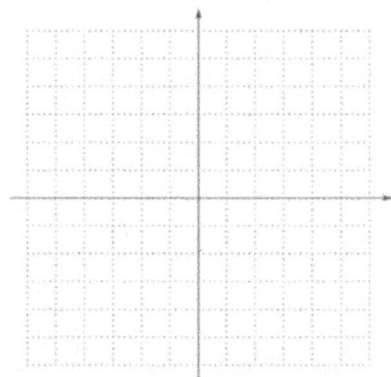

c) Explain what happened to the graph when the negative was added to the front of the function. Has the graph been flipped over the x-axis (upside down) or over the y-axis (right to left)?

d) Using Geogebra, graph this modified square root function and sketch it below.

$$f(x) = \sqrt{-x}$$

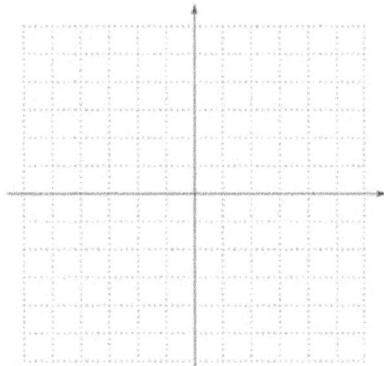

e) Here the negative sign was added inside the function, on the x. Has the graph been flipped over the x-axis (upside down) or over the y-axis (right to left)?

The following graphs have been transformed horizontally (right or left), vertically (up or down), and/or reflected (over the x-axis or over the y-axis). We want to provide an equation.

The order we do the transformations matters. Do right or left first. Then, reflections. Finally, up or down. Here's an example:

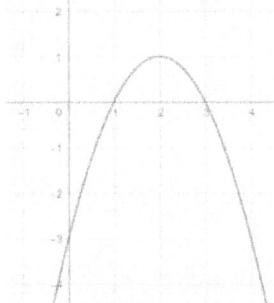

This has the shape of a quadratic equation. So, we are beginning with:

$$f(x) = x^2$$

It has moved two right. So, we have:

$$f(x) = (x - 2)^2$$

Then it has flipped upside down.

$$f(x) = -(x - 2)^2$$

Finally, it has been moved up one.

$$f(x) = -(x - 2)^2 + 1$$

Try these:

f) Give the equation of the function.

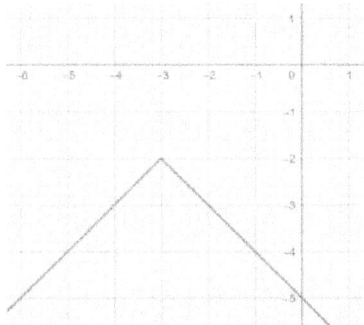

g)

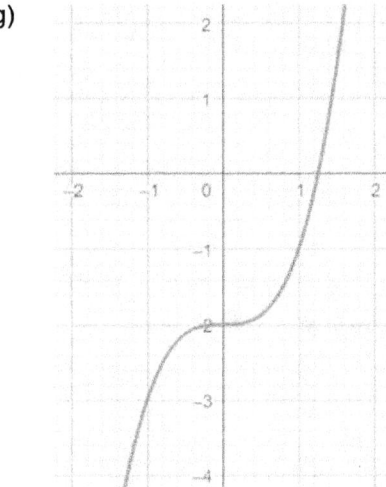

h)

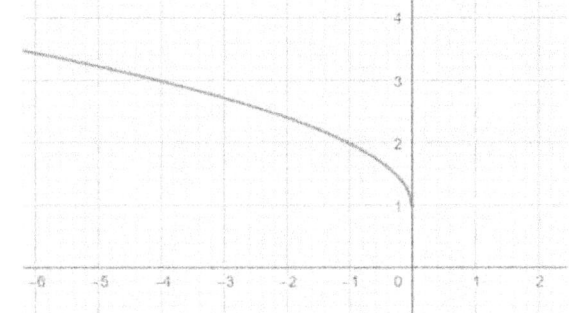

Finally, write an equation for the function based on the transformation which have been described. I have done the first one.

A quadratic function shifted 3 left and 2 down.

$$f(x) = (x+3)^2 - 2$$

i) A square root function flipped over the y-axis and shifted up 5.

j) An absolute value function flipped over the x-axis and shifted 2 right and 1 up.

A reciprocal function, $f(x) = \frac{1}{x}$, shifted 4 right and 5 down. (The changes on the x value happen inside the function. This gives students trouble. Here is what the answer would look like.)

$$f(x) = \frac{1}{(x-4)} - 5$$

k) A reciprocal square function, $f(x) = \frac{1}{x^2}$, shifted 3 left and 6 up.

Algebra II

Active Learning: 3.5c

Vertical Stretches and Shrinks

Using Geogebra, graph the absolute value function. Now, modify the graph and sketch it as indicated below.

a) $f(x) = .5|x|$

b) $g(x) = 2|x|$

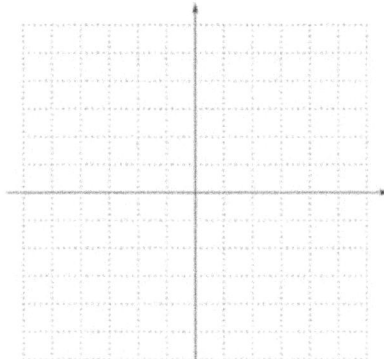
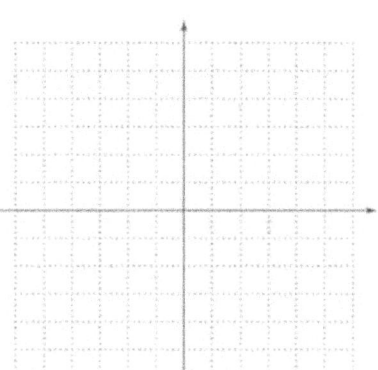

c) Which function appears to be a vertical stretch of the original absolute value function (taller and thinner): $f(x)$ or $g(x)$?

d) Which function appears to be a vertical shrink of the original absolute value function (shorter and thicker): $f(x)$ or $g(x)$?

Using Geogebra, graph the quadratic function. Now, modify the graph and sketch it as indicated below.

e) $h(x) = 2x^2$

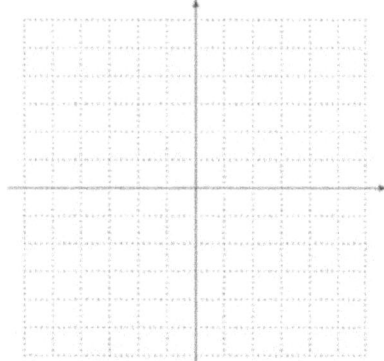

f) $r(x) = .5x^2$

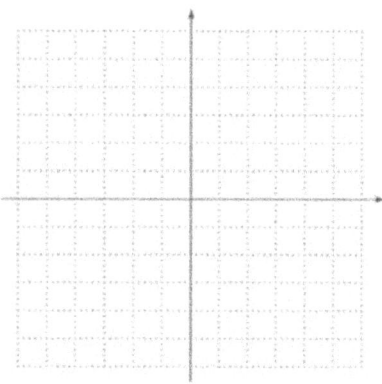

g) Which function appears to be a vertical stretch of the original absolute value function (taller and thinner): $h(x)$ or $r(x)$?

h) Which function appears to be a vertical shrink of the original absolute value function (shorter and thicker): $h(x)$ or $r(x)$?

The change for these functions has been on the outside, which makes these vertical stretches and shrinks (compressions).

Horizontal Stretches and Shrinks

Using Geogebra, graph the absolute value function. Now, modify the graph and sketch it as indicated below.

i) $f(x) = |.5x|$

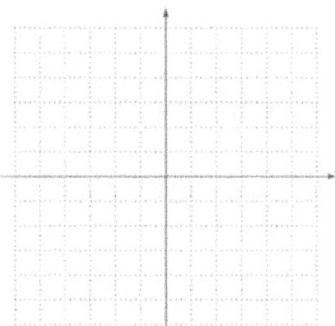

j) $g(x) = |2x|$

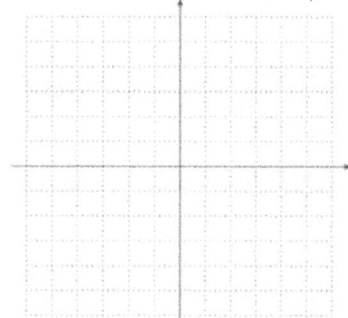

k) Which function appears to be a horizontal stretch of the original absolute value function (shorter and thicker): $f(x)$ or $g(x)$?

l) Which function appears to be a shrink of the original absolute value function (taller and thinner): $f(x)$ or $g(x)$?

Using Geogebra, graph the quadratic function. Now, modify the graph and sketch it as indicated below.

m) $h(x) = (2x)^2$ n) $r(x) = (.5x)^2$

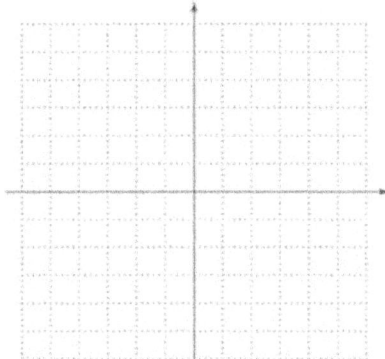
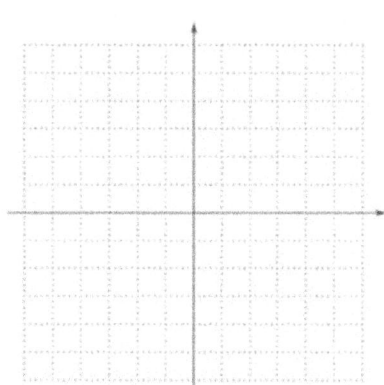

o) Which function appears to be a horizontal stretch of the original absolute value function (shorter and thicker): $h(x)$ or $r(x)$?

p) Which function appears to be a horizontal shrink of the original absolute value function (taller and thinner): $h(x)$ or $r(x)$?

The change for these functions has been on the inside with the x, which makes these horizontal stretches and shrinks (compressions). Students find horizontal stretches and compressions difficult. Anything that happens on the horizontal is the opposite of what you would expect.

Combining Transformations

When we combine the different transformations, it must follow the order of operations and work from the inside out.

1. Horizontal Shift—This is inside a grouping symbol and must come first.
2. Reflection – A reflection is multiplication of the original function.
3. Stretch or Shrink – A stretch or a shrink is also multiplying the original function.
4. Vertical Shift – Lastly, addition and subtraction on the outside.

Try graphing these on your own. Use Geogebra to check your answer if you aren't sure.

q) $f(x) = -|x + 2| - 1$

r) $g(x) = -2\sqrt{x + 4} + 1$

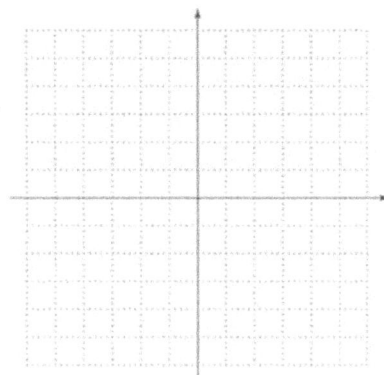
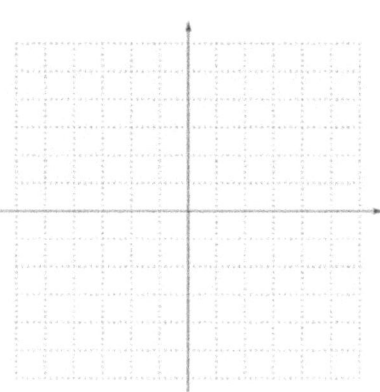

Write the function with the following transformations.

s) The cube function has been shifted right 2, reflected upside down, and the shifted vertically up by 3.

t) The absolute value function has been vertically compressed by .25 and shifted vertically down by 4.

u) The cube root function has been shifted left 1, stretched vertically by 3, and reflected upside down.

Even/Odd Functions

Below, I'm evaluating the quadratic function at the value of $-x$.

$$f(-x) = (-x)^2$$

It is the same as this:

$$f(-x) = (-1 \cdot x)^2$$

v) Finish simplifying this:

w) If we started with the function $f(x) = x^2$ what did we end with?

x) When this happens, we have something called an even function. Try another, evaluate this function at the value of $-x$.

$$f(x) = (x)^4 + (x)^2$$
$$f(-x) = (-x)^4 + (-x)^2$$

y) How did what we ended with compare to the original function?

Here are the graphs of $f(x) = x^2$ and $f(x) = x^4 + x^2$.

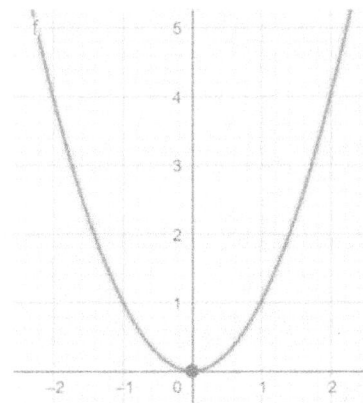 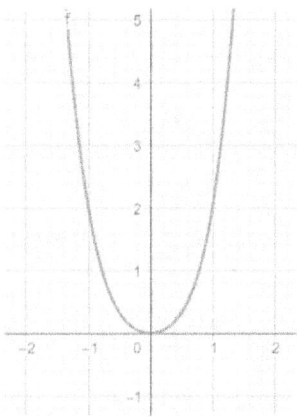

z) What happens if we fold these graphs over the y-axis?

Now, let's evaluate the cube function at $-x$.

$$f(-x) = (-x)^3$$

Which is the same as:

$$f(-x) = (-1 \cdot x)^3$$

If we finish simplifying, we end up with:

$$f(-x) = -x^3$$

aa) How did what we ended with compare to the original function?

bb) When this happens, we have something called an odd function. Next, evaluate this function at $-x$.

$$f(x) = x^3 + x$$
$$f(-x) = (-x)^3 + (-x)$$

cc) How did what we end with compare to the original function? (If you aren't sure, pull a GCF of -1 out of your final answer.)

Here are the graphs of $f(x) = x^3$ and $f(x) = x^3 + x$.

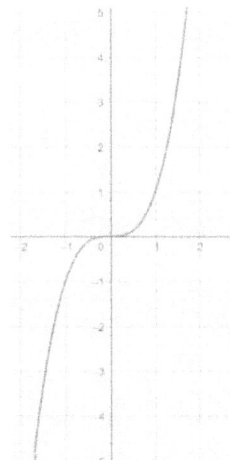 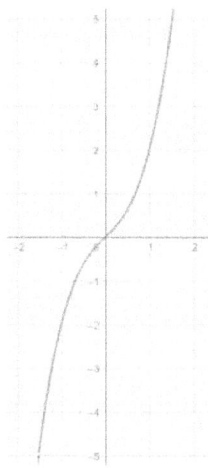

dd) What would happen if you folded these graphs, first over the y-axis and then over the x-axis?

To determine if you have an even or odd function, evaluate it at $-x$. If you get back to what you started with, you have an even function. If you could pull out a GCF of -1 and then you have -1 times what you started with, you have an odd function. Even functions are always symmetrical if you fold them over the y-axis. Odd functions are called symmetrical over the origin, which means folding it over the y-axis and then folding it over the x-axis.

Be careful, students often think they can determine these in their head. It can be a bit harder than you think and it is better to evaluate them at $-x$.

Are the following functions even, odd, or neither. (If they don't fit the rules for even or odd, they are neither.)

ee) $f(x) = x^5 + x$

ff) $f(x) = x^2 + 1$

gg) $f(x) = x^6 + x^2$

hh) $f(x) = x^3 + 2$

Algebra II

Active Learning: 3.6

Math is a language. It is a language that simplifies. The absolute value function, $f(x) = |x|$, asks the question, "how far away is a number from zero?" We don't care if a number is positive or negative; all that matters is how many steps we've gone from zero.

The textbook doesn't go into much detail on this, but a horizontal shift changes the fundamental question.

$$f(x) = |x - center|$$

Instead of asking how far we are from zero, this now asks how far away we are from the center number. For instance:

$$f(x) = |x - 5|$$

Would ask, "how many steps are we away from 5?" And:

$$f(x) = |x + 2|$$

Would ask, "how many steps are we away from -2?" Remember, a negative is a shift to the right and a positive is a shift to the left.

In the last section we already learned about all the transformations which apply to all functions. They, of course, apply to absolute value functions too. Write a function based on the description of the transformations:

a)   An absolute value function, flipped over the x-axis, shifted left 4 and up 7.

b)   An absolute value function, vertically stretched by 2 and shifted down 5.

c)   An absolute value function, horizontally stretched by $\frac{1}{2}$. (The textbook might say, horizontally stretched by a factor of 2.)

Graph the following transformed functions:

d) $f(x) = -|x + 3| - 2$

e) $g(x) = 2|x - 1| + 1$

f) Next, give the equation of the absolute value function by reading the graph.

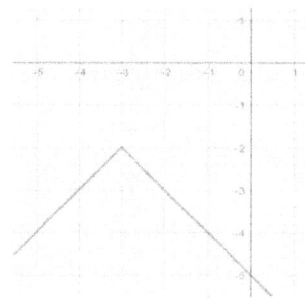

On this problem, the absolute value function has been stretched. The stretch is unknown so we call it a.

$$f(x) = a|x - 1| + 2$$

To find the value of a, we plug in any point from the graph other than the vertex of the absolute value. A clear point is $(2, 5)$. Remember, $f(x)$ is just y.

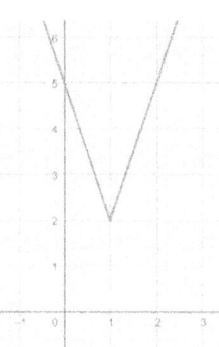

g) Once you find the value of a, rewrite the function in generic form with the value of a added. Give the equation in the space below.

h) Next, describe the transformations based on the graph. You don't need to write the equation; put it in words.

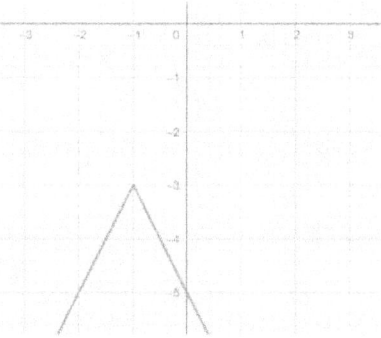

i) The only difficult portion of this is problem involves a vertical stretch. Use the same approach as the last problem; create the entire function with a variable a in front. Then, use a clear point from the graph, like $(-2, -5)$, to find the value of the vertical stretch.

Finally, we want to solve absolute value functions. Because an absolute value is about the distance away from a number (typically zero), there will be two answers. One answer coming from the right side of the number line and one answer coming from the left side of the number line.

The key to solving many equations in Algebra II involves getting the difficult portion of the equation alone. Here, we will isolate the absolute value:

Find the value of x such that $f(x) = 0$. (In other words, solve.)

$$f(x) = |x + 4| - 5$$
$$0 = |x + 4| - 5$$

First, get the absolute value function alone:

$$5 = |x + 4|$$

j) When the absolute value is isolated, we now need to work the problem twice. We work it once for $+5$ and once for -5. Solve for the two values of x:

$$x + 4 = 5 \text{ and } x + 4 = -5$$

If there is a stretch in front of the absolute value, we still begin by isolating the absolute value.

$$f(x) = 2|x - 4| - 6$$

Treat the absolute value like it is a giant variable:

$$0 = 2X - 6$$

You would first add 6 and then divide by 2. We do the same to isolate the absolute value. Isolate the absolute value and then work it twice. Solve:

k) $0 = 2|x - 4| - 6$

Try these on your own:

Find the value of x such that $f(x) = 0$.

l) $f(x) = 3|x + 6| - 12$

m) $f(x) = -2|x - 3| + 22$

There is something wrong with this statement:

$$|x - 3| = -3$$

This can never happen. Because of the x, we don't know exactly what the value is on the left. However, because of the absolute value symbol, we know that it is a positive. There is no way that a positive number on the left can equal a negative number on the right. Therefore, this is no solution.

An absolute value problem only becomes no solution when you are left with nothing but the absolute value on one side and a negative on the other. This problem is not "no solution." When you get down to the absolute value, the difficulty has been fixed.

Solve:

n) $2|x - 3| - 24 = -8$

Show why the following problem does, in fact, end with "no solution."

Find the value of x such that $f(x) = 0$.

o) $f(x) = 4|x + 6| + 24$

Algebra II

Active Learning: 3.7a

Find the composite functions:

$$f(x) = x^3 + 6$$
$$g(x) = \sqrt[3]{x - 6}$$

a) Given the two functions find $f(g(x))$.

b) Given the two functions find $g(f(x))$.

We learned earlier that a composite function is an assembly line.

$x \longrightarrow \boxed{f(x)} \longrightarrow \boxed{g(x)} \longrightarrow g(f(x))$

But on the problem that you worked above, an odd thing occurred.

$x \longrightarrow \boxed{f(x)} \longrightarrow \boxed{g(x)} \longrightarrow x$

At the end of the assembly line, we had the same thing we started with. That must mean that the second function was just the reverse of the first function. When this occurs, the two functions are called inverse functions.

c) Test to see if these functions are inverses. Make them a composite function. It doesn't matter what order that you do the composite; it will work either way.

$$f(x) = x^2$$
$$g(x) = \sqrt{x}$$

d) If you create a composite function, but you don't get back to x, the two functions aren't inverses. Show that the following two functions are not inverses.

$$f(x) = \sqrt[2]{x} - 1$$
$$g(x) = (x-1)^2$$

When we first worked with functions, we used the vertical line test. It works because for a particular x there cannot be two y's. Today we begin to look at inverse functions. They are function machines that work in reverse. And so, an inverse flips the domain and range. To check to see if a function can have an inverse, we use a horizontal line. (We previously saw the horizontal line test to check if a function is one-to-one. For a function to have an inverse, it must be one-to-one.)

If a horizontal line touches the graph twice, it cannot have an inverse. (Because the inverse function would fail the vertical line test.) If a horizontal line touches the graph only once, it can have an inverse.

e) Circle any graphs below which can have an inverse.

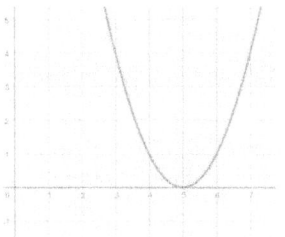 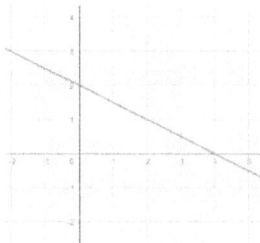 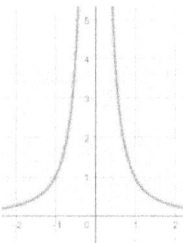

Because inverse functions work in reverse, all the x values become the y values and vice versa (the domain and the range are exchanged). As a result, all inverse functions "flip" over a certain line. Below are graphs of functions and their inverses. See if you can determine the equation of the line which inverse functions flip over.

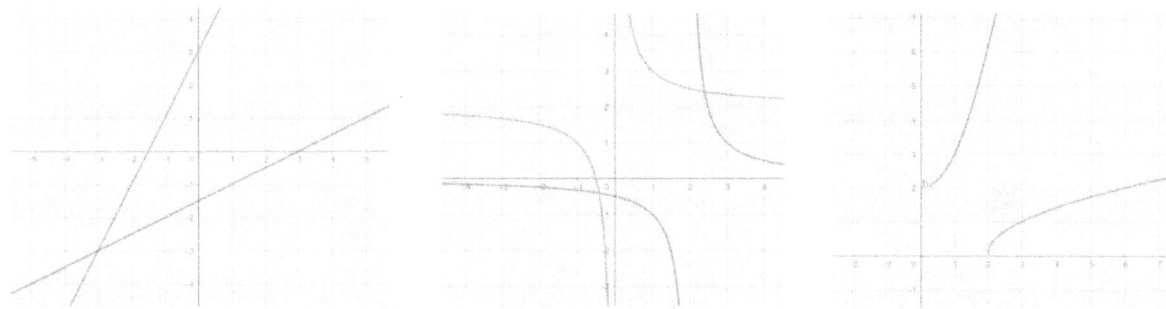

f) What is the equation of the line which inverse functions flip over? (Remember, the equation of a line is $y = mx + b$.)

If the domain of a function f is $(2, \infty)$ and the range of f is $(-\infty, 4)$, what is the domain and range of the inverse function g?

g) Domain of g:

h) Range of g:

Below is a graph of the function f:

i) Find the value of $f(1) =$

Inverse functions have the notation $f^{-1}(x)$. If the graph is $f(x)$ then $f^{-1}(x)$ would reverse the x's and the y's. To find $f^{-1}(2)$ we would start at $y = 2$ and find the value of x which goes with it. So:

$f^{-1}(2) = 1$

j) 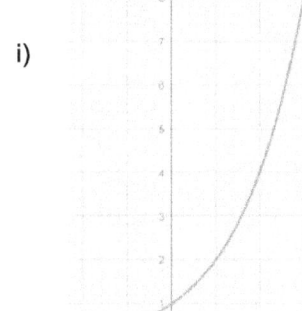 What would be the value of $f^{-1}(8)$?

$f^{-1}(8) =$

Try another. The graph is the function $f(x)$. Evaluate the following:

k) $f(2) =$

l) $f^{-1}(-8) =$

m) $f^{-1}(-2) =$

The same idea can be extended to a table. This is the function $f(x)$ in table form.

x	1	2	3	4
f(x)	4	8	12	16

n) Evaluate $f(2) =$

The inverse function $f^{-1}(x)$ goes in reverse, so the x's are the y's and the y's are the x's. To evaluate the inverse function, simply go backwards.

Evaluate $f^{-1}(12) = 3$

Use the same table to evaluate the following:

o) $f(4) =$

p) $f^{-1}(4) =$

q) $f^{-1}(8) =$

Algebra II

Active Learning: 3.7b

Solve the following equations for y:

a) $x = \frac{1}{2}(y - 9)$

b) $x = \frac{1}{y-3}$ Treat $(y - 3)$ as a group.

c) $x = 5 + \sqrt{y - 2}$

Since an inverse function switches the domain and range (x values and y values), to find an inverse function involves making the same switch. Suppose we want to know the inverse function of $f(x)$.

$$f(x) = 3x + 2$$

First, switch the x's and the y's:

$$y = 3x + 2$$
$$x = 3y + 2$$

Now, solve for the new y.

$$x - 2 = 3y$$
$$\frac{x - 2}{3} = y$$

Finally, add the inverse notation:

$$f^{-1}(x) = \frac{x-2}{3}$$

Try some on your own. Find the inverse of the following functions:

d) $f(x) = \frac{1}{2}(x-9)$

e) $f(x) = \frac{1}{x-3}$

f) $f(x) = 5 + \sqrt{x-2}$

g) $f(x) = \sqrt[3]{x+3}$

h) $f(x) = \frac{2}{x-6} + 8$

In this next problem, there will be two values of y. To get them alone, we will need a trick. Get all the y terms alone and pull out y as a greatest common factor. I've shown one below.

$$f(x) = \frac{x}{x+3}$$

$$x = \frac{y}{y+3}$$

$$x(y+3) = y$$

$$xy + 3x = y$$

$$3x = y - xy$$

$$3x = y(1-x)$$

$$\frac{3x}{1-x} = y$$

$$f^{-1}(x) = \frac{3x}{1-x}$$

Try a couple on your own.

i) $f(x) = \dfrac{x}{5-x}$

j) $f(x) = \dfrac{x+2}{x-3}$

Algebra II

Active Learning: 4.1a

Chapter four is going to take a look at the concept of lines as functions. We learned previously that a function is a machine in which something goes in and only one thing goes out. Lines are functions.

$$y = 3x + 2$$
$$f(x) = 3x + 2$$

Here was an example of putting 2 into a function.

$$f(2) = 3(2) + 2$$

2 → $3(2) + 2$ → 8

Remember, $f(x)$ is exactly the same thing as y. So, the ordered pairs (x, y) and $(x, f(x))$ are the same thing. What we will see next is primarily a review of lines within the context of function notation.

First, slope is the key to determining if a line is increasing, decreasing or constant. Indicate if the following lines are increasing, decreasing or constant. A positive slope is increasing. A negative slope is decreasing.

a) $f(x) = 115$

b) $f(x) = 65x - 320$

c) $f(x) = -.05x + 1.50$

d) $f(x) = -15 + 1.25x$

e) $f(x) = -200$

f) $f(x) = 102 - 15x$

Create linear functions for each of the following scenarios. Remember, changing values are slopes and constant values are y-intercepts.

g) A lawncare company charges $100 to start a service agreement and then $50 per visit. The input is the number of visits and the output is the total cost.

h) A tutoring company uses a system of credits. The basic package includes 150 credits. A student uses 5 credits per visit. The input is the number of visits and the output is the remaining credits.

i) A mobile phone service offers an unlimited plan for $25 a month. The input is the number of days of service and the output is the total cost per month. (Be careful here. The number of days of service has nothing to do with the total cost and doesn't need to be represented in the function.)

Calculating slopes of linear functions is the same as what we've always done. Here is the formula written in function notation.

$$m = \frac{y_2 - y_1}{x_2 - x_1} = \frac{f(x_2) - f(x_2)}{x_2 - x_1}$$

Find the slope in this applied problem.

j) The population of deer in a national park grew from 1,200 in 2005 to 2,800 in 2013. If we assume the growth was linear, find the average rate of growth for the deer per year. (Population will be our y value.)

k) Use the graph below to write the equation of the linear function. Use the two points to find the slope and then use that slope and one of the points in the point slope formula. (Do this by the math, even though you could do it visually.)

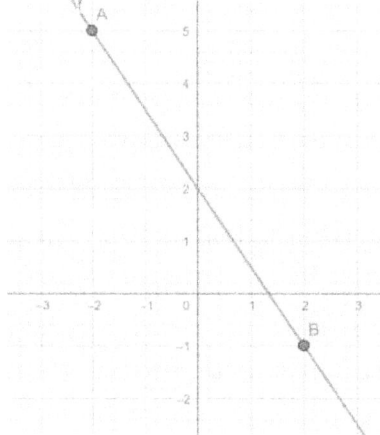

Write your final equation in function notation.

l) Here you are given two points in function notation. Remember $f(x) = y$. Find an equation for the linear function. First, find the slope. Then, use the point-slope formula.

$$f(2) = 4$$
$$f(-2) = -6$$

The cost of a product can often be written as a linear function.

$$C(x) = mx + b$$

Again, flat fees are b, the y-intercept, and changing fees are m, the slope. Write the following situations as cost functions.

m) To join a fitness center, you must pay $100 plus $10 per month.

n) A discount cellphone carrier charges $10 for their basic plan and then $5 per Gigabyte of data.

o) Use your cost function for the fitness center to determine how much you would pay if you were a member for 24 months.

p) Use your cost function for the cellphone carrier to determine how much someone would pay for 15 Gigabytes of data.

Finally, find a linear function that describes this commission-based salary.

q) A saleswoman's salary is based on commission. Last month, she sold three cars and received $2700 in salary. This month, she will sell five cars and will receive $2920 in salary. Find a function $S(x)$ which shows her salary based on how many cars she sells. (First, find the slope. Then, use the point-slope formula.)

r) What salary would she make if she sold 7 cars in one month?

Algebra II

Active Learning: 4.1b

We are continuing to look at linear functions and most of what we will cover is not new. The key is simply remembering that $f(x) = y$. First, we want to graph a line by plotting points.

$$f(x) = -\frac{3}{2}x + 2$$

a) Any values can be chosen for x. However, choosing values which simplify the fractions make the math easier. Finish the chart below by evaluating the function at the given values of x. Then graph the line.

x	f(x)
0	
-2	
2	

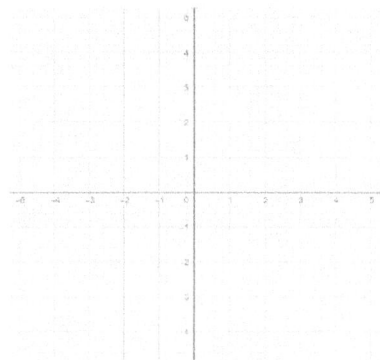

b) Next, graph the following linear function by using the slope and y-intercept. Recall that the y-intercept here is -1 and the slope is $\frac{1}{3}$.

$$f(x) = \frac{1}{3}x - 1$$

When a linear function is in slope-intercept form, we know that the y-intercept is the value of b. To find the x-intercept, we need to evaluate the function when $f(x) = 0$. Find the x-intercept of the following function. Remember, a x-intercept is a point, so give your answer accordingly.

$$f(x) = -\frac{3}{4}x + 2$$

c) x-intercept:

Now, we want to give the equation for horizontal and vertical lines. Recall, a horizontal line is a y line. So, the line is always equal to one value of y.

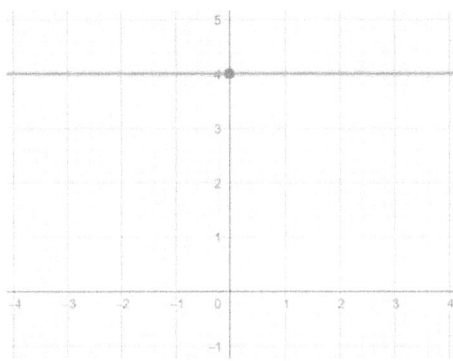

d) This is a linear function, so give the equation of this line in terms of $f(x)$.

e) Vertical lines are x lines, where the line is always equal to one value of x. Give an equation for the following vertical line.

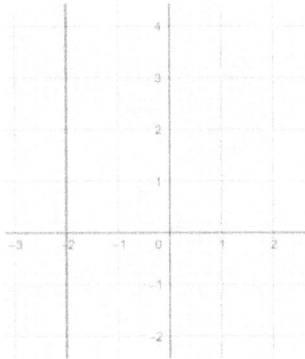

Finally, we previously looked at the ideas of transformations. Although we don't typically think of it like this, there is really one basic line.

$$y = x \text{ or } f(x) = x$$

Adding b is a vertical shift.

$$g(x) = x + 3$$

This is shifted vertically up 3.

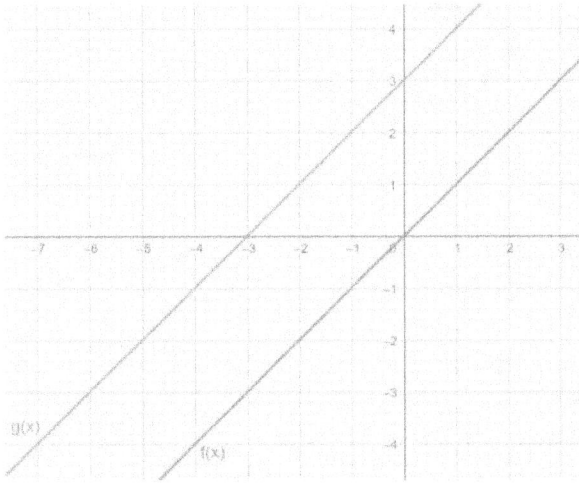

And, a slope is a vertical stretch or compression. This is a vertical compression of $\frac{1}{2}$.

$$h(x) = \frac{1}{2}x$$

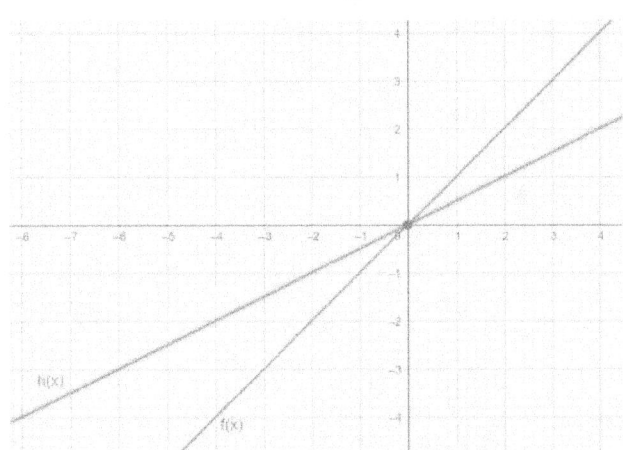

Below, draw a line connecting the transformation of $f(x) = x$ with its proper graph:

f) $f(x) = x - 2$

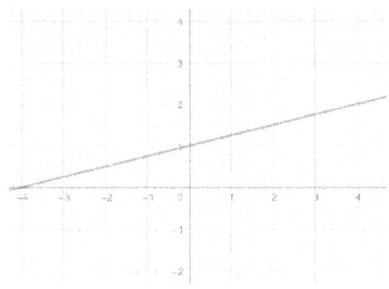

g) $f(x) = 3x$

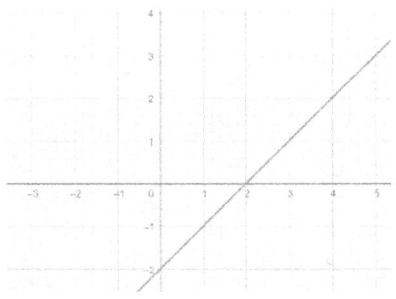

h) $f(x) = \frac{1}{4}x + 1$

Algebra II

Active Learning: 4.1c

Once again, we are examining linear functions. And, once again, the ideas have been examined previously.

Parallel lines are those which have the same slope.

$$f(x) = \frac{3}{2}x + 2$$

$$g(x) = \frac{3}{2}x - 3$$

The same slope causes each of the lines to be "built" in the same way. Perpendicular lines are those who have a reciprocal slope with an opposite sign.

$$h(x) = -\frac{4}{5}x + 1$$

$$p(x) = \frac{5}{4}x + 4$$

Identify if the following sets of lines are parallel, perpendicular or neither.

a) $f(x) = -\frac{3}{2}x + 1$
   $g(x) = \frac{2}{3}x - 7$

b) $h(x) = -\frac{5}{7}x - 2$
   $p(x) = -\frac{7}{5}x + 3$

c) $t(x) = \frac{4}{5}x + 5$
   $r(x) = \frac{4}{5}x + 2$

To find a line parallel to a given line, we want to use its slope. Then, use the point-slope equation to find our parallel line.

$$y - y_1 = m(x - x_1)$$

d) Find a line parallel to $f(x) = -\frac{3}{2}x + 1$ and passing through the point $(-4, 4)$. (Remember to use function notation for the new line.)

Follow the same procedure to find the equation for a line perpendicular to a given line. This time, take the negative inverse of the given line.

e) Find a line perpendicular to the $f(x) = \frac{1}{2}x + 3$ and passing through the point $(2, -3)$.

Finally, suppose we know two points on a line and wanted an equation for a line perpendicular. Find the slope through the two points and then use the negative inverse (and the new point) to find our line.

f) A line passes through the points $(-5, 2)$ and $(5, 6)$. Find the equation of a line perpendicular to this line and passing through the point $(-2, 7)$.

Algebra II

Active Learning: 4.2

If a mathematician finds that a linear function represents the real world, it provides them with valuable information. They can use this relationship to make predictions. Here's an example.

A stock price has been increasing in a linear relationship. In 2002, the stock price was $25. In 2016, the stock price was $60.

a) Find a linear function which represents this situation. (Years are x values. Stock prices are y values.) Assume that the stock price in 2002 is the y-intercept.

b) Use the linear function to predict what the stock price was in 2010. (2002 is the starting year. So, 2010 is year 8. Evaluate the linear function at $x = 8$.)

c) Use the linear function to predict what the stock price would be in 2030. (2002 is the starting year. So, 2030 is year 28. Evaluate the linear function at $x = 28$.)

Linear models are also helpful for comparing cost functions.

Hefty Freight Service charges 15 dollars for an item of freight plus .45 per mile. Speedy Freight Service charges 20 dollars for an item of freight plus .35 per mile.

d) Find Cost Functions for each shipping service. (x represents the number of miles.)

e) At what number of miles would the two services cost the same? (Set the two equations equal to each other and solve for x.)

f) How much would it cost to use Hefty Freight Service if you needed to ship the item 250 miles?

g) How much would it cost to use Speedy Freight Service if you needed to ship the item 250 miles?

These two types of problems are the most common applications of linear functions. We will look at one more, which is more of a traditional word problem and far less common in real application.

You are walking 5 miles per hour east from your house. Your friend is riding her bicycle and is moving 12 miles per hour south from your house. How far apart will you be after 2.5 hours?

h)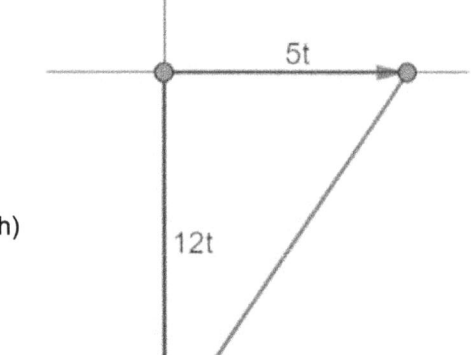

In this situation, two lines are forming a right triangle. The distance apart could be found with the Pythagorean Theorem:

$$(5t)^2 + (12t)^2 = d(t)^2$$

Find $d(t)$. Square each part of the left side and then combine like terms. To find $d(t)$, square root both sides.

i) $d(t)$ is a linear function which tells the distance between you and your friend. Find how far apart you are by evaluating $d(t)$ when $t = 2.5$.

Algebra II

Warm Up: 4.3a

Next, we want to look at a more complex mathematical model involving lines. It is called linear regression. First, however, we need to set the stage with something called a scatterplot.

A scatterplot relates two variables as a set of ordered pairs. Here, a speech pathologist wants to compare the age of a toddler (in months) with how many vocabulary words they know.

Age (Months)	Vocabulary (Word Count)
20	16
22	26
24	65
26	41
28	55
30	60
32	74
34	77

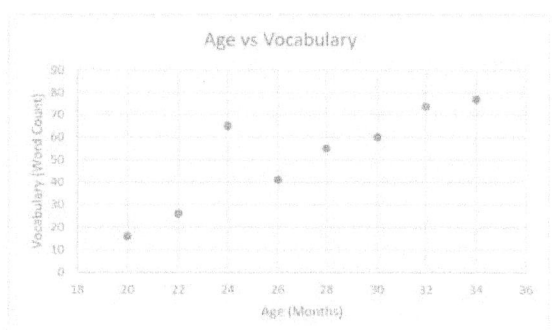

Scatterplots are described by their form, their direction and their strength. For our purposes, form is just one of two choices: linear (a line) or non-linear (not a line.) If you stare at the scatterplot and it has the general appearance of a line, it is linear.

a) What is the form of the scatterplot of age vs vocabulary?

b) Imagine a line is running through the dots on the scatterplot. The direction is positive if your line is running uphill (right to left) or negative if your line is running downhill (left to right). What is the direction of the scatterplot age vs vocabulary?

The strength of a scatterplot can be determined by a mathematical computation named correlation. Calculating correlation by hand is time consuming and I would permit it to be done on a calculator. This is what you need to know about a correlation value. Correlations range from -1 to 1. With a score of 1, all the dots form a perfect line going uphill. With a score of -1, all the dots form a perfect line going downhill. A score of 0 means there is no correlation at all. (The picture would look like a shotgun blast of random dots.)

The strength of a scatterplot is how good of a line the dots make. You can choose between: strong, moderate, or weak. The scale doesn't work as you might expect. Although it depends a bit on who you ask, a strong correlation would be .95 or above (or -.95 to -1). A moderate correlation would be above .85. And, a weak correlation would be between 0 and .85.

Here are four possible correlations for our graph of age vs vocabulary. Choose the correlation which best fits our graph. (Be careful, our graph has one dot which doesn't fit the pattern. It is called an outlier. An outlier will negatively impact the strength of our correlation. It won't destroy our correlation, but it will decrease it.)

c) $r = .98$ $r = .9$ $r = -.9$ $r = .5$

Now, we want to calculate the value of a correlation. Again, because the calculation is very tedious, you can use a calculator. Here is a walkthrough of how to find correlation on a TI calculator.

- First, a setting needs to be turned on. Push the "Catalog" button. (Hit "2nd" and then "0".)
- Us the down arrow to scroll down the list of choices until you get to "DiagnosticOn". Select it. Then hit "Enter." Your screen should now say "Done."
- Next, you need to enter the values into a table. Select "STAT". There is now an "EDIT" button. Select it and you are looking at a table. Enter the x values in L_1 and y values in L_2.
- (When you need to erase the lists, go to "STAT" and down to an option "CLRLIST". Select it then add "2nd L_1, 2nd L_2." Hit Enter. When it says "Done" you have emptied the lists.)
- To find a correlation, return to "STAT". Move to the right one and you are now under the "CALC" menu. Go down until you find a choice called "LINREG $(a + bx)$" and select it.
- If your calculator gives you a menu make sure $x: L_1$ and $y: L_2$.
- You will get a lot of information, but the value of r should be included.

An ice cream vendor wants to find the correlation between the outdoor temperature and the sale of ice cream. Enter the information in your calculator and find the value of r.

d)

Temperature (F)	Ice Cream Sales (Dollars)
82	176
84	175
86	220
88	250
90	265
92	268
94	275
96	310
98	291
100	360

What is the value of r?

e) Is there a positive or negative correlation between the outdoor temperature and the sale of Ice Cream?

f) What is the strength of the correlation?

Make a scatterplot of this data on the graph below. Label the x-axis as Temperature (F) and the y-axis as Ice Cream Sales (Dollars). Start your x-axis at the number 80 and go by twos up to 100. Start your y-axis at 170 and go by twenties up to 360. Remember, these are ordered pairs so (x, y) is $(Temp, Sales)$.

g) With the difficult scale, you won't be able to get your points exactly right, just do your best.

Correlation imagines a line running through the points. Suppose we wanted to put a real line. Choose two points which are well centered on an imaginary line through the dots. Find the slope between the points. This would be the slope of a "fit-line".

h) $m =$

Algebra II

Active Learning: 4.3b

In our last lesson, we looked at scatterplots. We imagining that a line ran through the middle of the points. Now, we want to put an actual line. It is called a "least-squares regression line" or a "line of best fit." Mathematically, it is literally finding the line which best averages its way through the points of data.

$$\hat{y} = a + bx$$

Mathematicians have changed the order of the "best fit" line, but it is still just a line. The a is the y-intercept. The b is the slope. The y gets a "hat" to show that values of x which go in this line are giving us "best fit" y's and not our actual y's.

As with correlation, the calculation of a best fit line is complicated. For this course, it is permissible to let the calculator find it for you. Most of the instructions are the same as finding correlation. Here are the instructions:

- First, you need to enter the data points. Select "STAT". There is now an "EDIT" button. Select it and you are looking at a table. Enter the x values in L_1 and y values in L_2.
- (When you need to erase the lists, go to "STAT" and down to an option "CLRLIST". Select it then add "2nd L_1, 2nd L_2." Hit Enter. When it says "Done" you have emptied the lists.)
- Return to "STAT". Move to the right one and you are now under the "CALC" menu. Go down until you find a choice called "LINREG (a+bx)" and select it.
- If your calculator gives you a menu make sure $x: L_1$ and $y: L_2$.
- After you calculate, the output will show a value for a. That value is the y-intercept of your line. It will also show a value for b. That value is the slope of your line.

Here is the data from the last lesson regarding age of a toddler and vocabulary.

Age (Months)	Vocabulary (Word Count)
20	16
22	26
24	65
26	41
28	55
30	60
32	74
34	77

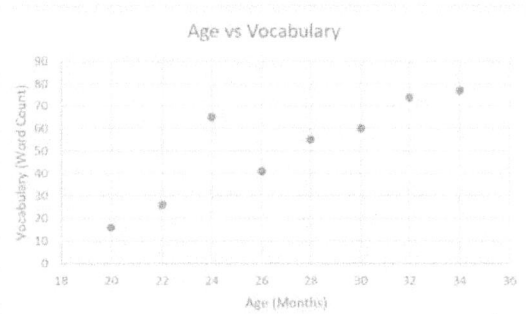

a) Find a least square regression line for this data. (Remember, it is a line in the form of $\hat{y} = a + bx$.)

A least square regression line is mathematical model. Mathematicians use it to make predictions about values for which they had not collected information.

b) For instance, suppose we wanted to predict the vocabulary score of a child who was 25 months old. We don't have a data point at that value of x. However, we have our least square line, which was created based on our data. Predict the vocabulary score of a child at 25 months by evaluating the least square regression line at $x = 25$.

c) This regression line was based on data points with x values between 20 and 34. It can be mathematically dangerous to try to use this line to predict values outside of that range. (The line wasn't built for that.) Doing so is called extrapolation. Use the least square regression line to predict the vocabulary score of a child at 2 months.

d) This prediction was extrapolation. What is odd about the predicted vocabulary score we obtained for a child who is 2 months old?

Here is the data for the ice cream vendor interested in studying the relationship between temperature and ice cream sales.

Temperature (F)	Ice Cream Sales (Dollars)
82	176
84	175
86	220
88	250
90	265
92	268
94	275
96	310
98	291
100	360

e) Find a least square regression line for predicting the amount of ice cream sales from the temperature.

f) Predict the ice cream sales if the outdoor temperature is 87 degrees.

Algebra II

Active Learning: 5.1a

One of the key functions of algebra is the quadratic equation.

$$f(x) = x^2$$

Like any function, quadratic equations can be transformed. When transformed, the Vertex (the bottom of the u) where the quadratic rests will move. Give the vertex of the following quadratic equation. It is a point, so give your answer as an ordered pair.

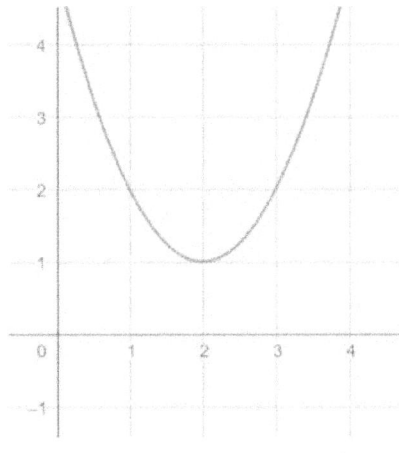

a)   Vertex:

Using Geogebra, we want to have a better understanding of how transformations impact the vertex. Enter the following quadratic equations in Geogebra. Then, graph the result below, and give the vertex.

b) $f(x) = (x-3)^2 + 3$

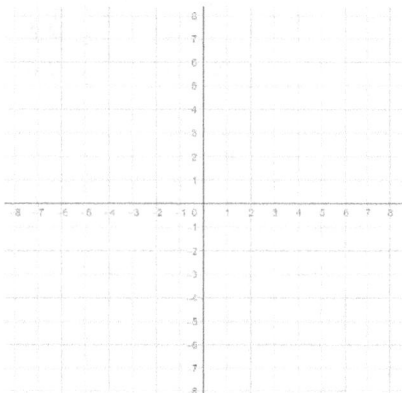

Vertex:

c) $g(x) = (x-1)^2 - 2$

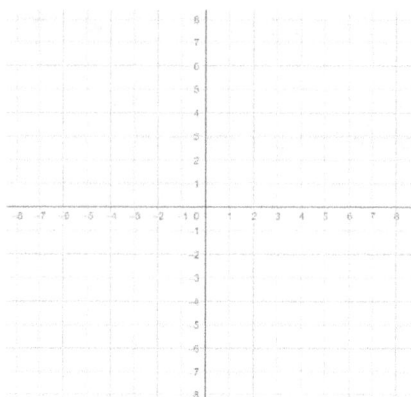

Vertex:

d) $h(x) = 2(x+2)^2 + 1$

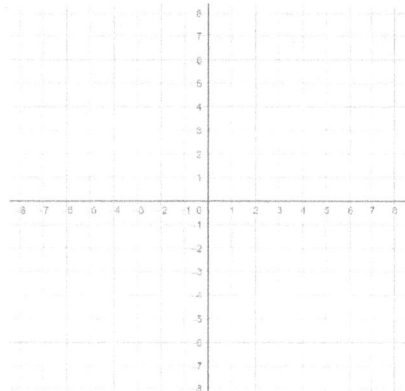

Vertex:

e) $n(x) = \frac{1}{2}(x+3)^2 - 1$

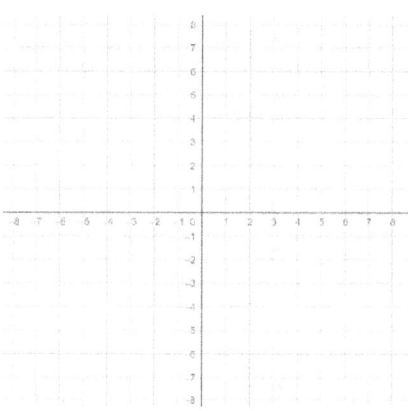

Vertex:

A quadratic function can be written in what is called vertex form.

$$f(x) = a(x-h)^2 + k$$

Using the variables from that equation, what are the coordinates of the vertex.

f) Vertex:

Here is the standard form of the quadratic equation:

$$f(x) = ax^2 + bx + c$$

Converting between standard form and vertex form will always create the same value of h.

$$h = -\frac{b}{2a}$$

g) Given the following quadratic function, find the value of h.

$f(x) = 2x^2 + 8x + 5$

$h =$

To find the y-coordinate of the vertex, we need to remember that x and y are an ordered pair. So, put h into the function to find k.

$$k = f(h)$$

h) Find the value of k for the function above.

$k =$

i) Work another. Find the vertex.

$g(x) = 3x^2 - 24x + 2$

Next, let's look at the domain and range of a quadratic function. Below are the graphs of two quadratics. Give their domain and range. Remember, for the domain, imagine a vertical line moving alone the x-axis. If you are hitting graph, that is part of the domain. For the range, imagine a horizontal line moving along the y-axis. If you are hitting graph, that is part of the range. Give the domain and range in interval notation.

j)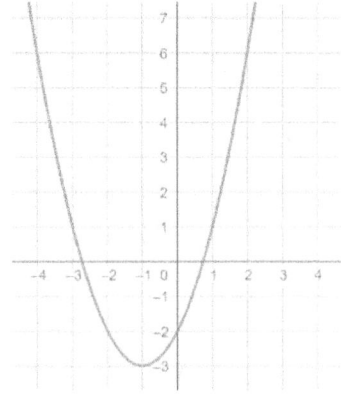

Domain:

Range:

k)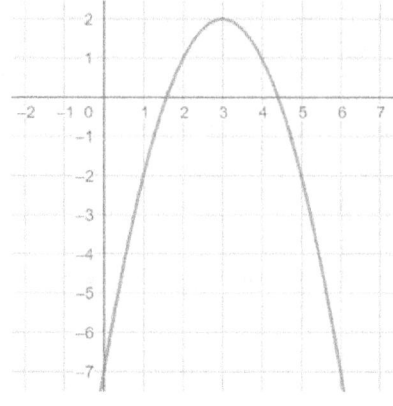

Domain:

Range:

Because a quadratic is the shape of a parabola, the domain will always be the same as these two graphs. The range depends on the value of k and whether or not the function is upside down or right side up. If it is right side up, the range will always be:

$$(k, \infty)$$

If it is upside down, the range will always be:

$$(-\infty, k)$$

Given the following quadratic equation, give the domain and range without graphing. (Find the value of k by the procedure we followed previously.)

$$f(x) = -x^2 + 4x - 2$$

l) Domain:

m) Range:

Finally, we want to write an equation for a quadratic from the graph. First, find the vertex.

n) 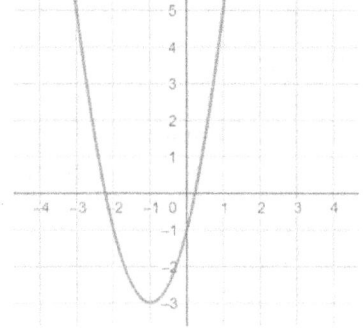   Vertex:

The vertex form of a quadratic is:

$$f(x) = a(x-h)^2 + k$$

We now can enter the values of the vertex into the equation. The only thing remaining would be to find the value of a. To find it, we simply need any point from the graph. We can then substitute that point in for x and y. The point $(1, 5)$ is a clearly on the graph. Substitute it into the standard form and then solve for a.

o) $a =$

p)   Give the full equation with a, h and k. (It should now go back to a generic x and a generic $f(x)$.)

q)   Try one more. Find the equation of the quadratic from the graph.

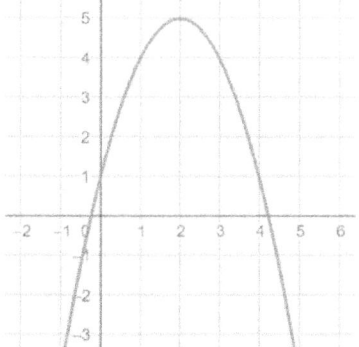

Algebra II

Active Learning: 5.1b

Next, we want to find the x and y intercepts of a quadratic function.

$$f(x) = 3x^2 + 10x - 8$$

It is done in the same way as with lines. The y-intercept occurs when $x = 0$.

$$f(0) = 3(0)^2 + 10(0) - 8$$

a) Finish evaluating the function at $x = 0$. (Remember, the y-intercept is a point, so it should be a coordinate.)

$y - intercept$:

The x intercept occurs whey $y = 0$. In a function, y is $f(x)$. So:

$$0 = 3x^2 + 10x - 8$$

b) We have solved quadratics like this before. Solve for x by either factoring or using the quadratic formula. Give your answers as coordinates to show they are x-intercepts.

c) Try another. Find the x and y intercepts.

$$f(x) = 5x^2 + 3x - 2$$

Look at the following quadratic functions.

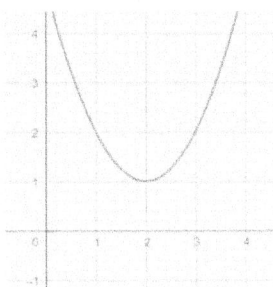
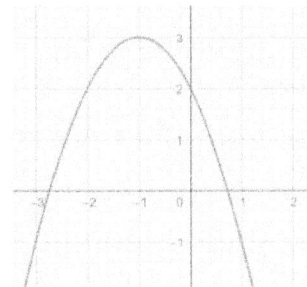

If you have a quadratic then the vertex must be either the highest point (a maximum) or the lowest point (a minimum). The following word problems will use this idea.

A homeowner wants to add a large garden up against her house. She purchased 100 feet of fencing.

a) Find a formula for the area of the garden.

$$Area = l \cdot w$$

To find the area, we would need the length and width, but we don't have them. The best we can do is start with the perimeter.

$$100 = 2l + w$$

Solving for width, we get:

$$100 - 2l = w$$

If we put this in the area formula, we have an equation dependent only on length.

$$Area = l(100 - 2l)$$

Multiplying this out, we get:

$$A(l) = 100l - 2l^2 \text{ or } A(l) = -2l^2 + 100l$$

b) Find the maximum area for this garden.

We are using different variables, but l is like x and $Area$ is like y. First, use the area formula to find the vertex. Find h using the equation below.

d)

$$h = -\frac{b}{2a}$$

e) Find k by substituting the value of h back into the Area equation.

$$A(h) =$$

The vertex here is the maximum. The value of k is the maximum area.

Finally, let's look at a problem involving a company's revenue.

Revenue is the amount of money which a company makes. It makes sense that the amount of money that a company makes depends on the price of the item and how many they sell.

$$Revenue = Price \cdot Quantity$$

Or:

$$Revenue = p \cdot Q$$

Research has shown that an educational website sells 500 classes at $100 per class and 420 classes at $120 per class. These are simply points on a line. Since price is directly impacting the number of classes sold, we will make x the price.

$$(100, 500) \text{ and } (120, 420)$$

f) Find the slope between them.

g) Use that slope and either point to find an equation for the Quantity. Since price is x, we will put a p (for price) into the equation where x would typically be.

$$Q = mp + b$$

h) The equation you just found is *quantity*. Put that into the revenue formula and then multiply the p.

$$Revenue = p(mp + b)$$

i) You now have a quadratic equation. You would have the maximum *price* if you find the value of h from the vertex. And, you would have the maximum *Revenue* if you find k from the vertex. Find those values.

231

Algebra II

Active Learning: 5.1c

Once again, we want to find the x and y intercepts of a quadratic function.

$$f(x) = x^2 + 4x - 2$$

a) The y-intercept occurs when $x = 0$. Evaluate the function at $x = 0$ and give the y-intercept. (Remember, the y-intercept is a point.)

The x-intercept occurs when $y = 0$.

$$0 = x^2 + 4x - 2$$

b) However, this quadratic can't be factored. Use the quadratic formula to find the x-intercepts. (They will look unusual because they will involve roots, but they still can be written as points.)

c) Try another. Find the x and y intercepts.

$$f(x) = 2x^2 + 3x - 3$$

In the last activity, we saw that the vertex of a parabola will be the maximum or minimum. If you recall our lessons on transformations, the quadratic will be right side up (a smile) if the leading term is a positive. The vertex of this quadratic will be a minimum because it is at the bottom of the smile.

$$f(x) = 2x^2 + 3x - 3$$

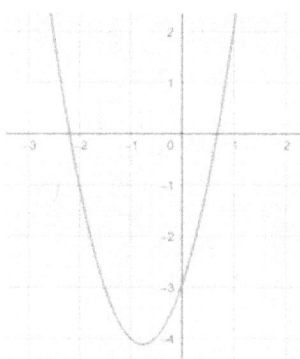

The quadratic will be upside down (a mountain) if the leading term is a negative. The vertex of this quadratic will be a maximum because it is at the top of the mountain.

$$f(x) = -3x^2 + 2x + 2$$

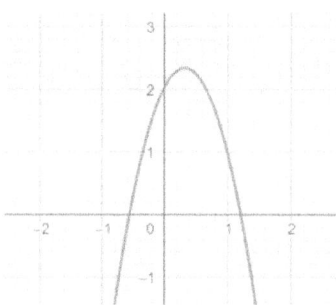

We will use this idea again. This time in an application regarding a projectile, which is common in physics.

A ball is thrown off a 20-foot-high roof. The following quadratic shows the height of the ball based on the time the ball is in the air.

$$H(t) = -16t^2 + 40t + 20$$

The function begins with a negative, so it is upside down and the vertex would be a maximum.

- When does the ball reach the maximum height?

This is asking for the time at which the ball is at the vertex. Time (t) is acting as x, so to find the x-
d) component of the vertex, we need to find h. Find the time of the maximum height.

$$h = -\frac{b}{2a}$$

- What is the maximum height of the ball?

e) The maximum height of the ball would be the y-component of the vertex. So, we need to find k. Find the maximum height of the ball.

$$k = H(h)$$

-When does the ball hit the ground?

f) The x-axis would be the ground. Therefore, the time the ball hits the ground would be the x-intercepts. Use the quadratic formula to find the x-intercepts. Since we want the time, use your calculator to approximate your intercepts.

g) You get two answers. One of the answers is negative. Since this is about time, it can't be negative and so we disregard the negative answer. At what time does the ball hit the ground?

Algebra II

Active Learning: 5.2a

The key to this section will be something called polynomial functions. However, we first need to mention a type of function named a power function.

$$f(x) = ax^n$$

a) Any function which is only one term and the variable can be written with an exponent is a power function. The exponent could be a positive, a negative, or a fraction. Circle any of the following which are power functions:

$$f(x) = 5x^3$$

$$f(x) = 3x$$

$$f(x) = \frac{1}{x}$$

$$f(x) = \sqrt[3]{x}$$

$$f(x) = 3x^2 - 2x$$

Each of the above functions are power functions, except the last. A power function has only a single term but it has two. The last function is a polynomial function. Polynomial functions have multiple terms and the exponents on each term must be positive. In Algebra I, we discussed polynomials. Here, all the ideas are the same except extended to make them functions. (Remember, a function is a machine where for any value of x which goes in there is only one value of y which comes out.) Below, I have taken the Algebra I material regarding polynomials and updated it to reflect polynomial functions.

A polynomial function is made up of different groups (called terms) each separated by addition or subtraction, where the exponents must be positive.

$$f(x) = 4x^3 + 2x^2 + 4x - 3$$

This polynomial has four terms.

There are special names for some smaller polynomials. One term is called a monomial: $g(x) = 3x$. Two terms is called a binomial: $h(x) = 12x^4 - 2x^2$. Three terms is called a trinomial: $r(x) = x^3 - 5x^2 + 8x$. And if it exceeds three terms, we simply called it a polynomial.

Name each of the following polynomials based upon its number of terms:

b) $f(x) = 15x^7 - 30x$

c) $g(x) = 12x^3 - 20x^2 + 6x - 50$

d) $r(x) = 3x^3$

e) $t(x) = x^2 - 6x + 15$

A polynomial function can also be classified by its highest exponent.

$$f(x) = 4x^3 + 2x^2 + 4x - 3$$

The highest exponent here is three, so this is a polynomial of degree three. Give the degree of the following polynomial functions:

f) $f(x) = 15x^7 - 30x$

g) $h(x) = 12x^3 - 20x^2 + 6x - 50$

h) $r(x) = 3x^3$

i) $t(x) = x^2 - 6x + 15$

It can also be important to identify a polynomial functions leading term and leading coefficient.

$$f(x) = 5x - 3x^2 + 4$$

The leading term is the one which has the highest degree. The leading coefficient is the number in front of that leading term. This function is out of order. Rearranged from highest degree down we have:

$$f(x) = -3x^2 + 5x + 4$$

Leading Term: $-3x^2$

Leading Coefficient: -3

Give the leading term and leading coefficient for the following polynomial functions.

$$f(x) = 8 + 17x^2 - 12x^5 - 2x^3$$

j) Leading Term:

k) Leading Coefficient:

$$g(x) = x^4 - 21x^3 + 5x^2 + 16x - 2$$

l) Leading Term:

m) Leading Coefficient:

Algebra II

Active Learning: 5.2b

Polynomial functions are those which have only x's raised to powers. There are no denominators and no special features like square roots or absolute value signs. For instance:

$$f(x) = 3x^4 - 2x^3 + 15x^2 - 5x + 12$$

When graphing polynomial functions, no matter how complicated, they follow a pattern. I'll give you three examples of each. Look at the graphs and the formulas to find the pattern. (Hint: It has to do with the coefficient and degree of the first term.)

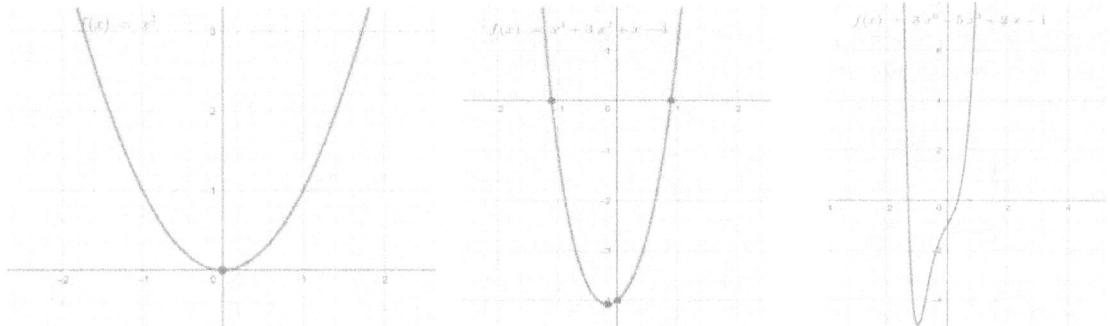

a) This type always opens upward. (Starts in quadrant II and ends in quadrant I.) What two things do they have in common? (If you aren't sure, look ahead at the next groups. It will help.)

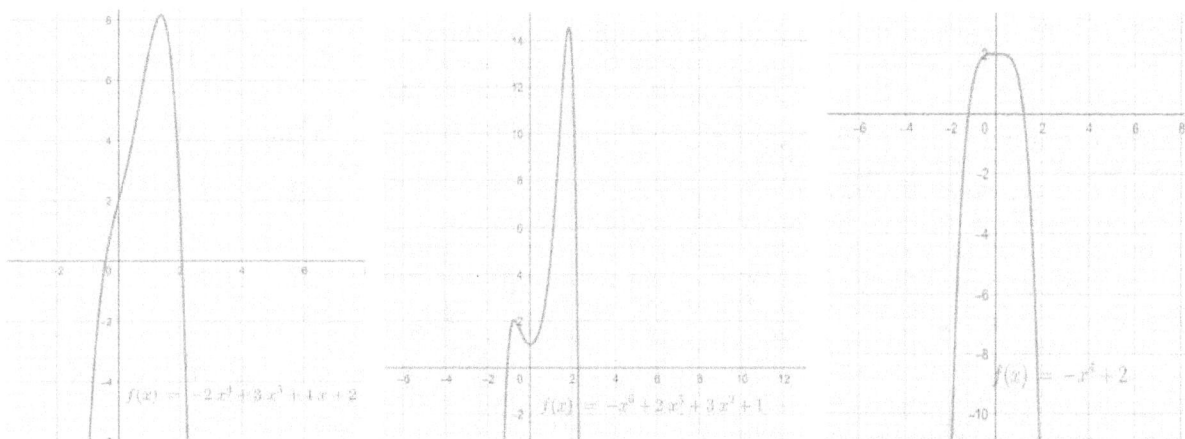

b) This type always opens downward. (Starts in quadrant III and ends in quadrant IV.) What two things do they have in common?

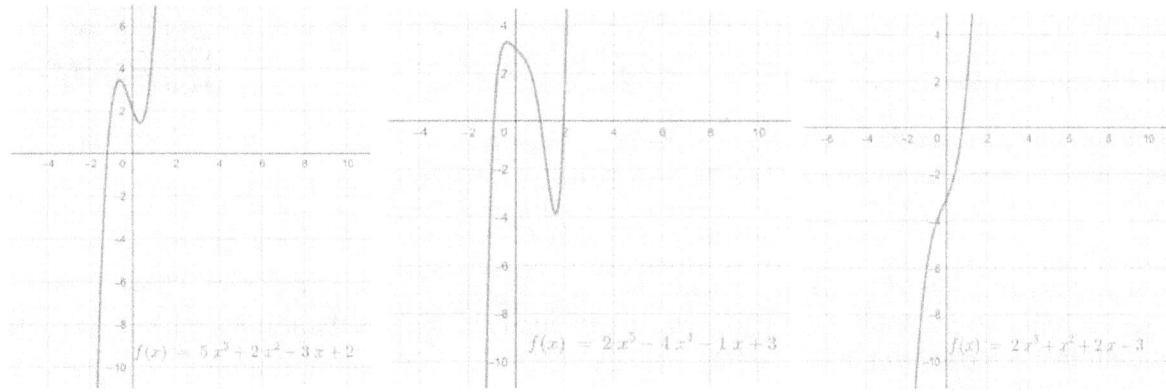

c) This type always goes diagonally upward. (Starts in quadrant III and ends in quadrant I.) What two things do they have in common?

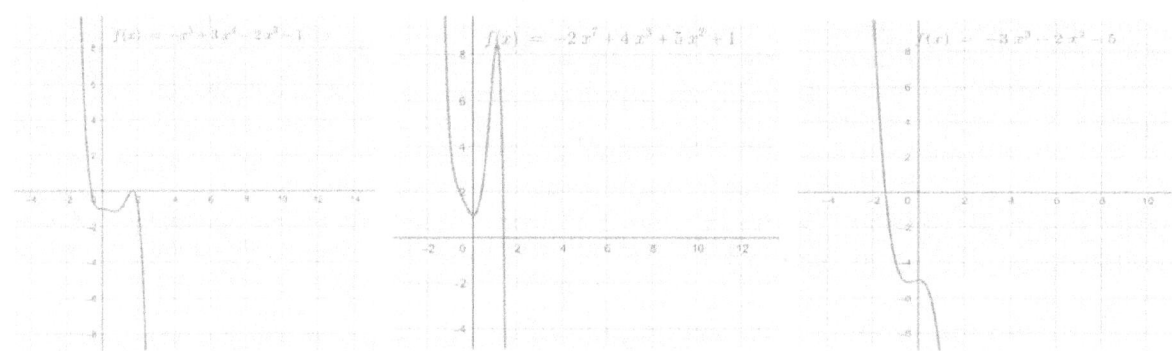

d) This type always goes diagonally downward. (Starts in quadrant II and ends in quadrant IV.) What two things do they have in common?

Finally, look at the following equations and indicate their "end behavior." Your choices are:

- Up Left, Up Right (Opening Upward)
- Down Left, Down Right (Opening Downward)
- Down Left, Up Right (Diagonally Upward)
- Up Left, Down Right (Diagonally Downward)

e) $f(x) = -4x^7 + 5x^6 - 12$

f) $g(x) = -14x^4 + 6x^3 - 2x^2 + 6x - 12$

g) $h(x) = 300x^{20} - 150x + 6$

h) $p(x) = x^5 + 4x^4 + x^3 + x^2 + x + 1$

Each of the choices in the last exercise has special notation.

- Up Left, Up Right (Opening Upward)

$$x \to -\infty, f(x) \to \infty$$
$$x \to \infty, f(x) \to \infty$$

It means: As x becomes negative, y becomes positive. As x becomes positive, y becomes positive. In other words, it opens upward.

- Down Left, Down Right (Opening Downward)

$$x \to -\infty, f(x) \to -\infty$$
$$x \to \infty, f(x) \to -\infty$$

It means: As x becomes negative, y becomes negative. As x becomes positive, y becomes negative. In other words, it opens downward.

Finish the notation for the final two choices.

- Down Left, Up Right (Diagonally Upward)

$$x \to -\infty, f(x) \to$$
$$x \to \infty, f(x) \to$$

- Up Left, Down Right (Diagonally Downward)

$$x \to -\infty, f(x) \to$$
$$x \to \infty, f(x) \to$$

If a polynomial function is in factored form, we can still follow the pattern that we discovered earlier.

$$f(x) = -4x^2(x-2)(x+3)$$

Add up the total number of x's.

- $-4x^2$- Two
- $(x-2)$- One
- $(x+3)$- One

If we multiplied this polynomial out, the highest degree of x would turn out to be 4. And since the leading coefficient would be negative. We have an even degree and a negative coefficient. It would open downward.

Identify the end behavior of the following factored polynomials.

$$f(x) = 3x^3(x+5)(x-2)$$

i) End Behavior:

$$f(x) = 5x^2(2x-3)(x+6)$$

j) End Behavior:

$$f(x) = -x(7x-5)(3x+6)$$

k) End Behavior:

In factored form, it is easy to find the x and y intercepts.

$$f(x) = -4x^2(x-2)(x+3)$$

The x-intercept occurs when $y = 0$.

$$0 = -4x^2(x-2)(x+3)$$

Because it is factored, we can use the zero-product property. Find the three x-intercepts. Be sure to give them as coordinates.

l) $\qquad -4x^2 = 0 \text{ or } x - 2 = 0 \text{ or } x + 3 = 0$

The y-intercept occurs when $x = 0$.

$$f(0) = -4(0)^2(0-2)(0+3)$$

m) Find the y-intercept. Be sure to give it as a coordinate.

n) Find the x and y intercepts for the following polynomial function.

$$f(x) = 5x^2(2x-3)(x+6)$$

For unfactored polynomial functions, check to see if they can be factored to find the x and y intercepts.

$$f(x) = x^4 - 13x^2 + 36$$

Even though this is higher order, the factoring is basic. What two numbers multiply to get 36 and add to get -13?

$$f(x) = (x^2 - 4)(x^2 - 9)$$

o) The factoring isn't complete. Each of the terms is a perfect square binomial. In the space below, factor them.

p) Give the x and y intercepts.

When factoring a polynomial function, it is possible that you get complex solutions. Because complex solutions involve imaginary numbers, they will not show up on the cartesian plane and therefore they won't be x-intercepts.

To find x-intercepts, we set a polynomial function equal to zero.

$$0 = x^4 - 13x^2 + 36$$

An important algebra theorem, called the Fundamental Theorem of Algebra, says that there will be one solution per degree. Here, the degree is 4, so there should be four solutions. However, some solutions could be complex and therefore not x-intercepts. So, we can know the maximum possible number of x-intercepts. Give the maximum possible number of x-intercepts in each of the following:

q) $f(x) = 3x^7 - 15x^4 + 21$ r) $g(x) = -25x^5 + 17x^4 - 121x^3$ s) $h(x) = (x^3 - 27)(x^2 - 16)$

Finally, we want to look at the idea of "turning points."

Here is the graph of the function $f(x) = x^4 - 13x^2 + 36$.

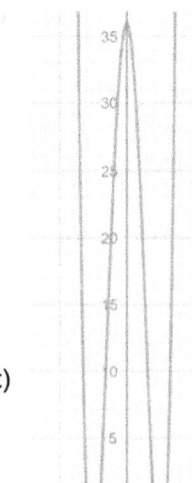

We know that there is a maximum of four x-intercepts, and the graph actually shows that there are four.

Graphs go left to right up the x-axis. A turning point is literally a place where the graph turns from down to up or up to down.

t) How many turning points does this polynomial function have?

Every turning point occurs because the graph needs to go through a pair of x-intercepts. Because of this, it will always cause the maximum number of turning points to be one less than the degree.

Find the maximum number of turning points for the following polynomial functions.

u) $f(x) = 3x^7 - 15x^4 + 21$ v) $g(x) = -25x^5 + 17x^4 - 121x^3$ w) $h(x) = (x^3 - 27)(x^2 - 16)$

Algebra II

Active Learning: 5.3a

The graphs of polynomial functions will be smooth curves. (No sharp points of any kind.) And they will be continuous. (They never have holes or breaks.)

a) Circle any of the following graphs which are polynomial functions.

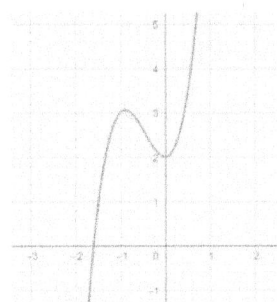
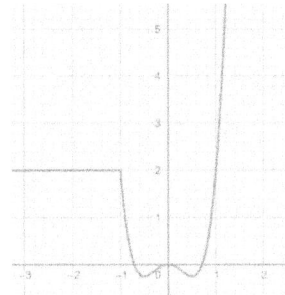

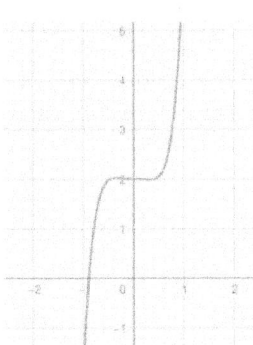
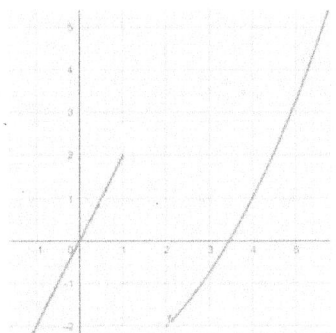

In the last activity, we learned how to find the x-intercepts of a polynomial function which could be factored. We will do the same here with some new techniques.

$$f(x) = x^5 - 5x^3 + 6x$$

When factoring, always start with a GCF. The GCF here is x.

$$x(x^4 - 5x^2 - 6) = 0$$

Now that the GCF has been removed, the right can be factored as we did last time.

$$x(x^2 - 3)(x^2 - 2) = 0$$

b) Finish finding the x-intercepts. There will be five intercepts, and four of them will involve square roots. (Remember, taking a square root creates a $\pm$.)

$$x(x^2 - 3)(x^2 - 2) = 0$$

c) Try another. Factor to find the x-intercepts.

$f(x) = x^6 - 9x^4 + 20x^2$

d) Next, we have a variation which involves factoring by grouping. Pull a GCF from the first two terms and the last two terms. (On the back, pull a -4.)

$$g(x) = \underline{x^3 - 4x^2} \; \underline{-4x + 16}$$

e) Done properly, you should have a matching term of $(x - 4)$. Pull that term out as a GCF.

f) With the function now factored, find the x-intercepts.

Next, use the graph below to find the x-intercepts of the function $h(x) = x^3 - 7x^2 + 12x$.

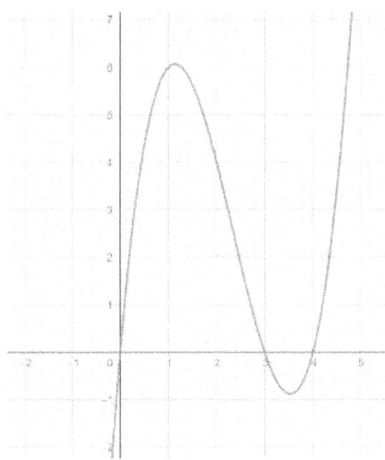

g) x-intercepts:

h) Test those intercepts by evaluating the function at those values of x. (If you have the correct intercept, the function will equal zero.)

Our final idea will be an important one for the next activity. It begins by examining a polynomial function in factored form.

$$f(x) = x^3(x - 4)(x + 5)^2$$

The x-intercepts are 0, 4 and −5. But there are actually repeats. Technically our factored function is:

$$f(x) = x \cdot x \cdot x(x - 4)(x - 5)(x - 5)$$

So, the intercept at 0 occurs three times and the intercept at 5 occurs twice. When an x-intercept repeats it is called *multiplicity*. Here is how I would write my x-intercepts (also called roots) noting the multiplicity.

$$(0, 0) \times 3$$
$$(0, 4) \times 1$$
$$(0, 5) \times 2$$

Find the x-intercepts (roots) and multiplicities of the following polynomial functions. Write the roots in such a way as to indicate the multiplicities.

i) $f(x) = x^2(2x+4)^3(x-1)$

j) $g(x) = x(x+12)^2(3x-2)^2$

When we graph polynomial functions, roots with multiplicities do different things at the x-intercepts. Here is a graph of the function:

$$f(x) = x^2(4x-4)(x+1)^3$$

k)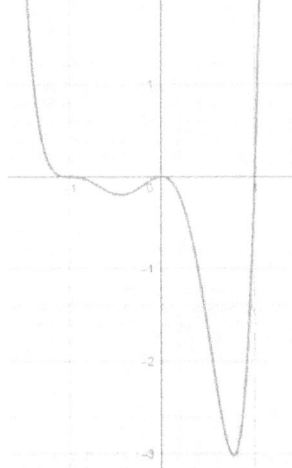

Write the roots and multiplicities.

Notice how the multiplicities have different behavior at the x-intercepts. A multiplicity of 1 always goes straight through the intercept. A multiplicity of 2 bounces at the intercept and returns in the direction from which it came. A multiplicity of 3 goes through the intercept but it flattens out both before and after it crosses.

Using the graphs below, identify the x-intercepts (roots) and note the multiplicities based on the behavior of the graph.

l)

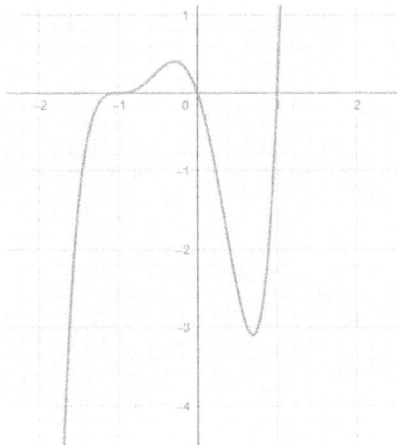

m)

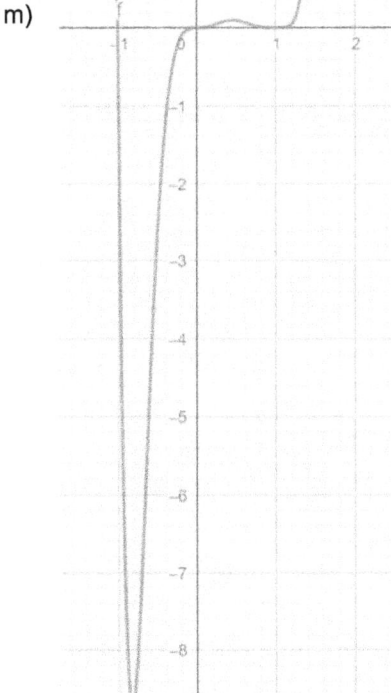

Algebra II

Active Learning: 5.3b

In this activity, we will put together the ideas from the last few sections in order to create a graph of a polynomial function. First, we will need the end behavior.

$$f(x) = x^3(x-4)(x+5)^2$$

a) There are a total of 6 x's in this function and the leading term is positive. Which of the following would be true about the end behavior of the graph?

    a) Up Left, Up Right (Opening Upward)
    b) Down Left, Down Right (Opening Downward)
    c) Down Left, Up Right (Diagonally Upward)
    d) Up Left, Down Right (Diagonally Downward)

Give the end behavior of the following functions:

b) $g(x) = -x^2(x+7)(2x-4)^2$

c) $h(x) = -x(x-1)^3(x+3)^2$

The book briefly returns to the idea of the maximum number of turns for the polynomial function. In truth, you don't need this to make the graph. However, the maximum possible turns is one less than the degree of the polynomial.

$$f(x) = x^3(x-4)(x+5)^2$$

There are 6 x's involved in this polynomial, so the maximum possible turns would be 5.

Give the maximum possible turns for the two polynomial functions above.

d) $g(x)$ e) $h(x)$

Possible Turns: Possible Turns:

We will also need the roots (x-intercepts) and multiplicities to graph a polynomial function. Give the roots and multiplicities of each of the three functions we've used.

f) $f(x)$ g) $g(x)$ h) $h(x)$

We put that information together to make a graph. Here is the process for $f(x)$. First, we mark the end behavior.

$$f(x) = x^3(x-4)(x+5)^2$$

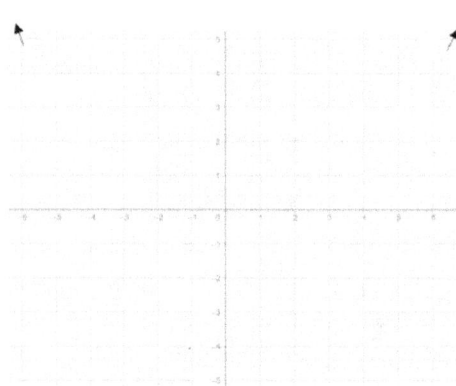

Then we mark the roots. (I've exaggerated my points to help them show up.) You could find and add the y-intercept as well, however I don't typically require it.

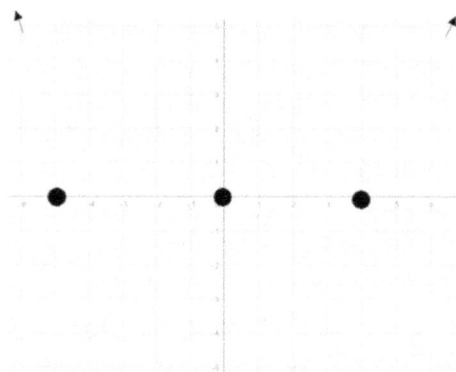

Now, since a graph goes from left to right, and with the knowledge of the multiplicities, there will only be one way in which the graph could be built.

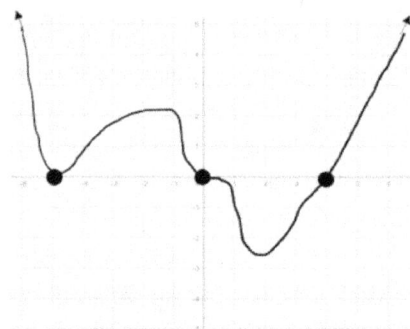

We bounce at -5 because of a multiplicity of 2. We pass in and out flat at 0 because of a multiplicity of 3. And, we go directly through at 4 because of a multiplicity of 1.

Use the same process to create the graphs of the following polynomial functions. (You do not need an accurate y-intercept.)

i) $g(x) = -x^2(x+7)(2x-4)^2$

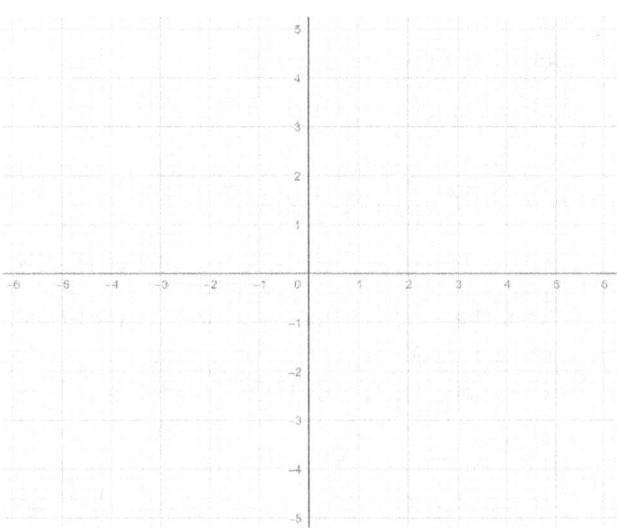

j) $h(x) = -x(x-1)^3(x+3)^2$

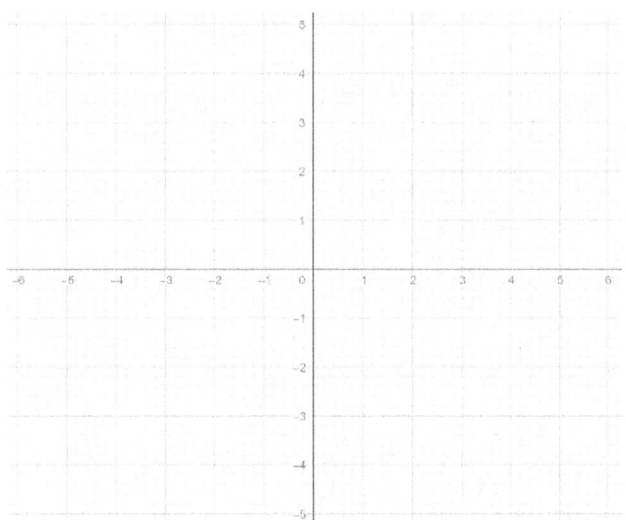

Finally, we will start with the graph and find the equation for a polynomial function.

Let's begin with this graph.

We have three roots, and our function would look like this:

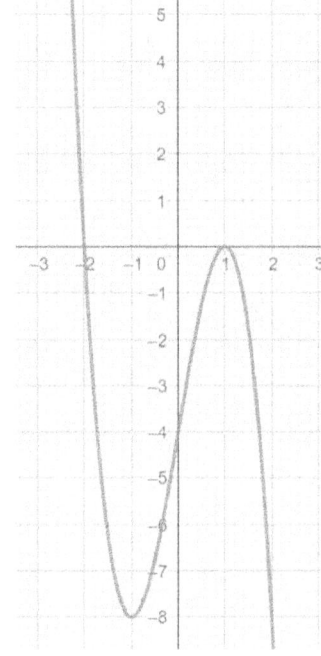

$$f(x) = a(x - 1)(x + 2)$$

Remember that the root at 1 actually became the $x - 1$ term. And, the root at -2 became $x + 2$.

The a must be put in front due to the possibility of a stretching factor.

Next, the action at the roots can tell us the multiplicities. We went straight through at -2 and so it has a multiplicity of one. We bounced at 1 and so our multiplicity there must be two.

$$f(x) = a(x - 1)^2(x + 2)$$

We need to solve the stretching factor, a. To do so, we need to enter a clear point (other than an x-intercept) from the graph into the equation. This will allow us to have only a as a variable. A clear point would be $(-1, -8)$. Plug this point into our function and solve for a.

k)

l) Give the final equation.

Algebra II

Active Learning: 5.4a

(The concept of dividing polynomials was covered in Algebra I. This activity was borrowed from that course.)

The idea in this section will follow something you know, but which you may not have done for a while. We are going to do long division. Divide the following. If you remember how, go ahead. If not, I will walk you through it below.

a)   12 | 2 5 9

First, can 12 go into the 2. No. So, try going into the first two numbers.

```
12 | 2 5 9
   - 0
     2 5
```

Now, 12 goes into 25 two times. 2 times 12 is 24. Check for any leftovers by subtracting. There is one left over.

```
       0 2
12 | 2 5 9
   - 0
     2 5
   - 2 4
         1
```

Bring down the 9 to join the 1.

```
        0  2  1
    12 | 2  5  9
       - 0
         2  5
       - 2  4
            1  9
          - 1  2
               7
```

You can't divide any further and so your final answer is: 21 R7. (Remember the R is the remainder.) Now we are going to do it with polynomials. But it will follow the exact same process. Try this if you can. If not, I will walk you through.

b) $\quad x \,|\, x^4 + x^3 - x^2 + x + 1$

We will follow the exact same steps as regular long division. How many times does x go into x^4? It can go x^3 times. x^3 times x is x^4. Check for any leftovers by subtracting. There aren't any here. Bring down the next number to continue.

$$\begin{array}{r} x^3 \\ x \,|\, x^4 + x^3 - x^2 + x + 1 \\ - x^4 \\ \hline 0 + x^3 \end{array}$$

Play the game again. How many times does x go into x^3? It can go x^2 times. x^2 times x is x^3. Check for any leftovers by subtracting. There aren't any here. Bring down the next number to continue.

```
              x³  +  x²
         _____
    x  |  x⁴  +  x³  -  x²  +  x  +  1
       -  x⁴
       _____
           0  +  x³         ↓
              -  x³
              _____
                 0  -  x²
```

Play the game again. How many times does x go into $-x^2$? It can go $-x$ times. $-x$ times x is $-x^2$. Check for any leftovers by subtracting. There aren't any here. Bring down the next number to continue.

```
              x³  +  x²  -  x
         _____
    x  |  x⁴  +  x³  -  x²  +  x  +  1
       -  x⁴
       _____
           0  +  x³                ↓
              -  x³
              _____
                 0  -  x²       ↓
                    -  x²
                    _____
                       0  +  x
```

Play the game again. How many times does x go into x? It can go 1 time. x times 1 is x. Check for any leftovers by subtracting. There aren't any here. Bring down the next number to continue.

```
              x³  +  x²  -  x  +  1
         _____
    x  |  x⁴  +  x³  -  x²  +  x  +  1
       -  x⁴
       _____
           0  +  x³
              -  x³
              _____
                 0  -  x²
                    -  x²
                    _____
                       0  +  x
                          -  x
                          _____
                             0  R  1
```

You can't play again. x can't go into 1, so it is your remainder. In algebra, we will take the remainder and put it over the divisor. Here is what our answer would look like:

$$x^3 + x^2 - x + 1 + \frac{1}{x}$$

Try one on your own:

c) $\quad x \overline{\smash{\big)}\, x^5 \;-\; x^4 \;+\; x^3 \;+\; x^2 \;-\; x \;+\; 2}$

Unfortunately, the problems aren't this easy. We are typically dividing by more than a simple monomial, but the plan stays the same.

$\quad x+1 \overline{\smash{\big)}\, 2x^4 \;+\; 5x^3 \;-\; 3x^2 \;+\; x \;+\; 6}$

Divide the first term of the binomial into the first term of the polynomial. How many times can x go into $2x^4$? It can go $2x^3$. But this time, we will multiply $2x^3$ times the binomial out front. $2x^3(x+1) = 2x^4 + 2x^3$. Check for any leftovers by subtracting. (Be careful here, you are changing the signs of both terms.) This time there are $3x^3$. Bring down the next number to continue.

```
                    2x³
x+1 | 2x⁴   +   5x³   -   3x²   +   x   +   6
   -  (2x⁴  +   2x³)
                3x³   -   3x²
```

Play again. How many times can x go into $3x^3$? It can go $3x^2$. Multiply $3x^2$ times the binomial out front. $3x^2(x + 1)$. Check for any leftovers by subtracting. (Again, the negative sign distributes.) There are $-6x^2$. Bring down the next number to continue.

```
                    2x³  +  3x²
         ┌─────────────────────────────────────
    x+1  │  2x⁴  +  5x³  -  3x²  +  x  +  6
       - (2x⁴  +  2x³)
         ─────────────
                3x³  -  3x²
             - (3x³  +  3x²)
               ─────────────
                      -6x²  +  x
```

Play again. How many times can x go into $-6x^2$? It can go $-6x$. Multiply $-6x$ times the binomial out front. $-6x(x + 1)$. Check for any leftovers by subtracting. (Again, the negative sign distributes.) There are $7x$. Bring down the next number to continue.

```
                    2x³  +  3x²  -  6x
         ┌─────────────────────────────────────
    x+1  │  2x⁴  +  5x³  -  3x²  +  x  +  6
       - (2x⁴  +  2x³)
         ─────────────
                3x³  -  3x²
             - (3x³  +  3x²)
               ─────────────
                      6x²  +  x
                   - (6x²  -  6x)
                     ─────────────
                             7x  +  6
```

Play one more time. How many times can x go into $7x$? It can go 7. Multiply 7 times the binomial out front. $7(x + 1)$. Check for any leftovers by subtracting. (Again, the negative sign distributes.) There is -1.

```
                    2x³  +  3x²  -  6x  +  7
         ┌─────────────────────────────────────
    x+1  │  2x⁴  +  5x³  -  3x²  +  x  +  6
       - (2x⁴  +  2x³)
         ─────────────
                3x³  -  3x²
             - (3x³  +  3x²)
               ─────────────
                      6x²  +  x
                   - (6x²  -  6x)
                     ─────────────
                             7x  +  6
                          - (7x  +  7)
                            ─────────
                                    -1
```

Write your remainder as a fraction and your final answer is: $2x^3 + 3x^2 - 6x + 7 - \frac{1}{x+1}$.

Try one on your own.

d) $x-2 \overline{\smash{\big)}\ 3x^4 - 5x^3 - 4x^2 + 4x - 2}$

When you do long division, you need terms of all the degrees. For instance, the problem below is missing a term in x^3 position.

$2x+1 \overline{\smash{\big)}\ 8x^4 - 4x^2 - 7x + 4}$

Since a term is missing and we need one, we add a placeholder with a zero. The process works then works the same. Finish the problem now that it has the placeholder.

e) $2x+1 \overline{\smash{\big)}\ 8x^4 + 0x^3 - 4x^2 - 7x + 4}$

Algebra II

Active Learning: 5.4b

a) The following quadratic function is in factored form. What are the roots (the x-intercepts)?

$$f(x) = (x+3)(x-1)$$

b) Use long division to divide the following:

$$(2x^3 - 6x^2 + 8x + 5) \div (x-1)$$

In this activity, we will learn an approach called synthetic division. Synthetic division starts with the root of the divisor. So, to divide $(2x^3 - 6x^2 + 8x + 5) \div (x-1)$ we will use 1. And we will only list the coefficients.

```
1 | 2  -6  8  5
  |_____
```

Next, bring down the first coefficient.

```
1 | 2  -6  8  5
  |_____
    2
```

Now, multiply the 1 by the 2.

```
1 | 2  -6  8  5
  |    2
    2
```

Add down.

```
1 | 2  -6  8  5
  |    2
    2  -4
```

Repeat the process. Multiply 1 times -4. Then add.

```
1 |  2   -6    8    5
  |       2   -4
  |_____
     2   -4    4
```

Continue until you run out of terms.

```
1 |  2   -6    8    5
  |       2   -4    4
  |_____
     2   -4    4    9
```

Now, return the x's to your numbers. Since you divided by $x - 1$, what remains has a degree which is one lower, and your final term is your remainder.

$$2x^2 - 4x + 4 + \frac{9}{(x-1)}$$

This process is simply a shortcut version of long division. By using the root, we have accounted for the subtraction in long division. However, synthetic division will only work for binomials of degree 1 with a leading coefficient of 1. Otherwise, the shortcut breaks down. Try one on your own.

c) Divide: $(6x^3 + 3x^2 - 8x + 2) \div (x - 2)$

Just as we saw with long division, we must add placeholders if we have a missing term. Since there is nothing in the x^3 place, we would add a zero.

d) Divide: $(2x^4 - 5x^2 - 3x + 9) \div (x + 4)$

```
-4 |  2    0   -5   -3    9
   |
   |_____
```

Try one more.

e) Divide: $(8x^4 + 4) \div (x - 2)$

Algebra II
Active Learning: 5.5a

a) Use either synthetic or long division to divide the following:

$$x^3 + 3x^2 - 4x - 12 \div (x + 3)$$

b) What is your remainder?

c) Now factor the following and simplify.

$$\frac{x^3 + 3x^2 - 4x - 12}{x + 3}$$

d) Use either synthetic or long division to divide the following:

$$x^3 - x^2 + 5x - 5 \div (x - 1)$$

e) What is your remainder?

f) Now factor the following and simplify.

$$\frac{x^3 - x^2 + 5x - 5}{x - 1}$$

g) If you divide a polynomial and there is no remainder, you have found a factor. Use long division or synthetic division to divide the following.

$$-6x^5 + 6x^4 - 2x^3 - 8x^2 + 6x - 2 \div (x - 1)$$

h) Is $x - 1$ a factor of $f(x) = -6x^5 + 6x^4 - 2x^3 - 8x^2 + 6x - 2$? How do you know?

If our goal was to find the x-intercept (or root) of $x - 1$, we would set it equal to zero.

$$x - 1 = 0$$

i) And the root would be $x = 1$. Evaluate the polynomial function $f(x) = -6x^5 + 6x^4 - 2x^3 - 8x^2 + 6x - 2$ at $x = 1$.

$f(1) =$

j) There is a connection between the number you got when you evaluated $f(1)$ and the remainder you found when you divided $f(x)$ by $x - 1$. What is the connection?

k) Next, divide $g(x) = 3x^4 - 2x^3 + 5x^2 - 8x + 12$ by $x + 2$. What is your remainder?

l) The root of $x + 2$ is -2. Evaluate $g(-2)$.

m) There is a connection between the number you got when you evaluated $g(-2)$ and the remainder you found when you divided $g(x)$ by $x + 1$. What is the connection?

We have two major concepts (known as theorems) in this section. The first concept was the factor theorem. If you divide a polynomial function by another polynomial function and the remainder is zero, you have found a factor. The second concept is called the remainder theorem. If you want to divide a polynomial function by a binomial, evaluating the root of the binomial will get you the same remainder you would get if you divided.

Use the remainder theorem to indicate what remainder you would get if you divided the following polynomials. (Don't divide. Just evaluate at the root.)

n) $(4x^3 + 6x^2 - 2x + 9) \div (x - 1)$ (Evaluate at 1.)

o) $(6x^3 + 4x^2 - 40x + 6) \div (x + 3)$ (Evaluate at -3.)

p) Did either of the problems involve factors? If so, circle the factor.

Algebra II

Active Learning: 5.5b

Earlier, we learned that the Fundamental Theorem of Algebra says that there is one root for each degree of a polynomial function.

$$f(x) = x^3 + 3x^2 - 4x - 12$$

Because this function is degree 3, it has three roots. However, the roots may or not all be rational numbers. It is possible some involve imaginary numbers, which would not show up on a Cartesian Plane as x-intercepts. In this activity, we want to learn a theorem which will help us find the rational roots. The theorem is called the Rational Zero Theorem.

$$\pm \frac{p}{q} = \pm \frac{factors\ of\ the\ constant}{factors\ of\ the\ lead\ coefficient}$$

The rational zeros must be among the combinations created by this formula. Looking at $f(x)$, we have the following:

$Constant: 12\ \ \pm 1, \pm 2, \pm 3, \pm 4, \pm 6, \pm 12$

$Lead\ Coefficient: 1\ \ \pm 1$

Now, we make all the possible combinations of $\pm \frac{p}{q}$.

$$\pm \frac{p}{q} = \pm \frac{1}{1}, \pm \frac{2}{1}, \pm \frac{3}{1}, \pm \frac{4}{1}, \pm \frac{6}{1}, \frac{12}{1} = \pm 1, \pm 2, \pm 3, \pm 4, \pm 6, \pm 12$$

This is a lot of numbers, but using the Remainder Theorem, we could quickly find which have a remainder of zero, and if it has a remainder of zero, it is a root. (In a TI calculator, you can enter the function and then evaluate the function at each of these possibilities.)

Evaluate $f(x)$ at the following:

a) $f(1) =$

b) $f(-1) =$

c) $f(2) =$

d) $f(-2) =$

e) $f(3) =$

f) $f(-3) =$

g) $f(4) =$

h) $f(-4) =$

i) $f(6) =$

j) $f(-6) =$

k) $f(12) =$

l) $f(-12) =$

m) Circle those above which are rational roots. (Remember, they will equal zero.)

n) Try one on your own. Use the Rational Zero Theorem to find the roots of the following polynomial function.

$$g(x) = x^3 + 4x^2 + x - 6$$

Although we can, it isn't necessary to use the Rational Zero Theorem to find all the roots. Upon discovering one, it is sometimes possible that we are left with something we can factor.

$$h(x) = x^3 + 2x^2 - 5x - 6$$

For $h(x)$, -3 is a factor. So, we could use division to divide out this factor. If we did, we would have:

$$(x + 3)(x^2 - x - 2)$$

o) Factor $x^2 - x - 2$ to find the other two roots.

p) What are the three roots of $h(x)$?

q) Use the Rational Zero theorem to find one root of the following polynomial function. Then, use division and factoring to find the other two.

$$f(x) = x^3 + 5x^2 - x - 5$$

The Fundamental Theorem of Algebra tells us that the following polynomial function has three roots.

$$r(x) = 4x^3 + 2x^2 + 3x - 9$$

However, only one of them is a rational root. It has a rational root at 1. Using division, we find $r(x)$ in factored form.

$$(x - 1)(4x^2 + 6x + 9)$$

r)   Now, use the quadratic formula on $4x^2 + 6x + 9$ to find the two complex roots.

s)   Work one final problem. Find the roots of the following polynomial function. Only one is rational. Use the Rational Zero Theorem to find that root. Then, divide. Finally, use the quadratic formula (or algebraic methods) to find the two complex roots.

$$f(x) = x^3 - 2x^2 + x - 2$$

Algebra II

Active Learning: 5.5c

Here is the quadratic formula.

$$\frac{-b \pm \sqrt{b^2 - 4ac}}{2a}$$

a) Use the formula to find the roots of $x^2 - 2x + 4$.

The two roots you found are called complex conjugates. We have seen the idea of conjugates before. Here are conjugates.

$$(x - 3) \text{ and } (x + 3)$$

$$(2 + \sqrt{6}) \text{ and } (2 - \sqrt{6})$$

Everything is the same except they have the opposite sign between them. Here is a pair of complex conjugates.

$$3 + 4i \text{ and } 3 - 4i$$

When finding roots, the $\pm$ in the quadratic formula will always ensure that we get complex numbers as a pair of conjugates.

The application of this is called the Complex Conjugate Theorem. If we know a complex number is a root of a polynomial function then we know its complex conjugate must be also.

Here's an application.

Find a polynomial function which has roots at 2 and $-3i$ such that $f(1) = 20$.

Because of the complex conjugate theorem, we know there is also a root at $3i$. Here is what we know about the equation:

$$f(x) = a(x - 2)(x + 3i)(x - 3i)$$

b) Use the knowledge that $f(1) = 20$ to find the value of a and complete the formula.

Try one on your own.

c) Find a polynomial function which has roots at 1 and $2i$ such that $f(2) = 16$.

So far, we have learned a number of theorems to help us find the roots of a polynomial function. However, in a real-world application, the degree of the polynomial function could be very high and finding the roots by the Rational Zero Theorem might be quite time consuming. Because of this, we have one more tool to provide us with additional information for finding roots. It comes from Rene Descartes and is called Descartes' Rule of Signs.

$$f(x) = -x^4 - 3x^3 + 12x^2 - 6x - 9$$

Descartes found that the sign changes could help find the possible number of rational roots. First, we can find the possible number of positive roots. Find the number of sign changes between the terms.

$$f(x) = -x^4 \underbrace{- 3x^3 + 12x^2}_{} \underbrace{- 6x - 9}_{}$$

As you go down the function, the sign changes from + to − twice. The maximum positive roots are then 2, *or it could continue to decrease by two until it gets to zero*. So, 2 or 0.

To find the possible negative roots, he put $-x$ into the function and simplified. Then he found the sign changes.

$$f(-x) = -(-x)^4 - 3(-x)^3 + 12(-x)^2 - 6(-x) - 9$$

$$f(-x) = -x^4 + 3x^3 + 12x^2 + 6x - 9$$

The maximum negative roots is 2, *or it could continue to decrease by two until it gets to zero*. So, 2 or 0. Therefore:

$$+: 2 \text{ or } 0$$

$$-: 2 \text{ or } 0$$

Armed with this information, we have an additional tool to reason our way through the possible roots.

Use Descartes Rule of Signs to list the possible number of positive and negative roots for the following polynomial functions.

d) $f(x) = 4x^4 - 8x^3 + 6x^2 - 5x + 1$

e) $g(x) = 3x^4 - 12x^3 - 8x^2 - 7x + 5$

Algebra II
Active Learning: 5.6a

Asymptotes are lines which impact the graph of a function. There are three types. The first are <u>vertical asymptotes</u>. I have graphed the function (called a rational function because it has a polynomial divided by a polynomial):

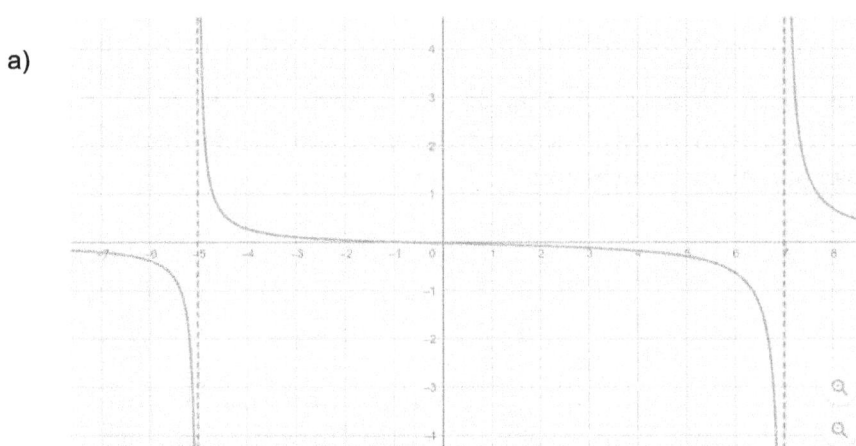

$$f(x) = \frac{x+1}{x^2 - 2x - 35}$$

a) Give me the equations for the lines of the two vertical asymptotes.

b) Factor the rational function in the space below.

c) Look at the equations for your two vertical asymptotes and your factoring of the function. Explain where vertical asymptotes come from.

<u>Horizontal Asymptotes</u>

The first term on the top and the first term on the bottom of a rational function indicate if there is a horizontal asymptote. There are two patterns we want to identify. Here are two examples of the first:

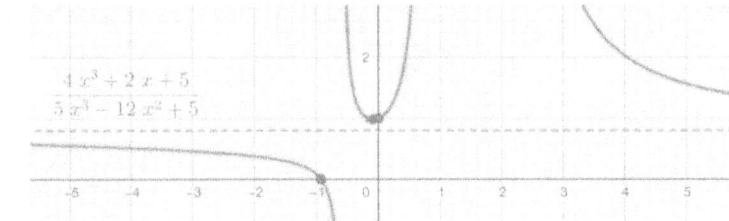

d) What is the equation of the horizontal asymptote?

e) 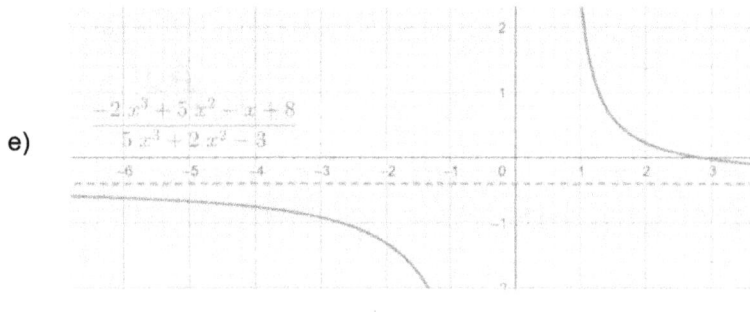 What is the equation of the horizontal asymptote?

f) Compare the equations of the horizontal asymptotes with the equations of the rational functions. (Remember, we only need to look at the first term on the numerator and denominator.) How does the rational function tell us the equation of the horizontal asymptote?

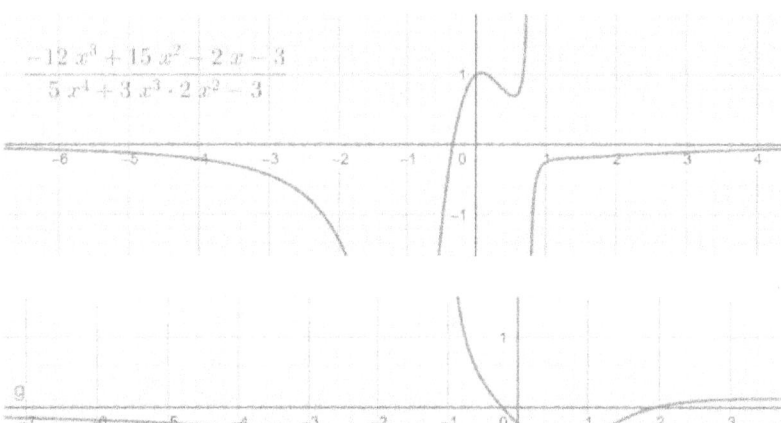 The following two graphs have a horizontal asymptote at $y = 0$. Examine the equations of the rational functions.

g) When do rational functions have a horizontal asymptote at $y = 0$? (Once again, you only need to look at the first two terms to find the pattern.)

Slant Asymptote

A third type of asymptote is called a slant asymptote. It limits the graph of the rational function diagonally. Below is a graph of the rational function $f(x) = \frac{2x^2+x-9}{x-3}$ along with its slant asymptote. Use the two marked points to find the equation for the slant asymptote. (Slant asymptotes are always lines.)

h) What is the equation of the slant asymptote?

Here is a zoomed-out graph of the function and its slant asymptote. (The function also has a vertical asymptote at $x = 3$.)

Slant asymptotes occur when the degree of the numerator (the polynomial on the top) is greater than the degree of the denominator (the polynomial on the bottom.)

i) When this occurs, we divided the bottom into the top. Divide the following:

$$(2x^2 + x - 9) \div (x - 3)$$

The slant asymptote is the equation of the line you get after dividing. However, we drop off the remainder. Here's why:

$$R: \frac{12}{x-3}$$

This is the remainder when you divided for your slant asymptote. Now, let's see what would happen if the values of x become very large or very small. First, I have randomly selected 1000 and -1000. Evaluate the remainder for the following: (Use a calculator.)

j) $x = 1000$

$R =$

k) $x = -1000$

$R =$

Now, evaluate the remainder at these values.

l) $x = 10000$

$R =$

m) $x = -10000$

$R =$

Finally, evaluate the remainder at these values.

n) $x = 100000$

$R =$

o) $x = -100000$

$R =$

p) What is happening to the value of the remainder as the values of x get larger (and smaller).

For this reason, when we find a slant asymptote, we can disregard the remainder because at the two extreme ends of the graph the remainder is irrelevant.

Removable Discontinuity

q) Finally, let's look at the rational function $f(x) = \frac{x-1}{x^2-1}$. Factor the denominator. Where would you expect to find vertical asymptotes? (Give the answers as vertical lines.)

Here is the graph.

r)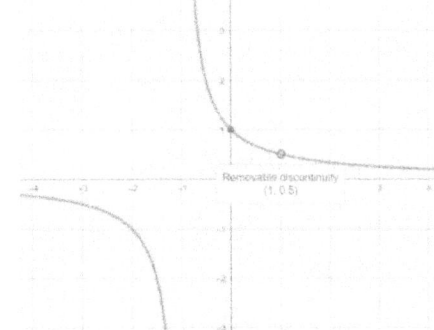

In place of one of the vertical asymptotes we have something called a removable discontinuity. You have already factored $f(x)$. Now, reduce the fraction by cancelling anything on the top and bottom which can be cancelled. Give your reduced function below.

s) Which root dropped out when you reduced?

A removable discontinuity is created when a vertical asymptote is "reduced-out" of the function. As we saw in Algebra I, any value which would have made the original problem undefined is not allowed. So, we still have a "hole" in the graph, but the vertical asymptote is removed.

t) The following function would have a removable discontinuity. Find the value of that discontinuity.

$$h(x) = \frac{x-2}{x^2 + 2x - 8}$$

Algebra II

Active Learning: 5.6b

a) Below is a graph of the rational function $f(x) = \frac{x-1}{x+5}$. I have marked the x and y intercepts as well as a number of other points. Remember that graphs will not cross vertical asymptotes. (They occasionally cross horizontal asymptotes, but that does not happen in this graph.) Use your knowledge of asymptotes and the points marked below to fill in the graph of the function.

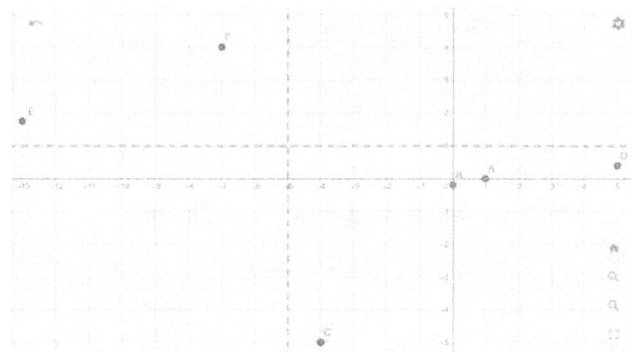

If we had not started with the information given, we could create the graph with five key pieces of information.

$$f(x) = \frac{x-1}{x+5}$$

1) Vertical Asymptotes

The denominator $x + 5$ causes the vertical asymptote. It occurs at values which make the denominator undefined. Here that would be $x = -5$.

2) Horizontal Asymptotes

The degrees of the numerator and denominator determine the horizontal asymptotes. Both are the same degree and so we take the coefficients of the leading terms: $y = \frac{1}{1} = 1$

3) x-intercepts

x-intercepts occur when $y = 0$.

$$0 = \frac{x-1}{x+5}$$

But the denominator will always multiply up to the other side and drop out, leaving:

$$0 = x - 1$$

$$x = 1$$

So, we have the x-intercept $(1, 0)$.

4) *y*-intercepts

y-intercepts occur when $x = 0$.

$$f(0) = \frac{0-1}{0+5} = -\frac{1}{5}$$

So, the *y*-intercept is the point $(0, -\frac{1}{5})$.

5) Does it cross the horizontal asymptote?

Occasionally, a graph will cross a horizontal asymptote. To test this, we check to see if a point exists at the value of the horizontal asymptote. The asymptote occurs at $y = 1$.

$$1 = \frac{x-1}{x+5}$$

If there is a value of *x* which occurs at this value of *y*, the graph crosses. If there is no value of *x* which occurs, the graph does not cross. Attempting to solve for *x*, we have.

$$x + 5 = x - 1$$
$$5 = -1$$

The *x*'s dropped out and 5 does not equal -1. So, no point exists, and the graph does not cross the horizontal asymptote.

Graphing the information above, we have:

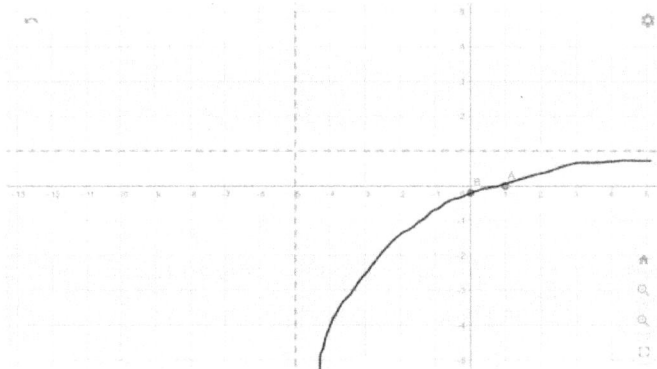

Knowing that the graph goes left to right, there is only on way to go through points B and A. To finish the graph, we need one more bit of information. Look at the vertical asymptote in the original equation.

$$f(x) = \frac{x-1}{x+5}$$

The concept is similar to multiplicity. If the degree (the exponent on the asymptote) is odd, on the opposite side of the asymptote, the graph will go in the opposite direction. If the degree (the exponent on the asymptote) is even, on the opposite side of the asymptote, the graph will go in the same direction. Ours is degree one, $(x + 5)^1$, so it is odd. The graph is going down on the right side of the asymptote. So, it will be going up on the left side.

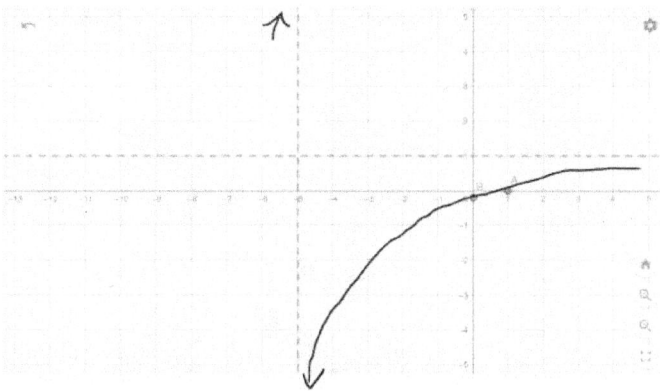

b) Knowing that we don't cross the asymptotes, finish the left side of the graph.

c) Next, we have the graph of the rational function $g(x) = \frac{-5x}{x^2-9}$. I've marked key points. Complete the graph. (Notice that it does cross the horizontal asymptote at the origin.)

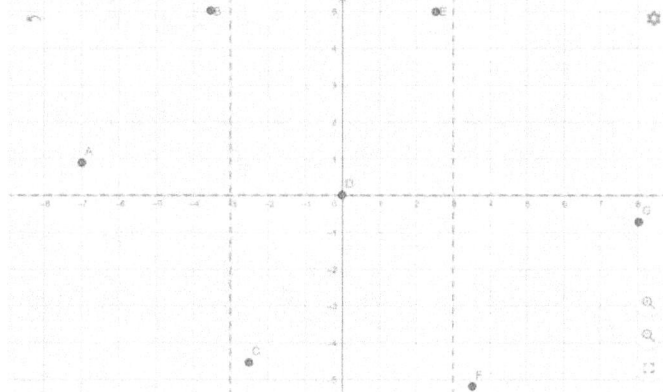

Let's recreate the graph without any prior knowledge of the points. Complete each step and then add it to the graph.

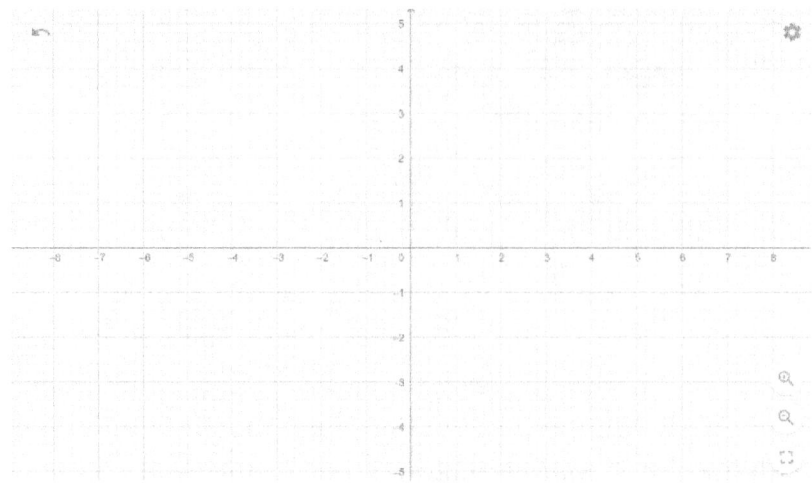

d) 1) Vertical asymptotes.

e) 2) Horizontal asymptotes

f) 3) x-intercepts

g) 4) y-intercepts

h) 5) Does it Cross?

(Notice that the exponent of both vertical asymptotes is one. Therefore, the graph will go in opposite directions on either side of those asymptotes.)

Lastly, we could also be asked to create the equation based on the graph.

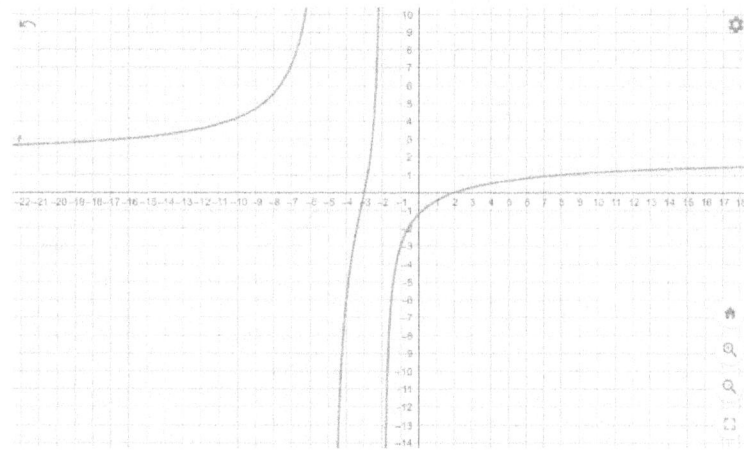

Here is the key information which we can take from the graph:

1) Vertical asymptotes at $x = -2$ and $x = -5$.
2) Horizontal asymptote at $y = 2$.
3) x-intercepts at 2 and -3.
4) We have opposite directions on each side of the vertical asymptotes, so they must be odd exponents. We will assume they are degree one.

With this information, we could build the following function.

$$g(x) = a\frac{(x-2)(x+3)}{(x+2)(x+5)}$$

To find the value of the stretch, a, we need a clear point from the graph. You must use a point other than the x-intercepts. The y-intercept is typically a good choice. Here it is a difficult to distinguish, but it

i) is the point $(0, -1.2)$. Plug those values into the function to find a.

j) In the space below, give the final equation.

k) Try another on your own. The y-intercept occurs at the point $(0, 4)$. Use the y-intercept to find the stretch factor. Give the equation for the complete function. (Assume degree one for the asymptotes.)

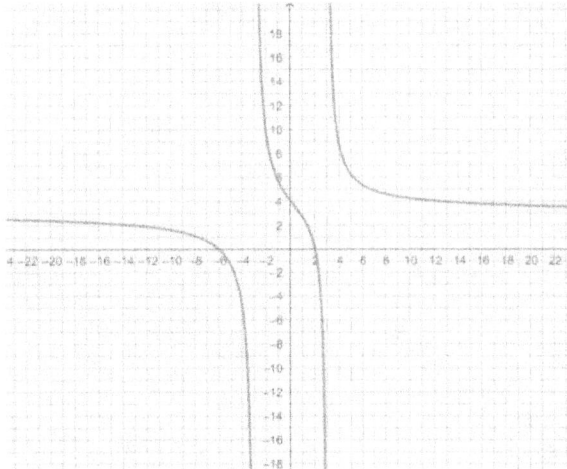

$$f(x) = \frac{6}{x+6} + 3$$

Algebra II

Active Learning: 5.7a

In Chapter Three, we learned about inverse functions. Functions are machines where x goes into the machine and only a single y comes out. Inverse functions are machines which go in reverse and have an output which is the same as the original function's input. This means that the domain and range are reversed for an inverse function. And, inverse functions must be one-to-one functions. A one-to-one function has a unique input for every output. For more details, review the ideas from the end of Chapter Three.

Composite functions are like assembly lines. One function directly follows another. If an inverse function and its partner function are composites, they will undo each other and they will give an answer of x. Here's a basic example:

$$f(x) = x - 1$$
$$g(x) = x + 1$$

Here's the composite function. (It doesn't matter which function comes first.)

$$f(g(x)) = (x + 1) - 1 = x$$

Because it equals x, the two functions are composites. Now, check to see if the following pairs of functions are composites. If so, write composite.

a)

$$f(x) = \frac{1}{x - 5}$$

$$g(x) = \frac{1}{x} + 5$$

(You could go either way, but working $g(f(x))$ would make the math easier.)

b) $f(x) = \dfrac{x - 3}{2}$

$g(x) = 2x + 5$

In Chapter Three, we also learned how to find the inverse of a function.

1) Replace the x and y in the equation.
2) Solve for the new y.
3) Give the inverse notation $f^{-1}(x)$.

Here we will find the inverse of a cubic function.

$$f(x) = 2x^3 - 3$$

Replace x and y. (Remember $f(x)$ is y.)

$$x = 2y^3 - 3$$

Solve for y.

$$x + 3 = 2y^3$$

$$\frac{x+3}{2} = y^3$$

$$y = \sqrt[3]{\frac{x+3}{2}}$$

$$f^{-1}(x) = \sqrt[3]{\frac{x+3}{2}}$$

Try these on your own.

c) $f(x) = \dfrac{x^3 + 2}{4}$

d) $f(x) = \sqrt[3]{x - 6}$

Algebra II

Active Learning: 5.7b

The graph below is the quadratic function $f(x) = x^2$. Give its domain and range in interval notation.

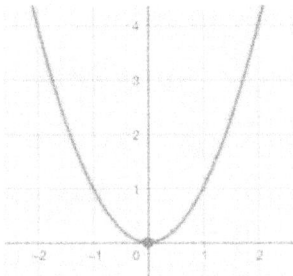

a) Domain:

b) Range:

c) Quadratic functions aren't one-to-one function. Explain why.

Since they aren't one-to-one functions, they can't have inverses. However, to solve that problem, mathematicians will cut the function. There can be multiple ways to make this cut. (This cut is called restricting the domain.) Below, I've show two ways to cut $f(x) = x^2$. Give the new domain and range of each.

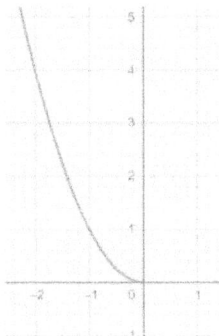 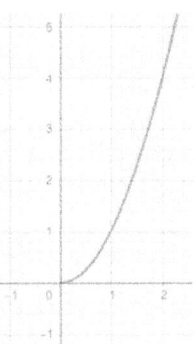

d) Domain: f) Domain:

e) Range: g) Range:

h) These cut functions can now have inverses because they are one-to-one. What will be true about the domain and range of the inverse functions?

Below is a quadratic function which has been both transformed and cut. Give its domain and range.

$$f(x) = (x - 3)^2 \text{ for } x \geq 3$$

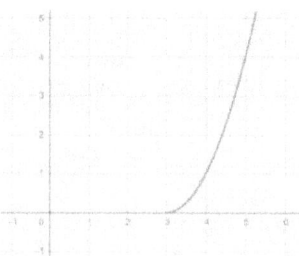

i) Domain:

j) Range:

Give the domain and range of the inverse function $f^{-1}(x)$

k) Domain:

l) Range:

Here is a formula for a transformed quadratic function. Without the graph, give a domain and range for the cut function.

$$g(x) = (x - 5)^2 \text{ for } x \geq 5$$

m) Domain:

n) Range:

Give the domain and range of the inverse function $g^{-1}(x)$

o) Domain:

p) Range:

Work another. (Hint: the vertical shift will only impact the range.)

$$h(x) = (x + 3)^2 - 2 \text{ for } x \leq -3$$

q) Domain:

r) Range:

Give the domain and range of the inverse function $h^{-1}(x)$

s) Domain:

t) Range:

Next, let's find the inverse function of a quadratic with a restricted domain.

$$g(x) = (x - 5)^2, \quad x \geq 5$$

Here are the steps:

1) Find the Domain and Range of the original function.
2) Find the inverse function.
3) For the inverse, reverse the Domain and Range from the original function.
4) Deal with the $\pm$ from any square roots by checking the necessary range for the inverse.

First, we find the domain and range.

Domain: $[5, \infty)$

Range: $[0, \infty)$ (I knew the range because it was a quadratic function opening upward.)

Next, we find the inverse in the typical fashion.

$$y = (x - 5)^2$$
$$x = (y - 5)^2$$
$$\pm\sqrt{x} = y - 5$$
$$\pm\sqrt{x} + 5 = y$$

So, we have:

$$g^{-1}(x) = 5 \pm \sqrt{x}$$

Now, we know that inverting will reverse the domain and range of the original. So, $g^{-1}(x)$ will have the following domain and range.

Domain: $[0, \infty)$

Range: $[5, \infty)$

Finally, if this is the domain and range which we need for $g^{-1}(x)$, our restricted inverse function would be:

$$g^{-1}(x) = 5 \pm \sqrt{x}, x \geq 0 \ (x \geq 0 \text{ because the domain requires it.})$$

We need a domain greater than or equal to 0. But we also need a range that is greater than or equal to 5, so we drop the negative sign in front of the square root. The negative sign would make values for the range that were less than 5.

$$g^{-1}(x) = 5 + \sqrt{x}, x \geq 0$$

Try one on your own.

Find the inverse function with a restricted domain. Start by giving the domain and range of the original function.

$$f(x) = (x-2)^2, x \geq 2$$

u) Domain:

v) Range:

w) Now, find $f^{-1}(x)$ in the typical manner.

What are the domain and range of the new inverse function?

x) Domain:

y) Range:

z) Finally, rewrite $f^{-1}(x)$ reflecting the restricted domain and by removing the sign which would generate the incorrect range.

Try another. This one includes a vertical shift. However, the process is the same.

$$h(x) = (x+3)^2 - 2 \text{ for } x \leq -3$$

aa) Domain:

bb) Range:

cc) Now, find $h^{-1}(x)$ in the typical manner.

What are the domain and range of the new inverse function?

dd) Domain:

ee) Range:

ff) Finally, rewrite $h^{-1}(x)$ reflecting the restricted domain and by removing the sign which would generate the incorrect range.

Here is a graph of a transformed square-root function. Give the domain and range.

$$f(x) = \sqrt{x-2}$$

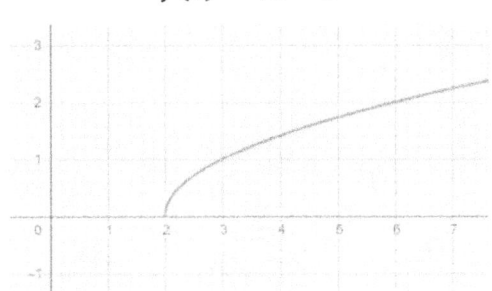

gg) Domain:

hh) Range:

Notice that the square-root function has a restriction on the range. Without it, the square-root wouldn't be a function. Find the inverse of this square-root function by following the same process we did for the quadratic.

ii) Find $f^{-1}(x)$. (Hint: There will not be a plus or minus because you will not take a square root. However, you will need to consider the acceptable values of the new domain.)

Algebra II

Active Learning: 5.7c

This section involves applications of inverse functions. In applications, the concept remains the same. The only difference is that we don't switch the variables. Instead, we simply solve for the new variable.

A pile of topsoil is shaped like a half-sphere. The volume can be expressed as a function of the radius.

$$V = \frac{4}{3}\pi r^3$$

a) Find the inverse function by solving for the radius. (Simply solve for r.)

b) Use the inverse function to find the radius if the volume measures 200 cubic feet. ($\pi = 3.14$)

c) The rational function $C = \frac{30+.5n}{100+n}$ expresses the concentration C of an acid solution after n mL of a 50% acid solution has been added to a 30% solution. Find an inverse function based on n. (Hint: Multiply $100 + n$ up to the other side and distribute the C. Then get all of the terms with an n over to one side. Finally, pull out the n as a GCF in order to solve for n.)

d) If the concentration of acid solution is 40%, find n. (This means $C = .40$) Use your new inverse formula.

Finally, we will look at the domain of a radical function with a rational expression inside.

$$f(x) = \sqrt{\frac{(x+3)(x-2)}{(x-4)}}$$

We know that what is under the root must be greater than or equal to zero.

$$\sqrt{\frac{(x+3)(x-2)}{(x-4)}} \geq 0$$

So, Geogebra only show where $f(x) \geq 0$. Give the domain of the following in interval notation. (Notice that there is a vertical asymptote at $x = 4$.)

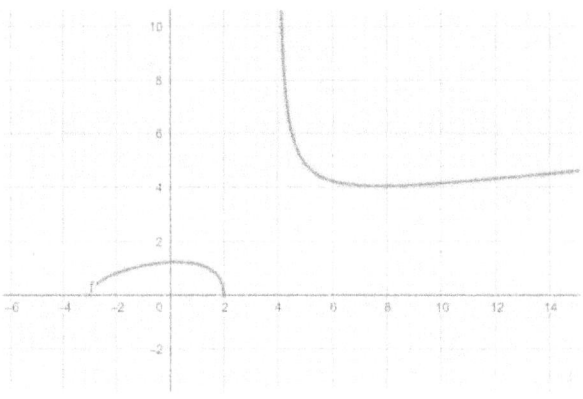

e) Domain:

If we were unable to graph the function, we could use an approach which we used earlier. The rational expression can change from positive to negative at the zeros and the asymptote. So, we could create the following regions.

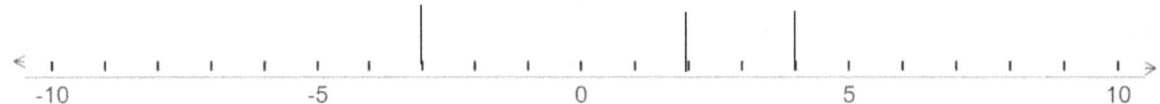

Test the regions to determine if they give positive or negative values. Below, I've chosen a value for each region, but any value in that region would work. Add a + or − to the number line above.

f) $f(-5) =$

g) $f(0) =$

h) $f(3) =$

i) $f(5) =$

Only regions which have a positive value would be permitted under a root, so only those values would be part of the domain. Give the domain based on the positive regions of your number line. (Remember that zeros can be included in your domain but asymptote values cannot.)

j) Domain:

If done correctly, the graphical approach and the test-point approach should yield the same domain.

Algebra II

Active Learning: 5.8

In Algebra I, we looked at the idea of direct variation.

$$y = kx$$

The idea is easier to understand when we look at a rearranged version of the formula.

$$\frac{y}{x} = k$$

The variables y and x always keep the same ratio, k. If y increases, x must also increase to keep the ratio the same. Likewise, if y decreases, x must also decrease to keep the ratio the same. Here is a typical problem.

If y varies directly with x when $y = 8$ and $x = 1.6$, find y when $x = 3$.

First, we need to find k, called the constant of variation. We do that in the same way we could for any equation. Plug in the x and y and solve for k.

$$\frac{8}{1.6} = k$$

$$k = 5$$

Now that we know k we can use it in the generic equation.

$$y = 5x$$

And, to find y when $x = 3$, we plug in.

$$y = 5(3) = 15$$

Work one on your own.

a) If y varies directly with x when $y = 6$ and $x = 1.25$, find y when $x = 7$.

Direct variation does not have to be limited to a degree of one. It can be extended for any degree.

$$y = kx^n$$

So, y is varying directly with some degree of x.

$$\frac{y}{x^n} = k$$

A problem will follow the same idea as the basic ones.

If y varies directly with the square of x when $y = 12$ and $x = 4$, find y when $x = 6$.

The set up would look like this.

$$\frac{12}{4^2} = k$$

b) Find the value of k and then find y when $x = 6$.

Try another.

c) If y varies directly with the cube of x when $y = 15$ and $x = 2$, find y when $x = 4$.

The concept of inverse variation is similar.

$$y = \frac{k}{x}$$

Rearranged we have.

$$yx = k$$

Here, if y increases, x must decrease to keep the ratio the same. Likewise, if y decreases, x must increase to keep the ratio the same. And, this can also be extended to higher degrees.

$$y = \frac{k}{x^n}$$

Working a problem follows the same procedure.

d)   If y varies indirectly with the square of x when $y = 4$ and $x = 3$, find y when $x = 2$.

Finally, we can extend this idea to something called joint variation. As the name implies, joint variation involves multiple variations within the same problem.

$$x = \frac{ky}{z}$$

Let's look at a problem.

A quantity x varies directly with the cube of y and indirectly with square of z when $x = 2, y = 2$ and $z = 3$. Find x when $y = 3$ and $z = 4$.

The concepts are the same, but the set up can be a little confusing. If x varies directly with the cube of y, we have the y^3 in the numerator.

$$x = ky^3$$

If y varies indirectly with z^2, we have z^2 in the denominator.

$$x = \frac{ky^3}{z^2}$$

e)   Plug the values in for $x = 2, y = 2$ and $z = 3$ to find k. Then, use k to find x when $y = 3$ and $z = 4$.

Try another.

f)   A quantity x varies directly with the square root of y and indirectly with cube of z when $x = 3, y = 9$ and $z = 2$. Find x when $y = 4$ and $z = 3$.

Algebra II

Active Learning: 6.1a

Exponential Functions

Suppose that you had $100 and you doubled your total amount of money every month for six months.
a) Fill in the table below with how much money you would have.

Month	Money
0	100
1	
2	
3	
4	
5	
6	

As you can see, this quickly becomes a lot of money. This kind of growth is called exponential growth. Here's the math:

$$100 * 2 * 2 * 2 * 2 * 2 * 2$$

You have the money you started with and then you doubled it six times. But in math language, there is a way to simplify repeated multiplication—exponents.

So, our formula becomes:

$$100(2)^6$$

There are many applications of exponential growth in the real world. And the idea is always the same. You have the number you started with (in our example $100). And then there is a rate of growth (for us 2). And it grows over some number of intervals (for us 6). And the equation for exponential growth looks like this:

$$f(x) = ab^x$$

$$a = amount\ you\ start\ with$$

$$b = rate\ of\ growth$$

$$x = the\ number\ of\ intervals$$

Although we usually talk about exponential growth, it is possible for exponential decay, where you are losing money (or whatever it is you started with). One of the following is **exponential decay**, one is **exponential growth**, and one is **no growth** at all. Which is which? **Explain what makes it the one you chose.**

b) $f(x) = 100(1)^x$

c) $f(x) = 100(1.1)^x$

d) $f(x) = 100(.90)^x$

Calculating exponential growth works like any other function. Simply put in the value of x, which is the number of intervals.

e) Find $f(5)$ if $f(x) = 12(1.2)^x$

f) Find $f(2)$ if $f(x) = 250(1.152)^x$

Setting up an exponential equation requires identifying the initial amount and adjusting the "no growth" value of 1.

A rabbit population in a park is growing exponentially. In 2010, there were 20 rabbits. Set up an exponential growth equation if the rabbits are growing by 12% per year.

$$f(x) = ab^x$$

The initial is 20. And, we change the rate to a decimal, adding it to 1 since we have growth.

$$f(x) = 20(1.12)^x$$

Try one on your own.

g) The deer population in the state park is growing exponentially. In 2000, there were 150 deer. Set up an exponential growth equation if the deer are growing by 6% per year.

Sometimes, the rate of growth isn't given. If we have an ordered pair (x, y) we can substitute in those values and solve for b.

A rabbit population in a park is growing exponentially. In 2010, there were 20 rabbits. In 2014, there were 120 rabbits.

So, we have:

$$f(x) = 20b^x$$

However, we don't have b. But we know that in the year 2020, which was 4 years after the initial year, we have 65 rabbits. This gives us the ordered pair $(4, 120)$. We can substitute in and solve.

$$120 = 20b^4$$

First, we isolate the difficult portion which is the b^4.

$$6 = b^4$$

To get rid of a fourth power, we can take the fourth root of each side.

$$(6)^{\frac{1}{4}} = (b^4)^{\frac{1}{4}}$$

$$1.565 = b$$

If we were asked the rate of growth, we would take the decimal away from the 1 and turn it into a percentage.

$$56.5\%$$

Work this problem.

h) The deer population in the state park is growing exponentially. In 2000, there were 150 deer. In the year 2005, there were 210.

Next, we will be asked to find the exponential equation from the graph.

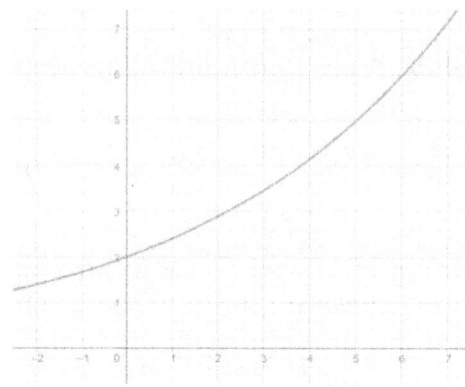

The initial value, a, would be the y value when $x = 0$. Reading the graph, we can see that is 2. So:

$$a = 2$$

Now, we have:

$$f(x) = 2b^x$$

i) And, with any point we could find the value of b. Reading the graph, there is a clear point at $(5, 5)$. Plug the point in to solve for b then give the final equation.

Try this problem.

j) Find the equation for the exponential function from the graph below. Use the point $(2, 4)$ to find the value of b.

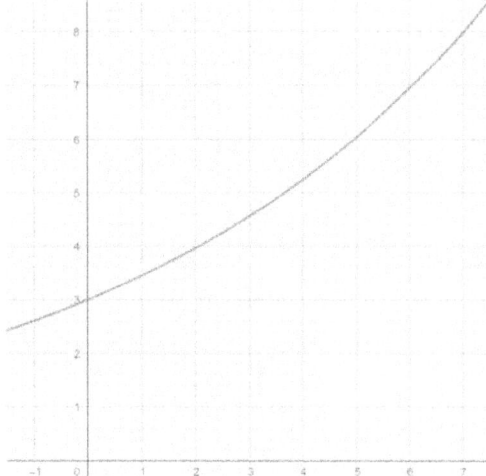

Next, we will find an equation for the exponential function given two points.

An exponential equation passes through these two points, find the equation: $(0, 4)$ and $(3, 6)$.

This works just as we did on the graph. We have the initial point a.

$$f(x) = 4b^x$$

k) Use the point $(3, 6)$ to solve for b. Then, give the final equation.

Finally, we need a systems approach to find the next equation.

An exponential equation passes through these two points, find the equation: $(-1, 3)$ and $(2, 4)$.

Without the initial point, we need to set up two equations.

$$f(x) = ab^x$$

Dropping in the point $(-1, 3)$.

$$3 = ab^{-1}$$

And the point $(2, 4)$.

$$4 = ab^2$$

We now have two variables and two equations. Just as we did with lines, we can use the substitution method.

$$3 = ab^{-1}$$

$$3 = \frac{a}{b}$$

$$3b = a$$

Now that we have an equation for a, we can substitute into the second equation:

$$4 = ab^2$$

$$4 = (3b)b^2$$

$$4 = 3b^3$$

$$\frac{4}{3} = b^3$$

$$\left(\frac{4}{3}\right)^{\frac{1}{3}} = (b^3)^{\frac{1}{3}}$$

$$b = 1.10$$

Now that we know b, we can find a.

$$3b = a$$
$$3(1.10) = a$$
$$a = 3.3$$

And our final equation is:

$$f(x) = 3.3(1.10)^x$$

I know this is not easy, but try one on your own.

I) An exponential equation passes through these two points, find the equation: $(-1, 4)$ and $(2, 6)$.

Algebra II

Active Learning: 6.1b

In our last lesson, we learned about exponential growth and the exponential function.

$$f(x) = ab^x$$

a) Suppose you started with $100 and your money grew by 10% every year. Why would $b = 1.10$ be the correct value? (Hint: Why does the 1 have to be there? We saw this in our last lesson.)

b) Now, calculate how much money you would have after five years.

Banks usually don't update your money only once per year. They split it up. But if they split it up you don't get more interest. They divide the interest rate. (Not all of b. Only the interest rate.) So, answer the following:

-If they pay out interest semi-annually, what is the value of b? (I've done this one for you.)

$b = (1 + \frac{.1}{2}) = 1.05$

(Again, they only divide the interest rate portion. We always need the 1.)

c) -If they pay out interest quarterly, what is the value of b?

d) -If they pay out interest monthly, what is the value of b?

But if the bank changes the value of b then they also need to change the value of x or it won't be right. The value of x needs to be the total number of times that they pay you. Answer the following:

-If your money is in the bank for 5 year and they are giving interest out semi-annually, what is the value of x? (I've done this one for you.)

$$x = 5 \cdot 2 = 10$$

e) -If your money is in the bank for 5 year and they are giving interest out quarterly, what is the value of x?

f) -If your money is in the bank for 5 year and they are giving interest out monthly, what is the value of x?

Now, let's find out how much money you would have in each of those situations. But remember, both b and x changed. I'll show you the first one.

g) -How much money do you have at the end of the five years if the bank pays you semi-annually?

$$Amount = 100(1 + \frac{.10}{2})^{2*5}$$

$$Amount = 100(1.05)^{10}$$

h) -How much money do you have at the end of the five years if the bank pays you quarterly?

i) -How much money do you have at the end of the five years if the bank pays you monthly?

Try a couple on your own.

j) If you invest $5000 compounded semi-annually at a rate of 8%, find the value of your investment at the end of 5 years.

k) If you invest $7500 compounded quarterly at a rate of 5%, find the value of your investment at the end of 6 years.

The compound interest formula gets different notation, but it is nothing more than the exponential function broken down as we did above.

$$A(t) = P\left(1 + \frac{r}{n}\right)^{nt}$$

$A(t)$ is the amount you have at the end of the investment.

P is the principal. (The starting amount.)

r is the interest rate.

t is the number of years.

n is the times per year you compound.

Again, the notation is different, but this is the exponential function.

Occasionally, a problem will require you to solve for P, the principal, instead of the final amount. In those cases, simply plug in all the values which you do have, simplify, and then use your algebra skills to solve for P. In this next problem, you have everything but the value of the principal.

l) At the end of 10 years, your investment has become $50,000. If you were getting a rate of 5% and the interest was compounded quarterly, find the initial amount of money which you invested. (Since this is money, round your answer to the nearest hundredth.)

Sometimes growth (or decay) isn't occurring at a set interval, but rather it is happening constantly. In such circumstances, our formula changes:

$$A(t) = Pe^{rt}$$

The letter *e* is not a variable. It is called Euler's number and it appears in a number of mathematical and scientific applications, including continuous growth. You can find the number *e* on your calculator, typically above the LN symbol. This is e^x, which will immediately prompt you for an exponent.

The e^r is essentially *b* from the exponential formula, and it covers the continuous nature of the compounding. When working problems, if you see "compounded continuously" then you know that you need to switch to the continuous compounding formula.

Try a problem.

m) You invest $2000 at a rate of 6% compounded continuously. How much money will you have at the end of 8 years?

Finally, this continuous formula applies in other settings beyond money. Here, I've changed the initial amount *P* to an *a*. The concept is the same, but the change takes us back to the idea of initial amount that we first saw in the exponential function.

$$A(t) = ae^{rt}$$

n) There are 200 grams of a certain bacteria which grows continuously at a rate of 12% a day. How many grams of the bacteria will there be after 15 days?

These problems can also represent continuous decay, where the value of r is simply entered into the equation as a negative rate.

o) An element decays at a continuous rate of 15% per day. Initially, there was 500 milligrams of the element, how much remains after 10 days?

I've set the problem up for you so that you could see the negative rate.

$$A(t) = 500e^{-.15(10)}$$

Algebra II

Active Learning: 6.2a

We have been looking at exponential functions. Here is what the graph of an exponential growth function looks like:

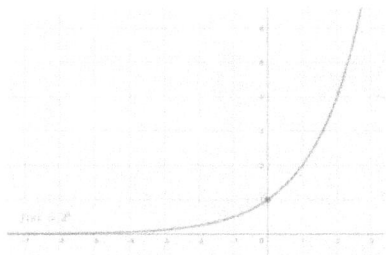

Transformations of exponential functions follow the same ideas we have seen earlier. Match the graph with the formula.

           1   2   3

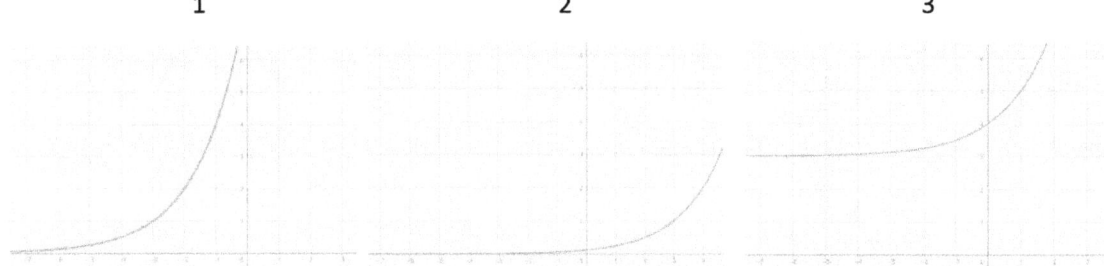

a) $f(x) = 2^x + 3$

b) $g(x) = 2^{x+3}$

c) $h(x) = 2^{x-3}$

The value of b in our exponential formula tells us if we have a growth or decay factor. We grow if b is greater than 1, and we decay if b is less than 1. However, the value of b can never be a negative. (Sometimes there is a vertical reflection, but that is out front of b and not b itself. So, when we are repeatedly multiplying the value of b, there is no exponent which can make b^x into a negative number. This makes a horizontal asymptote (a y line) which the graph won't cross.

Here is the graph of the exponential function $f(x) = 4^x$. It shows the horizontal asymptote at $y = 0$.

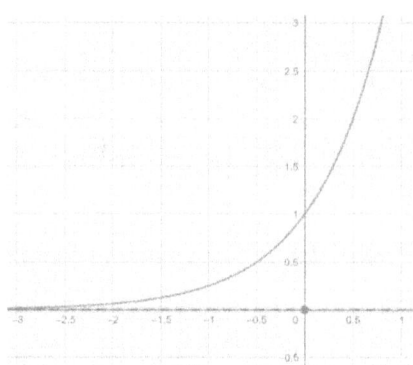

What are the domain and range of this function? (Give your answer in interval notation.)

d) Domain:

e) Range:

Here is the graph of $f(x) = 4^{x-2} + 1$. Provide the following.

f) Domain:

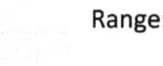

g) Range:

h) Asymptote: (Give as a y line.)

Here is the graph of $f(x) = 2^{x+3} - 2$. Provide the following.

i) Domain:

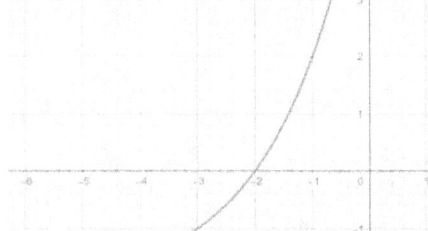

j) Range:

k) Asymptote: (Give as a y line.)

l) One of the transformations is changing the location of the asymptote, which transformation is it?

Give the domain, range, and asymptote for the following exponential functions:

m) $f(x) = 2^{x+5} + 4$

n) $f(x) = 3^{x-4} - 12$

Algebra II

Active Learning: 6.2b

In this activity, we want to graph exponential functions. The key is to find two points and the horizontal asymptote.

$$f(x) = 3^x$$

Because the value of x is an exponent, there are always two easy points we can find.

- No matter the value of b, we always have the point $(0, 1)$. This is because when you put $x = 0$ into the function, anything raised to the zeroth power is 1.
- Then, whenever we put $x = 1$ into the function, we always get the value of b, since anything raised to the first power is just the number itself.
- The horizontal asymptote is always the line $y = 0$, unless there has been a vertical shift.

So, with that information, we can graph the exponential function. Plotting these three pieces of information, we have:

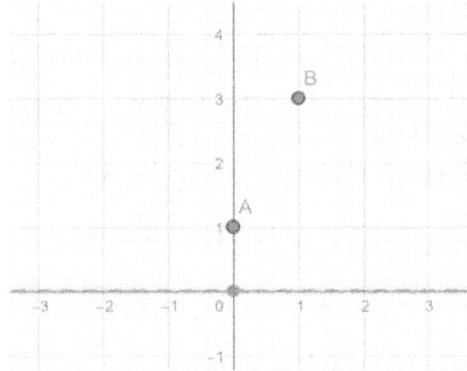

And, because we know it is exponential growth, we know the shape. We must go through points A and B, and we will curve down along the asymptote without crossing it.

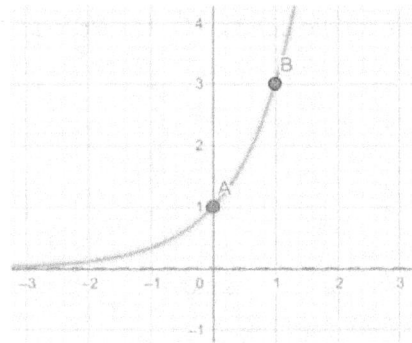

a) Try one on your own:

$$f(x) = 2^x$$

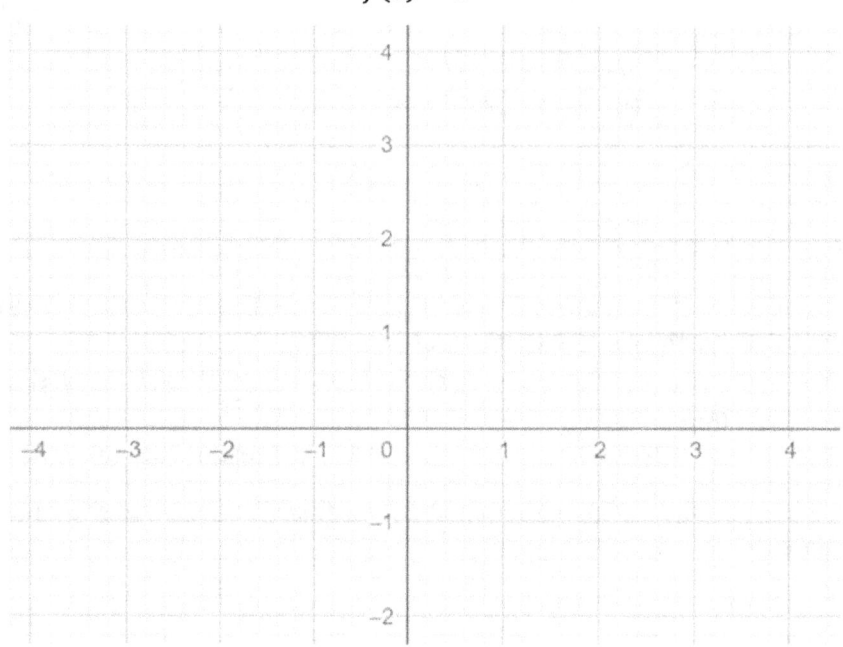

When an exponential function has transformations involved, we want to start by ignoring the transformations. First, we will look to the base function to find the two basic points and the asymptote. Then, we will perform the transformations on the asymptotes.

$$f(x) = 3^{x-2} + 1$$

We start with the basic, untransformed function, $f(x) = 3^x$. Our points and asymptote would be:

(0,1)

(1,3)

$y = 0$

Next, we perform the transformations on the points. The transformation requires shifting right 2 and up 1.

	Right 2		Up 1
(0,1) ⇨	(2,1)	⇨	(2,2)
(1,3) ⇨	(3,3)	⇨	(3,4)

Notice to move right, I changed the x values, and to move up, I changed the y values. And, because of the vertical shift up 1, I adjust the vertical asymptote.

$y = 1$

b) With the points transformed, we are ready to graph. use the transformed points and the new asymptote to graph the function below:

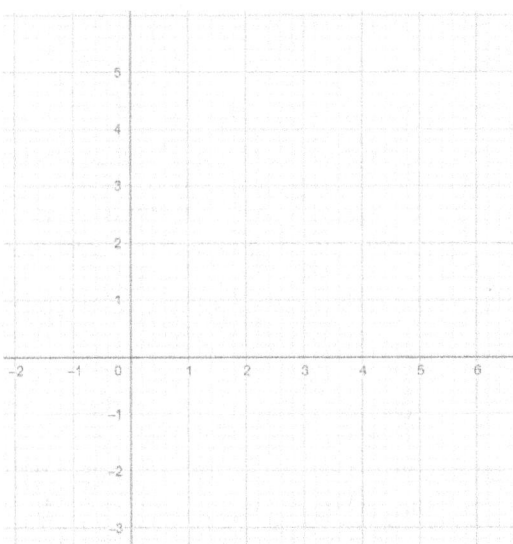

c) Try one on your own.

$$f(x) = 2^{x+1} - 2$$

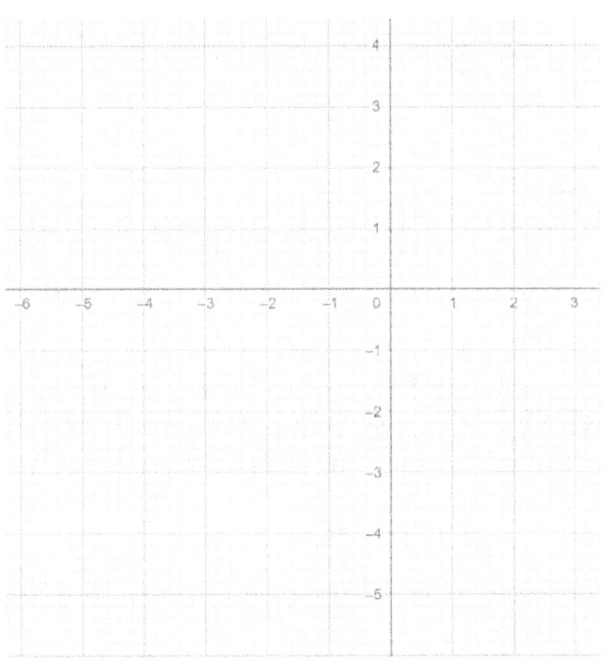

Work another. This one has a reflection.

$$f(x) = -2^{x-3} + 2$$

Here are our initial points:

$(0, 1)$

$(1, 2)$

The order of transformations requires that we start with horizontal shifts, then flips, and then vertical shifts. Here is the change to the points after the horizontal shift and the reflection over the x-axis.

d)
                  Right Three   Reflection

$(0, 1)$ ⇨ $(3,1)$ ⇨ $(3, -1)$ ⇨

$(1, 2)$ ⇨ $(4,2)$ ⇨ $(4, -2)$ ⇨

Reflecting over the x-axis flips the y values. Now, finish transforming the two points with the vertical shift. The horizontal asymptote is not impacted by the flip, but it is by the vertical shift. It will be $y = 2$. (Remember, we reflected the graph, so it will now be below the asymptote.)

e)
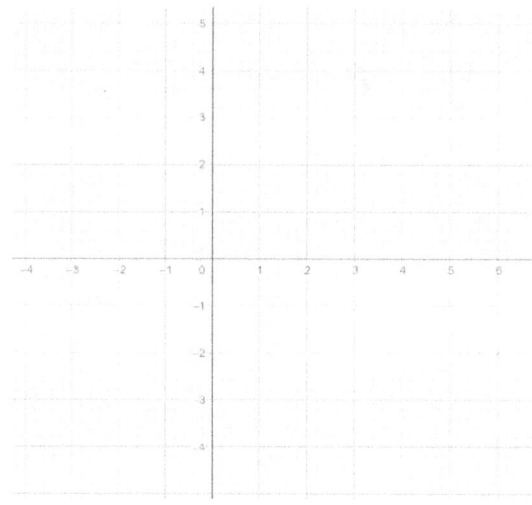

So far, we have graphed functions that did not have the a in front of the equation, $f(x) = ab^x$. In a classroom, I would not typically ask you to graph such a circumstance, but we'll look at one briefly. Anytime there is a value out front of the function it involves a vertical stretch or shrink. That is true for the a in this case too.

To graph an exponential function with a stretch, we will include the stretch in the shifting of points.

$$f(x) = 2(3)^{x-2}$$

Starting with our basic points, we will make the transformations:

        Right 2   Vertical Stretch

(0, 1) ⇨ (2, 1) ⇨ (2, 2)

(1, 3) ⇨ (3, 3) ⇨ (3, 6)

Notice that the vertical stretch involved multiplying the y-coordinates by 2. The asymptote did not shift, so it is still $y = 0$.

f) Use the two points and the asymptote to graph the stretched function.

Algebra II

Active Learning: 6.3

Logarithms

Answer the following questions:

a) 5 raised to some exponent turns into 125. What is the exponent?

b) 2 raised to some exponent turns into 32. What is the exponent?

c) 3 raised to some exponent turns into 81. What is the exponent?

A long time ago, some mathematician found themselves answering questions like that over and over again. And, as mathematicians like to do, they invented something in math language to simplify the idea. The came up with the logarithm.

The answer to a logarithm is an exponent. Let's look at one:

$$\log_5(25)$$

In English, this is asking a question. It is asking, what exponent can turn 5 into 25.

$$\log_5(25) = 2$$

The answer is 2. Try some: (I'll put the English next to them.)

d) $\log_4(16) =$ _____   (4 raised to what exponent equals 16.)

e) $\log_6(216) =$ _____   (6 raised to what exponent equals 216.)

f) $\log_2(16) =$ _____   (2 raised to what exponent equals 16.)

We call the small number the base. And, we read it $\log_2(16)$, log base 2 of 16. So, in generic form, $\log_2(x)$, would read log base 2 of x.

The answer to a logarithm can be a negative. For instance:

$$\log_2\left(\frac{1}{8}\right) =$$

Negative exponents create fractions. So, for a problem like this, simply ask what exponent turns 2 into 8 and then make it a negative.

$$\log_2\left(\frac{1}{8}\right) = -3$$

Give the answer to the following logarithms.

g) $\log_3\left(\frac{1}{27}\right) =$

h) $\log_2\left(\frac{1}{16}\right) =$

In math language, the following is true:

$$y = \log_b(x) \text{ is equivalent to } b^y = x$$

Again, this just says that there is some exponent (y) that turns b into x. Some problems will ask you to switch between the two.

Write the logarithm in exponential form.

$\log_2(8) = 3$

That just means: 2 to the third power equals 8.

$2^3 = 8$

Try these:

Write the following logarithmic equations in exponential form.

i) $\log_5(125) = 3$

j) $\log_{10}(10000) = 4$

These next problems are asking you to go the other direction.

Write the exponential equations in logarithmic form.

$$2^4 = 16$$

This just means that 4 is the exponent that turns 2 into 16.

$\log_2(16) = 4$

Try these:

Write the exponential equation in logarithmic form.

k) $3^4 = 81$

l) $4^3 = 64$

Finally, there are two special types of logarithms, those with base 10 and those with base e. Because base 10 is the most common type of logarithm it is called the common log and the base is often not shown. If you see a logarithm without a base, it is base 10.

$$\log_{10}(x) = \log(x)$$

And, because of the many natural applications of base e, it is given a special logarithm called the natural log.

$$\log_e(x) = \ln(x)$$

Use the log button on your calculator to answer the following:

m) $\log 45 =$

n) $\log \dfrac{1}{125} =$

Use the natural log button on your calculator, ln, to answer the following:

o) ln(15) =

p) ln(225) =

Algebra II

Active Learning: 6.4a

Below are the graphs of two exponential functions and their partnered logarithmic function.

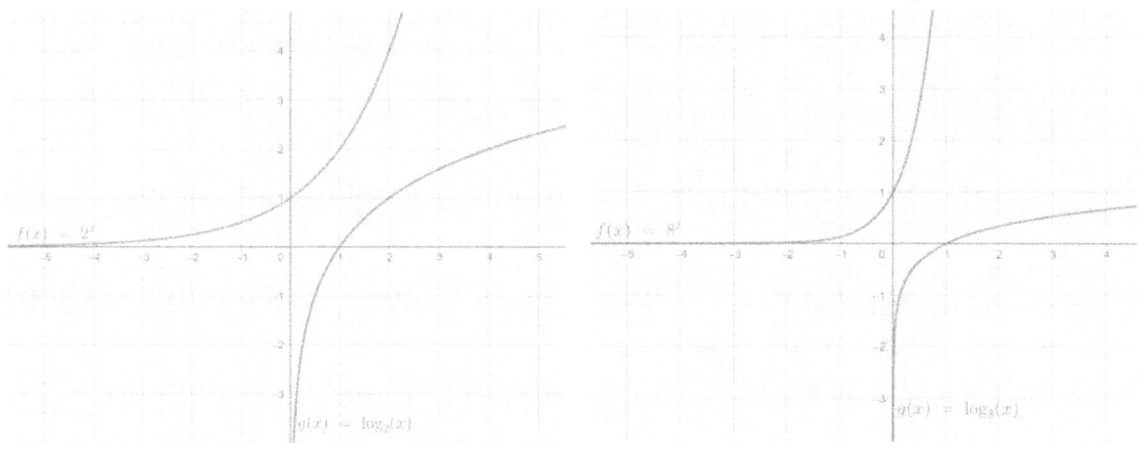

a) There is a special relationship between the two functions. Looking at the graph, the two functions are reflections over the line $y = x$. What is the name of this special relationship between these functions?

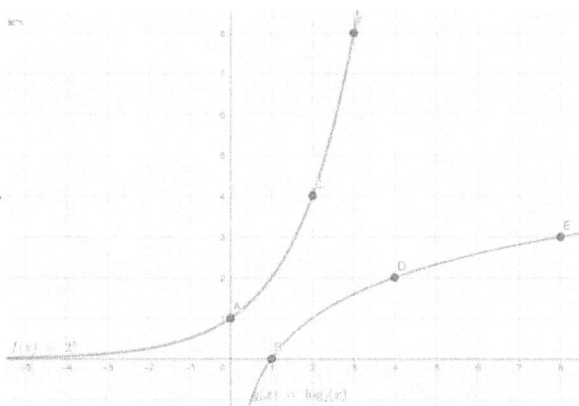

Complete the chart below with the ordered pairs of the points indicated.

b)

	Exponential		Logarithmic
A		B	
C		D	
F		E	

c) What is the relationship between the Domain and Range of the exponential function $f(x) = 2^x$ and the Domain and Range of the logarithmic function $g(x) = \log_2 x$?

d) Use the exponential function $f(x) = 3^x$ to graph the logarithmic function $g(x) = \log_3 x$. (Hint: Find the y-values for the exponential function and then use the special relationship between their Domains and Ranges to find the points for the logarithmic function.

	Exponential		Logarithmic
X	Y	X	Y
-1			
0			
1			
2			

e)
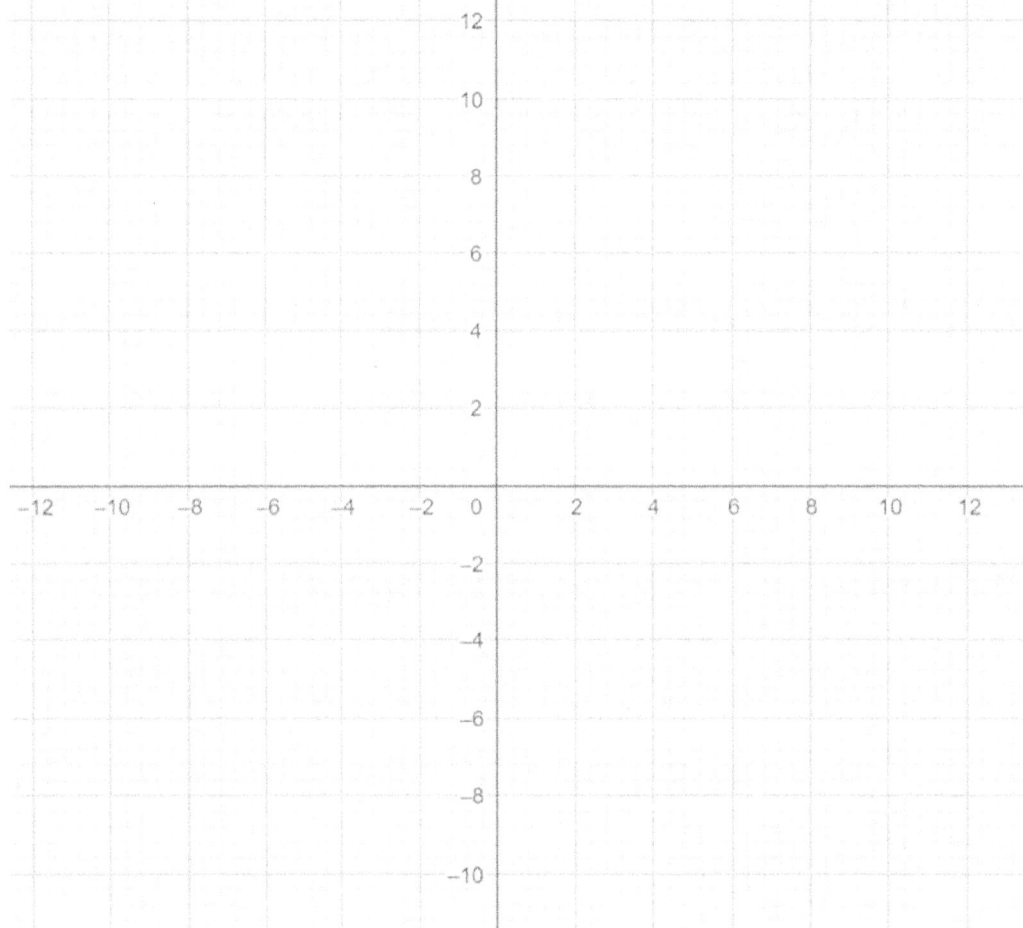

To graph a logarithmic function, the easiest way is to start with its partner exponential function.

$$f(x) = \log_2(x)$$

The partner exponential function would be:

$$f(x) = 2^x$$

As we saw in an earlier activity, we will use the two basic points:

$(0, 1)$

$(1, 3)$

But we don't want the exponential function, we want the logarithmic function. And the log function reverses the domain and range, which exchanges the x and y.

$(1, 0)$

$(3, 1)$

The exponential function has a horizontal asymptote at $y = 0$. Since the logarithmic function is the inverse, and the domain and range are reversed, the log function has a vertical asymptote at $x = 0$.

f) So, using this information, graph the log function.

$(1, 0)$

$(3, 1)$

$x = 0$

(Your graph will go left to right through the two points while starting along the vertical asymptote without crossing it.)

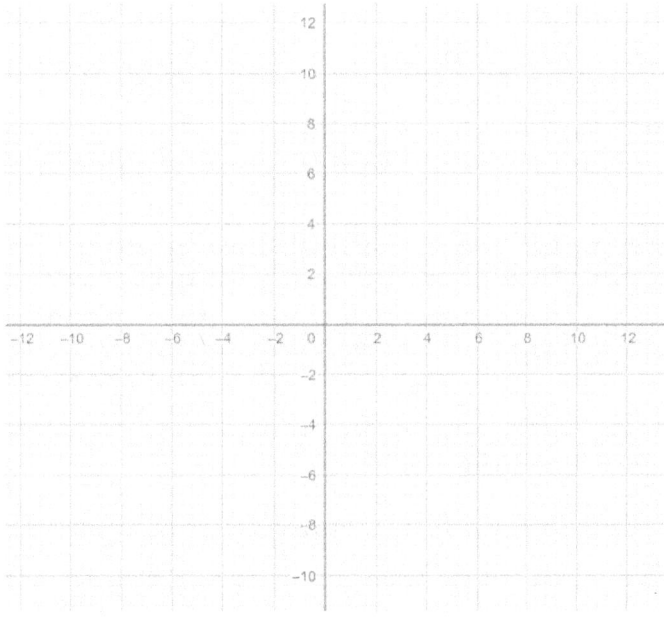

g) Try one on your own. Graph the following logarithmic function.

$$f(x) = \log_4(x)$$

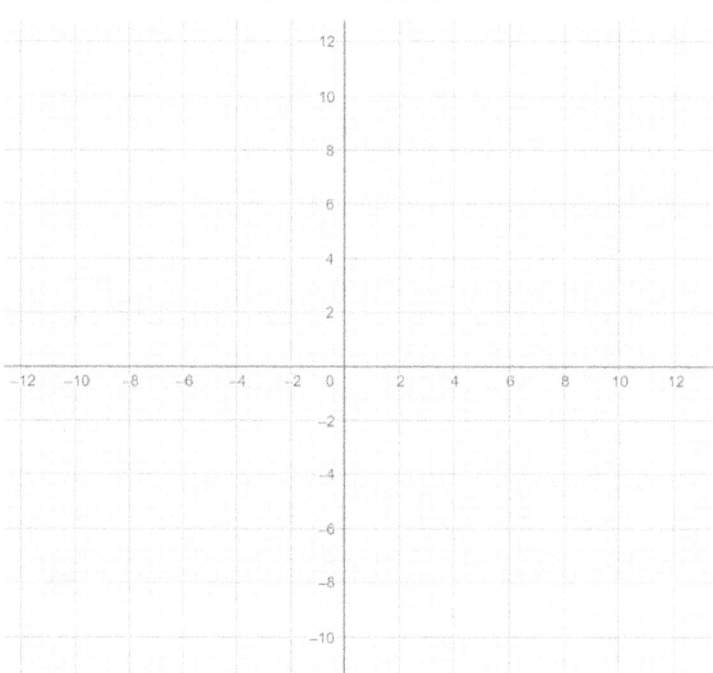

Give the domain and range of the function you graphed above. (Be sure to give them in interval notation.)

h) Domain:

i) Range:

Let's look at why a logarithm has a vertical asymptote at $x = 0$.

$$f(x) = \log_4(x)$$

A logarithm is an exponent. So, in this log, there must be some exponent which turns 4 into x. But there is no exponent which can turn 4 into zero. Likewise, there is no exponent which can turn 4 into a negative number. Therefore, we have the asymptote at $x = 0$. There are no acceptable x values equal to or less than 0.

Next, let's find the domain of a transformed logarithm.

$$f(x) = \log_4(x - 3)$$

The function has been transformed, but the issue is still the same. There is no exponent which can turn 4 into $(x - 3)$.

To find the domain, we need values which don't break the function. So, we need:

$$x - 3 > 0$$

Or:

$$x > 3$$

The domain would be:

$$(3, \infty)$$

Try some on your own.

Find the domain of the following functions. (The base of the logarithm doesn't matter.)

$$f(x) = \log_5(x - 9)$$

j) Domain:

$$f(x) = \log_2(3x - 12)$$

k) Domain:

$$f(x) = \log_3(x + 4)$$

l) Domain:

On this problem, the vertical shift has nothing to do with the domain. You still only need to set what is inside the logarithm to be greater than zero.

$$f(x) = \log_6(x - 6) + 12$$

m) Domain:

As you will see when you graph them, the range of a logarithmic function always turns out to be $(-\infty, \infty)$.

Horizontal shifts will change the location of the vertical asymptote.

$$f(x) = \log_4(x - 3)$$

Because the graph has shifted 3 units to the right, the vertical asymptote does too. It starts at $x = 0$ and therefore becomes, $x = 3$.

State the location of the vertical asymptotes for the following:

$$f(x) = \log_5(x - 9)$$

n) $x =$

$$f(x) = \log_3(x + 4)$$

o) $x =$

Here, the horizontal shift, $+12$, has nothing to do with changing the location of the vertical asymptote.

$$f(x) = \log_6(x - 6) + 12$$

p) $x =$

Finally, this problem has a horizontal stretch.

$$f(x) = \log_2(3x - 12)$$

To understand the horizontal stretch, you pull out a GCF of 3 on the inside of the parenthesis.

$$f(x) = \log_2(3(x - 4))$$

Now, the horizontal stretch can be clearly seen. It is 4 to the right and the vertical asymptote is $x = 4$.

Try this one.

$$f(x) = \log_3(4x + 16)$$

q) $x =$

Algebra II

Active Learning: 6.4b

To graph translations of logarithmic functions, we'll use the same ideas we've already seen.

$$f(x) = \log_2(x) + 2$$

We will start with the base points form the exponential function $f(x) = 2^x$. Those base points would be:

$(0, 1)$

$(1, 2)$

Now, we will reverse them since we are working with a logarithm.

$(1, 0)$

$(2, 1)$

Next, we transform them. In this problem, we have a vertical shift of up 2. This will change the y-values of the points:

Up 2

$(1, 0) \implies (1, 2)$

$(2, 1) \implies (2, 3)$

a) A logarithm has a vertical asymptote at $x = 0$. There are no right or left translations and so that asymptote won't change. Using our translated points and the asymptote graph our function:

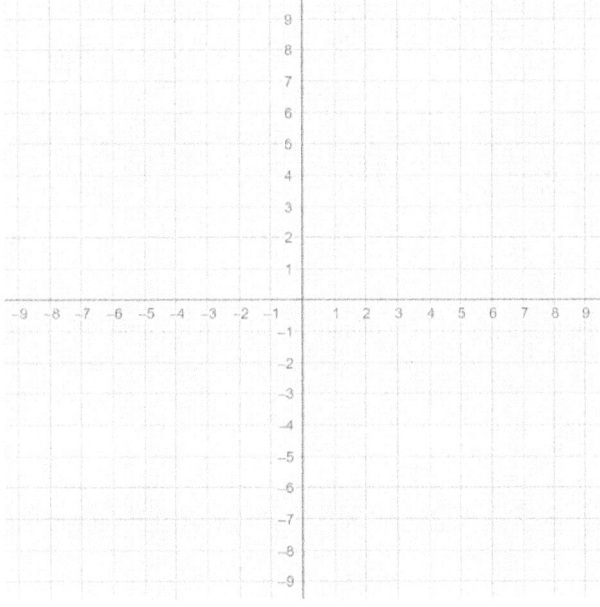

b) Try one with a horizontal shift. (The vertical asymptote will become $x = 2$.)

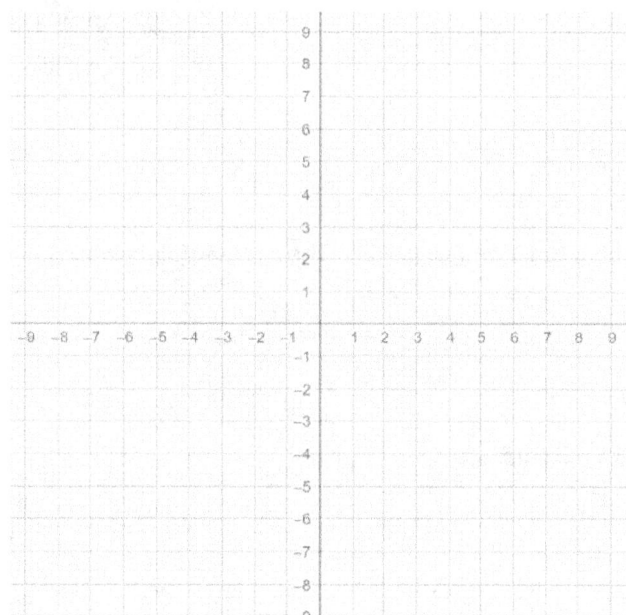

$$f(x) = \log_3(x - 2)$$

c) This problem has both a vertical and horizontal shift. The order of these shifts doesn't change the result, but remember to move the vertical asymptote too.

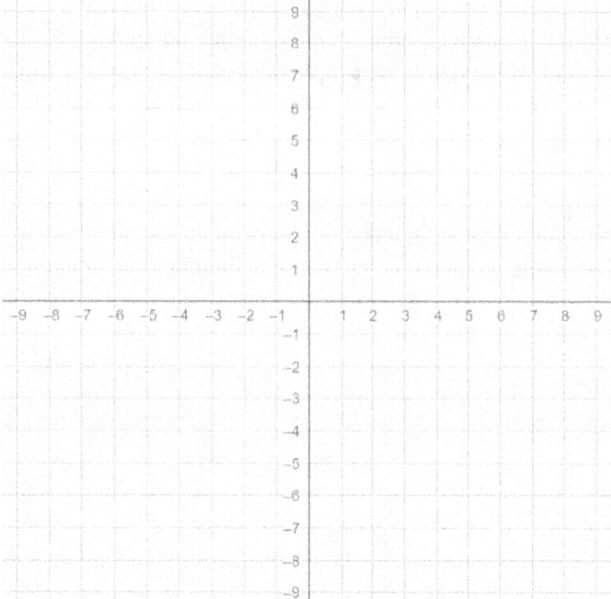

$$f(x) = \log_2(x + 1) - 2$$

Next, we look at vertical stretches.

$f(x) = 3\log_2(x)$

As always, we take our base points from $f(x) = 2^x$ and reverse them.

(0, 1) ⇔ (1, 0)

(1, 2) ⇔ (2, 1)

We have a vertical stretch because the entire function is being multiplied by 3. This will change the outcome of the function, so we multiply the y-values.

× 3

(1, 0) ⇒ (1, 0)

(2, 1) ⇒ (2, 3)

d) There was no horizontal shift, so the asymptote remains $x = 0$. Using our translated points and the asymptote graph our function:

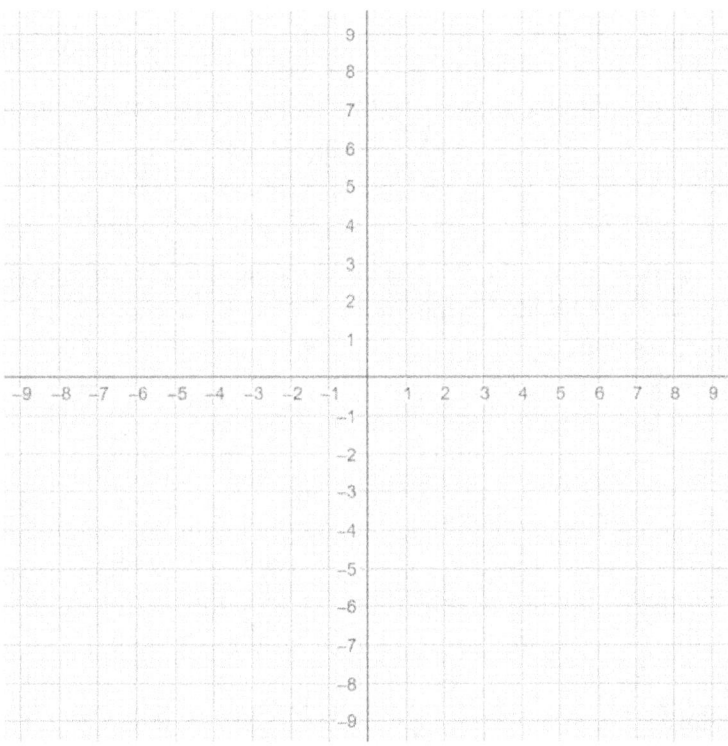

e) Now, try one that combines horizontal and vertical shifts, and a vertical stretch. The transformations must follow the order of operations. First, horizontal shift. Second, stretch. Third vertical shift. (Don't forget to adjust your asymptote.)

$$f(x) = 2\log_3(x-3) - 1$$

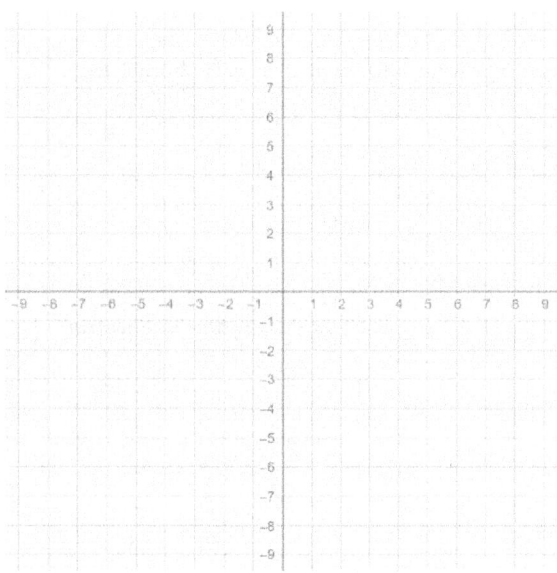

Finally, reflections work as they did before. A negative on the outside of the function reflects the graph over the x-axis. A negative on the inside, in front of the x, reflects the graph over the y-axis. Match the following functions to their graphs.

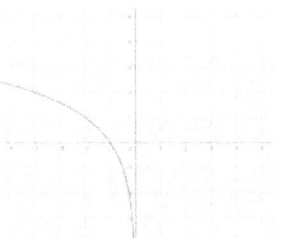   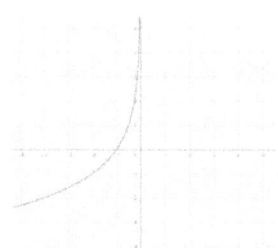   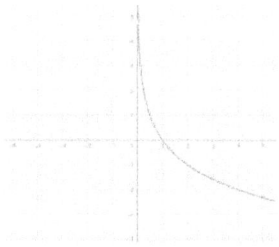

1   2   3

f) $f(x) = -\log_2(x)$

g) $f(x) = \log_2(-x)$

h) $f(x) = -\log_2(-x)$

342

Algebra II

Active Learning: 6.5a

Solve the logarithms:

a) $\log_2(32) =$

b) $\log_2(4) + \log_2(8) =$

(Notice that $4 \cdot 8 = 32$.)

Solve the logarithms:

c) $\log_3(81) =$

d) $\log_3(9) + \log_3(9) =$

(Notice that $9 \cdot 9 = 81$.)

A log with a product inside, for instance $\log_2(xy)$, is the same as the sum of the two individual logs: $\log_2(xy) = \log_2(x) + \log_2(y)$. With that in mind, expand the following logarithms. (These have variables, so you can't solve them. Just expand.)

e) $\log_5(7y) =$

f) $\log_3(2x) =$

This one takes the idea one step further.

g) $\log_7(3xy)$

Solve the logarithms:

h) $\log_2(\frac{64}{4}) =$ (Simplify what's inside the parenthesis and then solve.)

i) $\log_2(64) - \log_2(4) =$

Solve the logarithms:

j) $\log_3(\frac{81}{3}) =$ (Simplify what's inside the parenthesis and then solve.)

k) $\log_3(81) - \log_3(3) =$

A log with division inside is the same as subtraction.

$\log_2(\frac{x}{y}) = \log_2(x) - \log_2(y)$

With that in mind, expand the following logarithms.

l) $\log_5\left(\frac{3}{x}\right) =$

m) $\log_4\left(\frac{r}{s}\right) =$

Solve the logarithms:

n) $\log_2(64) =$

o) $3\log_2(4) =$

(Notice that $64 = 4^3$.)

Solve the logarithms:

p) $\log_5(625) =$

q) $2\log_5(25) =$

(Notice that $625 = 25^2$.)

Exponents in a log can be brought out front. For instance: $\log_2(x^3) = 3\log_2(x)$. With that in mind, expand the following:

r) $\log_3(y^7) =$

s) $\log_5(x^3) =$

(In the next problem, turn 8 into a number based on an exponent. Then use the exponent rule to simplify.)

t) $\log_3(8) =$

In this problem, move y^2 to the numerator and then use the exponent property to simplify.

u) $\log_2\left(\dfrac{1}{y^2}\right) =$

Here, recall that a root can be made into a fractional exponent. Then, use the exponent property.

v) $\log_3(\sqrt{x}) =$

Expand using the exponent rule. You will need a fractional exponent.

w) $\log_8\left(\sqrt[3]{x^2}\right) =$

Next, we want to combine the rules to expand more complicated logarithms. There are circumstances where the division property can cause problems when combined with other rules. For that reason, I bring denominators up as negative exponents and then use the exponent rule. Here is an example:

$$\log_2 \dfrac{\sqrt{x-2}}{y^3} = \log_2(\sqrt{x-2})y^{-3} = \log_2(\sqrt{x-2}) + \log_2(y^{-3}) = \log_2(x-2)^{\frac{1}{2}} + \log_2(y^{-3})$$

$$\dfrac{1}{2}\log_2(x-2) - 3\log_x(y)$$

The negative exponent on the y comes down to make subtraction.

Try some:

x) $\log_3 \dfrac{\sqrt{x}}{y^2} =$

y) $\log_2 \dfrac{xy^5}{z^3} =$

z) $\log_5 \dfrac{x^2}{\sqrt{y+3}} =$

aa) $\log_7 \dfrac{\sqrt[3]{x+1}}{\sqrt{y^2+5}} =$ (Here, the binomials under the roots are a group and can't be further simplified.)

Algebra II

Active Learning: 6.5b

Condensing Logarithms

Last time, we learned to use the property of logarithms to expand them. Now, we are going to go in the other direction. And, so, the steps simply reverse:

1. Exponents- put numbers out front of a log back as an exponent.
2. Multiplication- logs that are being added can now be put back together as being multiplied.
3. Division- logs that are being subtracted now go back to a single division.

Here is an example.

$2\log_3(x) + 3\log_3(y) - 5\log_3(z)$

1. Exponents

$\log_3(x^2) + \log_3(y^3) - \log_3(z^5)$

2. Multiplication

$\log_3(x^2 y^3) - \log_3(z^5)$

3. Division

$\log_3\left(\dfrac{x^2 y^3}{z^5}\right)$

As I mentioned before, you can run into some problems with division, so I avoid it. Instead, I make any subtraction into a negative exponent. Let me show you the same problem worked this way:

$2\log_3(x) + 3\log_3(y) - 5\log_3(z)$

First, I make everything an exponent:

$\log_3(x^2) + \log_3(y^3) + \log_3(z^{-5})$

Now, I turn everything into multiplication:

$\log_3(x^2 y^3 z^{-5})$

And, finally, I move the negative exponent to the bottom:

$\log_3\left(\dfrac{x^2 y^3}{z^5}\right)$

Try these. Condense the following logarithm.

a) $\log_3(x) + 2\log_3(y) + 3\log_3(z)$

On this problem, make a negative exponent and then move it down to the bottom.

b) $4\ln a + \ln b - 9\ln c$

c) $\dfrac{1}{2}\log(x-2) - 2\log y$

Here, all of the variables can go under one cube root sign.

d) $\dfrac{1}{3}\log_5(x) + \dfrac{1}{3}\log_5(y) + \dfrac{1}{3}\log_5(z)$

e) $5\log_7 x - 3\log_7 y - 2\log_7(z-1)$

The Change of Base Formula

Most calculators only have buttons for common logs, $\log(x)$, and natural logs, $\ln(x)$. So, some of the other bases can be a problem. The solution is the change of base formula.

Find the answer to the following:

f) $\log_3(27) =$

(Work each of these individual logs and then divide the answers.)

g) $\dfrac{\log_{10}(27)}{\log_{10}(3)} =$

h) $\log_8(64) =$

(Work each of these individual logs and then divide the answers.)

i) $\dfrac{\ln(64)}{\ln(8)} =$

This is the change of base formula.

$$\log_b(M) = \dfrac{\log_n(M)}{\log_n(b)}$$

You will have two logs being divided by each other. The numerator is the new base with what was inside the parenthesis. The denominator is the new base with the old base in the parenthesis.

Here is an example. Change the logarithm to base 10:

$$\log_3(12) = \dfrac{\log_{10}(12)}{\log_{10}(3)}$$

Your calculator could then easily work both the logarithm in the numerator and the logarithm in the denominator. It would be possible to calculate both of those and then divide your answer. I typically only ask students to show what I've shown above with the two logs over top of one another.

Try these.

Change the following to base 10.

j) $\log_3(51) =$

If it doesn't tell you what base to change it to, you can use common logs or natural logs and you will still get the same answer.

Use the change of base formula.

k) $\log_7(121) =$

l) $\log_2(37) =$

Algebra II

Active Learning: 6.6a

Using Like Bases to Solve Exponential Equations

Solve the following for x:

a) $\dfrac{x}{6} + \dfrac{1}{6} = \dfrac{3}{6}$

$x =$

In the same way that the denominators didn't matter, these bases don't matter. Since these have the same base, just set the exponents equal to each other.

b) Solve $3^{x+2} = 3^{2x-6}$

We only need to do:

$$x + 2 = 2x - 6$$

Work these problems.

c) $2^{5x-3} = 2^{2x+9}$

d) $5^{2x-7} = 5^{-6x+1}$

Solve the Equation by Creating Like Bases

Simplify the following:

e) $(2^3)^4 =$

f) $(3^2)^5 =$

g) $(5^3)^x =$

We can use this idea to manipulate an equation so we can get the same base on both sides.

Solve $8^{x+6} = 4^{x+2}$

Without the same base, it doesn't appear as if we can us our trick. However, both 8 and 4 can be made into exponents with a base of 2.

$(2^3)^{x+6} = (2^2)^{x+2}$

Notice that this is now a power raised to a power. And when you do that, you multiply the exponents.

$2^{3x+18} = 2^{2x+4}$

h) Now, both sides have the same base, and so the base doesn't matter. Finish solving it in the space below:

Try these:

i) Solve $3^{2x} = 9^{5x-12}$

j) Solve $25^{2x-5} = 125^{x-3}$

On this next problem, remember that a square root is a $\frac{1}{2}$ exponent.

k) Solve $\sqrt{5}^{2x+4} = 25^{3x-11}$

Solving Basic Exponential Equations with Logs

Solve the following:

l) $4\sqrt{x} = 16$

m) $2x^2 - 3 = 5$

In each of the problems above, our trick is to isolate the difficult portion of the problem and then use the opposite of what remains to help us. We can do the same with exponential equations.

$$25 = 5e^{2x}$$

The problem part of this equation is e^{2x}. First, we can get it alone by dividing both sides by 5.

$$5 = e^{2x}$$

Now, with the difficult part isolated, just take the ln of both sides. (Since ln is the opposite of e, they will cancel each other out.)

$$\ln 5 = \ln e^{2x}$$

$$\ln 5 = 2x$$

The key here is understanding that $\ln(5)$ is just a number. So, to get x alone, just divide both sides by 2. You get:

$$x = \frac{\ln(5)}{2}$$

And since $\ln(5)$ turns out to be a long decimal, mathematicians just leave it as is. Try these:

n) Solve $3e^{2x} - 3 = 15$

o) Solve $5e^{3x} + 20 = 45$

On this next problem, the opposite of base 10 is the common log.

p) Solve $400 + 10^{4x-1} = 63000$

Algebra II

Active Learning: 6.6b

Here again, we will be solving equations using logs.

Solving Equations Using the Definition of a Log

Our first strategy is one we've seen before. Look at this problem:

$$\text{Solve } 3\ln(x) - 5 = 7$$

The first thing to do is isolate the difficult part of the problem, which is the $\ln(x)$. To do this, just follow the typical algebra rules. We subtract 5 from both sides and then divide by 3.

$$3\ln(x) = 12$$
$$\ln(x) = 4$$

Now, here's where the definition of a log comes in. A natural log is really a log base e.

$$\ln(x) = \log_e(x) = 4$$

And, the answer to a log is an exponent. So, this is saying that e raised to the fourth power equal x. Let's write it like that.

$$e^4 = x$$

And although that looks strange, e^4 is just a number, so we are done.

Try a couple on your own.

a) Solve $5 + \ln(x) = 12$

On this problem, you'll have an extra step after using the definition of a logarithm. The 4 will need to be sent to the other side.

b) Solve $2\ln(x + 4) = 12$

Using the One-to-One Property of Logs to Solve Equations

Look at this problem.

Solve $\ln(x^2) = \ln(3x + 4)$

The trick to solving this is one which we saw for like bases. If both sides of the equation are both the same type of log than the logs have nothing to do with finding the answer. We drop the logs and just solve.

$x^2 = 3x + 4$

$x^2 - 3x - 4 = 0$

$(x - 4)(x + 1) = 0$

$x = 4 \text{ or } x = -1$

Try this one.

c)   Solve $\ln(x^2) = \ln 9$

Using the One-to-One Property with the Properties of Logarithms

Our next idea involves using the log properties in order to set up the One-to-One property. Here's what I mean.

$$\log(8x + 4) = \log(x) + \log(12)$$

If we have one log on the right and one log on the left, we can just ignore the logs altogether. Well, using the properties of logs, we can easily make that happen. Let's condense the log on the right.

$$\log(8x + 4) = \log(12x)$$

Now, the logs can be ignored and we solve:

$$8x + 4 = 12x$$

$$4 = 4x$$

$$x = 1$$

Try this one.

d) $$\log_5(4x) - \log_5(10) = \log_5(80)$$

Factoring using Substitution

Finally, look at this problem.

$$e^{2x} + e^x - 12 = 0$$

This may appear to be something impossible, but it is actually something which we already know how to factor. The key is to make a substitution: $a = e^x$. With that change we get:

$$a^2 + a - 12 = 0$$

Which we can easily factor and solve using the zero-product property.

$$(a - 3)(a + 4) = 0$$

$$a = 3; a = -4$$

But we aren't done. We made a substitution and need to undo our change.

$$e^x = 3; e^x = -4$$

These answers have one additional step. We need to take the ln of both sides.

$$x = \ln 3; x = \ln(-4)$$

However, -4 is outside the domain of the ln, so that answer isn't permitted. Our final answer is then $x = \ln 3$.

Work these problems using substitution.

e) $e^{2x} + 2e^x - 15 = 0$

f) $e^{2x} + 10e^x + 25 = 0$

Algebra II

Active Learning: 7.1a

In this activity, we want to understand systems of linear equations. A system of linear equations simply means that both lines exist in the same "universe." They exist together, and we are interested in knowing where (if anywhere) they intersect. The point where they intersect is the solution to the system of linear equations.

One way to find a solution to a system of linear equations is to graph the two lines and see where they intersect. What is the solution to the following systems of linear equations? (Hint: The solution is the point where the lines intersect. Give it as an ordered pair.)

$y = -x + 4$

$y = x - 2$

$2x + y = 0$

$3x + y = 1$

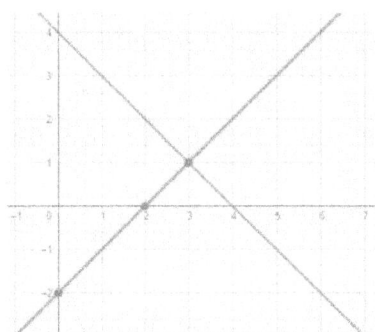
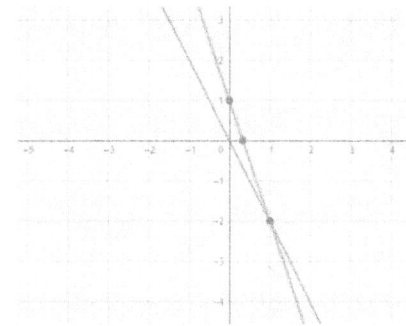

a) Solution:

b) Solution:

c) Graph the following system of linear equations and give the solution.

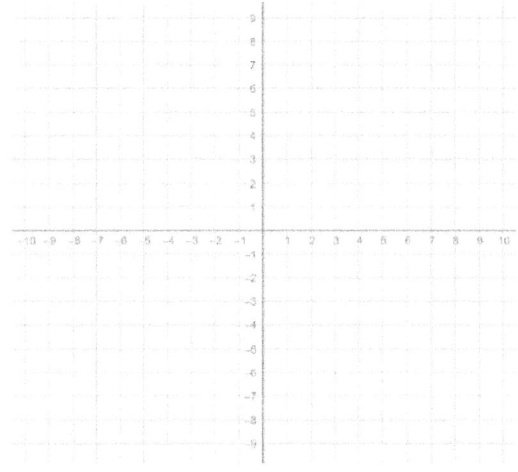

$x - y = -1$

$2x - y = -6$

d) Graph the following system of linear equations and give the solution.

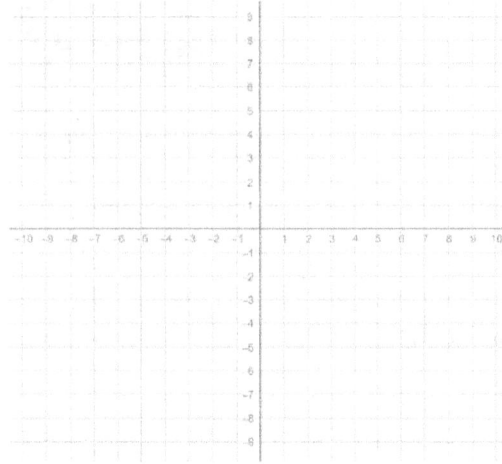

$$y = \frac{1}{2}x + 2$$

$$2x - 4y = -4$$

e) Something unusual has happened here. What is it?

f) Why has this happened?

g) Finally, one of the following two points is a solution to the system of linear equations. Test the two points. The point which gives a true statement in **both** equations is on both lines, and if it is on both lines, it must be where they intersect. And, intersection points are solutions. (Substitute in for x and y.)

$$y = \frac{1}{3}x + 2$$

$$2x + 8y = 2$$

(6, 4)   (−3, 1)

Algebra II

Active Learning: 7.1b

To understand the next approach that can be used to solve a system of linear equations, solve the following problems by substituting in the known value.

a) Solve for y if $3x + 2y = 45$ and $x = 5$

b) Solve for x if $2x - 9y = 82$ and $y = 2$

If we know x or y, solving these equations is easy. The concept for systems of linear equations is the same. We are going to solve for one variable and substitute it into the other equation. Look at this system:

$$2x + y = 0$$
$$3x + y = 1$$

I'm going to solve the top equation for y because it would be easy. (It can be x or y, it doesn't matter.)

$$y = -2x$$

Now, I know that y is equal to $-2x$. It isn't a simple number, but it will be enough to find what I need. Next, I substitute it into the next equation for y.

$$3x + (-2x) = 1$$

And I find $x = 1$.

The final step is to substitute the value of x back into either of the original equations. And you get the point where they intersect.

$$3(1) + y = 1$$

$y = -2$

$(1, -2)$ is the solution to the system of linear equations. (We graphed this in the last activity and got the same answer.)

Try one. (Choose the easiest variable to solve for. You also may start with either equation. It doesn't have to be the top one.) Since x is already solved, substitute it into the other equation.

$$x = -y - 1$$
$$2x - y = -5$$

c) Parenthesis are needed here. I've started it below. Finish it. Give the final answer as an ordered pair.

$$2(-y - 1) - y = -5$$

d) Try another. Solve the system of linear equations.

$y = 3x + 2$

$y = 2x$

e) I've purposely given easy equations. Try one more which is a bit harder.

$2x + y = 7$

$x - 2y = 6$

f) Do another. Something odd will happen here. Solve the system of linear equations by substitution.

$x - 5y = 15$

$5y = x + 20$

g) When you substituted your variables fell out. When what remains is false, like should have happened above, there is **no solution**. Why is there no solution? (Graph them if you aren't sure.)

h) Work one more. Solve the system of linear equations by substitution.

$2x + 4y = 8$

$4x + 8y = 16$

i) Again, when you substituted your variables fell out. When what remains is true, like should have happened here, there are **infinite solutions**. Why are there infinite solutions? (Graph them if you aren't sure.)

Your textbook will give fancy mathematical names to the three types of systems which you can encounter. When a system has one solution, it is called independent. When a system has no solution, it is called inconsistent. (This happens for parallel lines.) When a system has infinitely many solutions, it is called dependent. (This occurs when both lines are the same.)

Algebra II

Active Learning: 7.1c

So far, we've learned two methods for solving a system of linear equations: graphing and substitution. In this activity, we will learn a third called elimination. To understand the idea, we need to remember classic elementary school addition.

$$\begin{array}{r} 3 \\ +5 \\ \hline 8 \end{array}$$

a) Let's add some algebra to the idea. Add the like terms below:

$$\begin{array}{r} 2x \\ +5x \\ \hline \end{array}$$

b) We can solve linear equations like this too. Look at this system of linear equations lined up as an addition problem. We can add the like terms. (On the left, add the x's and the y's. Bring down the equal sign. On the right, add the numbers.)

$$\begin{array}{r} 4x + 2y = 14 \\ + x - 2y = 6 \\ \hline \end{array}$$

c) Now with only one variable, you can solve for x. Solve for x in the space below:

d) After you solve for x, you can put that value into either of the original equations in order to solve for y. Solve for y:

In this next system, adding the like terms won't get rid of a variable, but there is a simple trick.

$$2x + y = 7$$
$$+ x - 2y = 6$$

Any equation remains in balance if we multiply each term by the same number. So, if we multiply, the top equation by 2, we now create opposite terms for the y's.

$$2(2x) + 2(y) = 2(7)$$
$$+ x - 2y = 6$$

And we get:

$$4x + 2y = 14$$
$$+ x - 2y = 6$$

Adding now removes the y term and we could solve. (This system is actually the same as the first one.) It doesn't have to be the y term that is made to be the opposite. We could multiply through either equation by anything in order to make what we need.

$$2x + y = 7$$
$$+ x - 2y = 6$$

e) Multiply the bottom equation through by -2 and then add the terms. (You must multiply through every term in the equation, on both sides, to keep the equation in balance.) The x's will now drop out and you can solve for y. In the space below, show that this approach gets the same solution as before.

f) Try one on your own. You can get rid of either variable, but you must multiply through in order to create an opposite either with the x's or with the y's.

$$x - y = -1$$
$$+ 2x - y = -6$$

Sometimes you must multiply through both equations to create the opposites. Look at this system.

$$4x - 3y = 9$$
$$+ 7x + 2y = -6$$

g) Let's get rid of the y's. However, a single change won't be enough. The trick is to multiply the other equation by the coefficient in front of the y's. Finish the problem below.

$$2(4x) - 2(3y) = 2(9)$$
$$+ 3(7x) + 3(2y) = 3(-6)$$

Here is another that will require multiplying both equations. (You can eliminate either the x's or y's. However, if you choose the x's, you need to multiply through the top by 5 and the bottom by -3 in order
h) to create the opposites.) Solve the system of linear equations by elimination.

$$3x - 4y = -9$$
$$+ 5x + 3y = 14$$

i) Of course, the special cases of **No Solution** or **Infinite Solutions** can occur here too. Solve the following system of linear equations and determine if it should be "no solution" or "infinite solutions."

$$2x - 3y = -4$$
$$+\ 6x - 9y = -12$$

Finally, let's look at an application of systems of linear equations using cost and revenue functions. A candy company wants to sell a new inexpensive pack of chewing gum. It costs the company $1250 for the ingredients and .75 per pack to manufacture. Write a cost function. (Ingredient costs are the y-intercept and the per pack cost is the slope.)

j) $C(x) =$

The company will sell the packs of gum for $2 per pack. Write a revenue function. (There is only a slope here.)

k) $R(x) =$

You now have a system of equations. (Treat $C(x)$ and $R(x)$ as y.) The point of intersection for these two lines would be the break-even point for the company—when their costs and revenue are the same. Use any of our approaches for solving a system of equations to find the break-even point. (The value of x will be the number of packs of gum which the company will need to sell.)

l) What is the break-even point for the company?

Algebra II

Active Learning: 7.2

The concepts for solving systems of linear equations can be extended to three-dimensions. In three-dimensions there are now three variables, and we need one equation for each variable. Look at the following system:

$$x - 2y + z = 3$$
$$2x + y + z = 4$$
$$3x + 4y + 3z = -1$$

To solve a system with three equations, we must use elimination to simplify two equations down to the same two variables. We could eliminate any of the three variables, but here I will eliminate x.

$$-2(x - 2y + z = 3)$$
$$\underline{2x + y + z = 4}$$

$$-2x + 4y - 2z = -6$$
$$\underline{2x + y + z = 4}$$
$$5y - z = -2$$

Now, using the other two equations, we must once again eliminate x.

$$-3(2x + y + z = 4)$$
$$\underline{2(3x + 4y + 3z = -1)}$$

$$-6x - 3y - 3z = -12)$$
$$\underline{6x + 8y + 6z = -2}$$
$$5y + 3z = -14$$

a) With the x's eliminated this has become the same type of problem as we worked in the previous section. Use the two equations to find the values of y and z.

$$5y - z = -2$$
$$5y + 3z = -14$$

b) And once you have the values of y and z, you can plug those values into any of the original equations to find x. Plug your values of y and z into the equation below to find x.

$$x - 2y + z = 3$$

c) This approach isn't very enjoyable, and in a future section we will learn a much easier way to work these problems. For now, solve one more on your own. (Previously, we eliminated x's from two equations. Here, it would be easier to eliminate either y's or z's.)

$$5x + 2y + z = 5$$
$$-3x - y + 2z = 6$$
$$2x + 3y - 3z = 5$$

What is your solution to the system of equations?

These systems both had solutions, but as we saw before, it is possible that a variable drops out and we are left with a system that has **No Solution** or **Infinite Solutions**.

- $0 = 0$ Infinite solutions
- $0 = 4$ No solution

Algebra II

Active Learning: 7.3a

a) Solve the following system of linear equations by using the substitution method.

$$3x - y = 5$$
$$4x + 2y = 10$$

The concept of substitution doesn't need to be limited to linear equations. We can also find the intersection of non-linear systems.

$$y = x^2 + 2$$
$$y = x + 4$$

Below is the graph of this system of non-linear equations. Give the points of intersection. Notice that there are two.

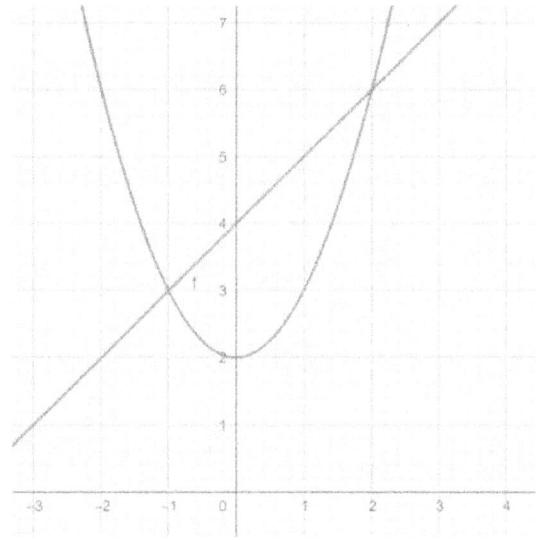

b) Points of Intersection

Point #1:

Point #2:

Let's use substitution to find the solutions. Both equations are solved for y, so we could use either. However, the math is a bit easier to substitute the second equation into the first.

$$y = x^2 + 2$$
$$y = x + 4$$

$$x + 4 = x^2 + 2$$

$$x^2 - x - 2 = 0$$

c) This is a quadratic equation. Factor it and solve for x. You will get two values.

d) Just as we did previously with substitution, put the values back into an original equation to find the values of y. There are two x's so we need two y's.

(The math is easier if we substitute back into $y = x + 4$)

Give your answers as the points of intersection.

e) <u>Points of Intersection</u>

Point #1:

Point #2:

f) Try another. Solve the following system of non-linear equations by substitution. (Hint: You will get a duplicate value of x, so there will only be one point of intersection.)

$$y = x^2 - 1$$
$$y = 4x - 5$$

Although the math will be a bit harder, you can use substitution to find the intersection of a line and a circle.

$$x^2 + y^2 = 25$$
$$y = x - 1$$

Here's the setup of the first step.

$$x^2 + (x - 1)^2 = 25$$

g) You will need to multiply out $(x - 1)^2$, and remember, to solve this you will need to get one side equal to zero. There are two points of intersection, find them.

In the two systems we've worked on above, elimination would not work. Both equations would need a x^2 and y^2. However, with the intersection of two circles (or ellipses), we could do elimination.

Look at the following system:

$$x^2 + y^2 = 25$$
$$2x^2 + y^2 = 36$$

Here is a graph.

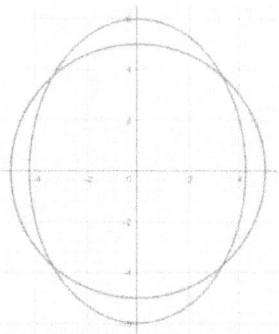

Notice that there are four points of intersection.

To use elimination, we would run a -2 through the top equation.

$$-2(x^2 + y^2 = 25)$$
$$\underline{2x^2 + y^2 = 36}$$

$$-2x^2 - 2y^2 = -50$$
$$\underline{2x^2 + y^2 = 36}$$

$$-y^2 = -12$$
$$y^2 = 12$$
$$y = \pm\sqrt{12}$$

h) Substitute the values of y back into the first equation. There are only two values of y, but you will create four points.

Point #1:   Point #2:

Point #3:   Point #4:

376

i) Try one on your own.

$$2x^2 + y^2 = 16$$
$$x^2 + y^2 = 12$$

Algebra II

Active Learning: 7.3b

We are going to find the intersection of non-linear inequalities. However, before we do, we need a refresher on graphing inequalities.

$$y < 2x - 1$$

To graph this inequality:

- Graph the line $y = 2x - 1$ but with a dashed line because we want *less than* and not greater than or equal to.
- The inequality says that the y's are less than the line we just made. So, shade below the line, where the lesser values of y are found on the y-axis.

$$y < 2x - 1$$

a) Graph the inequality below:

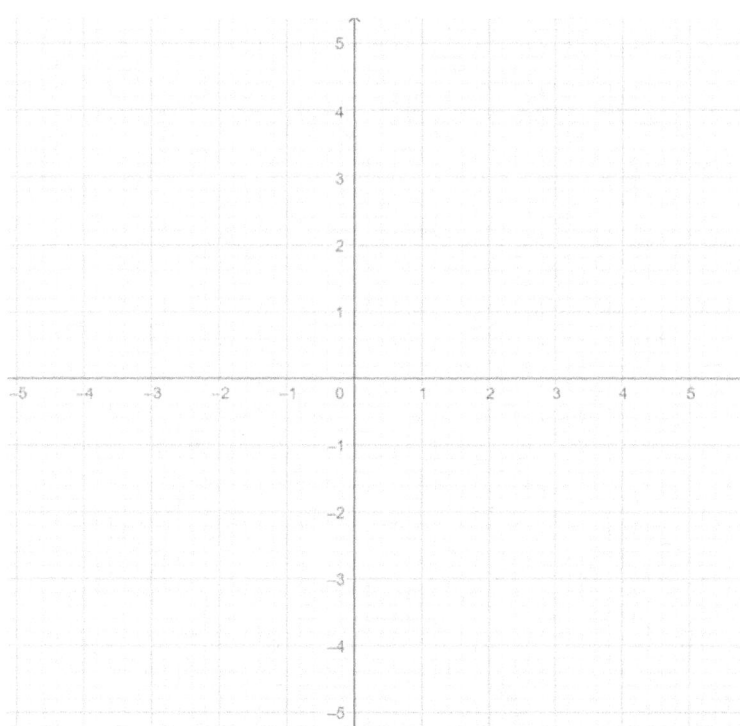

The same concept applies to parabolas.

$$y \geq x^2 - 1$$

To graph this inequality:

- Graph the parabola $y = x^2 - 1$. This will have a solid line because the inequality included equal to.
- To graph the parabola, use $h = -\frac{b}{2a}$ and $k = f(h)$ to find the vertex, or you can see that this is a basic parabola which has been shifted down 1.
- The inequality says that we want the values of y larger than the parabola. So, shade above the parabola. It is harder to see with a parabola, but you want all of the larger y values. This will be above and will go up the y-axis.

b) Graph the inequality below:

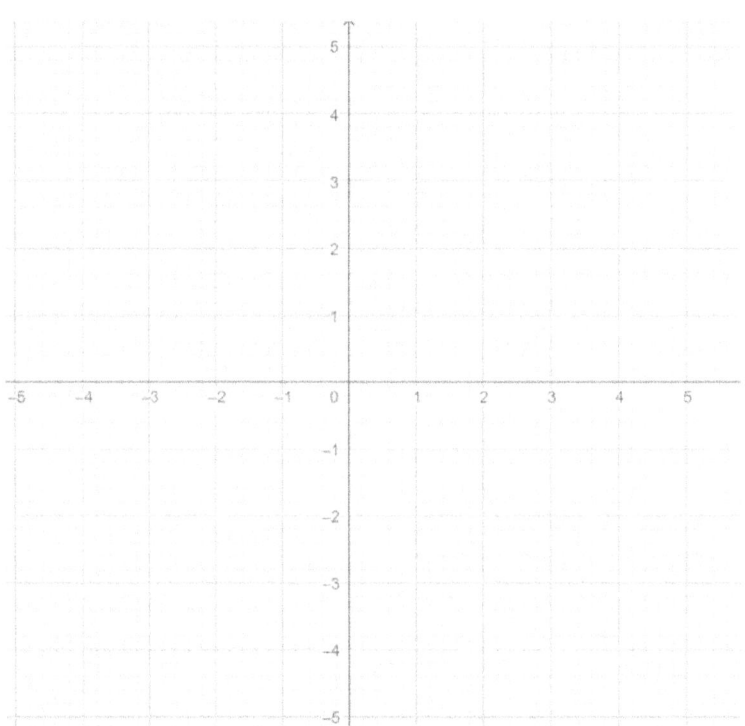

Next, we will put it together to examine a system of nonlinear inequalities.

c) We want to graph the following system:

$$y < 2x - 1$$
$$y \geq x^2 - 1$$

Combine the two previous graphs to create an overlap region. That overlap region would be the solution to the system of nonlinear equations.

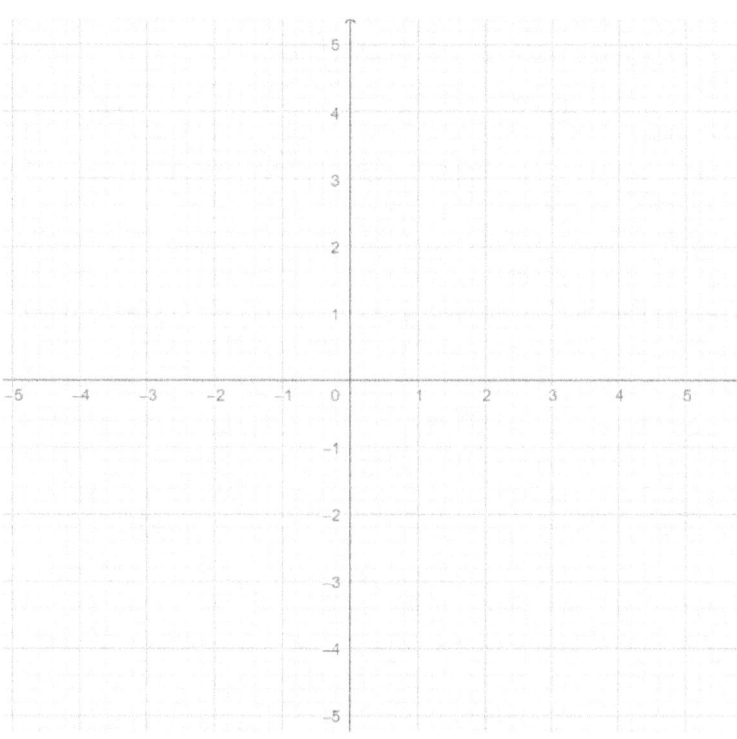

d) For more clarity, we can add the points of intersection. Use the concepts from the last lesson to find the intersection points of the two inequalities. (We will work with them as if they both had equal signs.)

$$y = 2x - 1$$
$$y = x^2 - 1$$

e) Add your points of intersection to the graph.

f) Following the same procedure, we could graph a system of non-linear inequalities where both involve parabolas. Graph the following and add to the graph the points of intersection:

$$2x^2 - y \le 3$$
$$x^2 + y \le 9$$

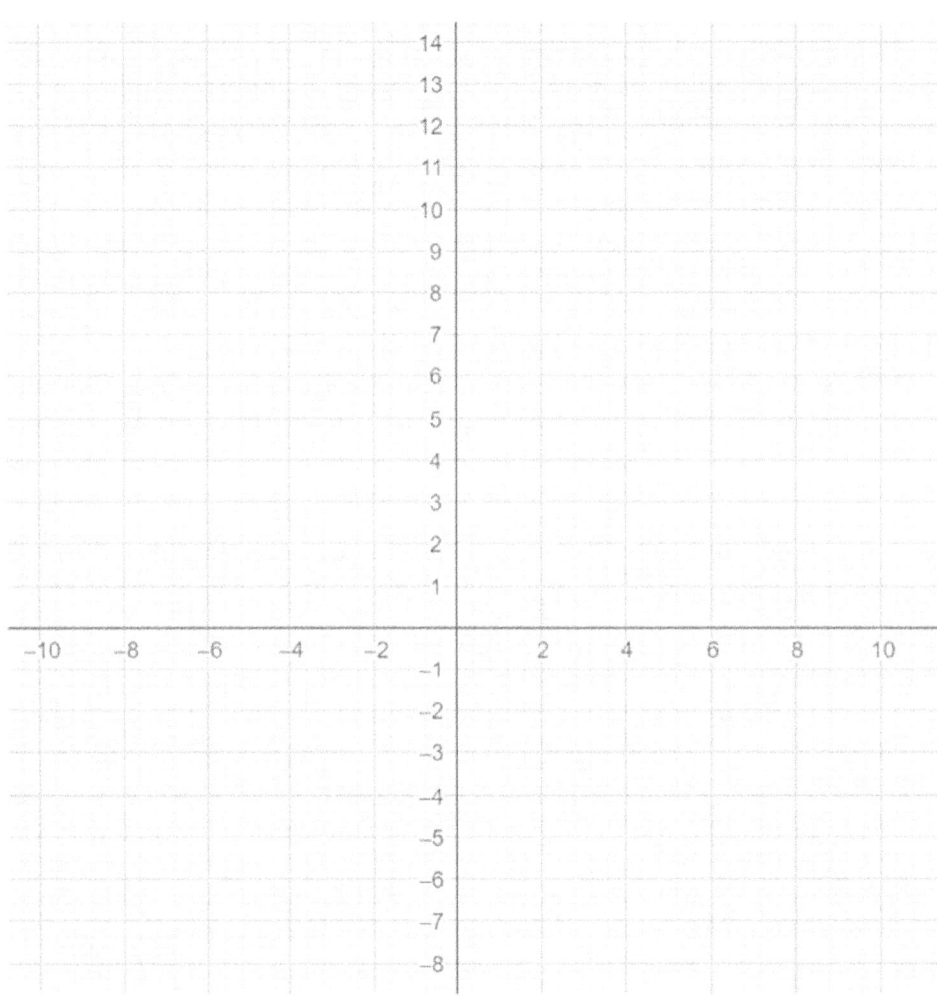

Algebra II

Active Learning: 7.4a

a) Simplify the following rational expressions. Remember, you must get a common denominator. And if you add something to the denominator of a fraction, you must add it to the numerator.

$$\frac{-4}{x-2} + \frac{6}{x+3}$$

In this section, we are going to learn how to decompose a fraction into a sum of fractions. Let's begin with the following:

$$\frac{2x}{(x-2)(x+3)}$$

We know it must turn into something of this form because we have the two factors in the denominator:

$$\frac{2x}{(x-2)(x+3)} = \frac{A}{x-2} + \frac{B}{x+3}$$

If we can find A and B, we will have broken the fraction into two pieces. To do this, let's get all three denominators to match and then the denominator will have nothing to do with solving the equation. If we add something to a denominator, we must add it to the numerator.

$$\frac{2x}{(x-2)(x+3)} = \frac{A(x+3)}{(x-2)(x+3)} + \frac{B(x-2)}{(x+3)(x-2)}$$

The denominators no longer matter.

$$2x = A(x+3) + B(x-2)$$

Expanded we have:

$$2x = Ax + 3A + Bx - 2B$$

The trick now is that x terms on the left must equal the x terms on the right. And, any coefficients on the left must equal any coefficients on the right. So, I will group the terms.

$$2x = Ax + Bx + 3A - 2B$$

$$2x = (A + B)x + (3A - 2B)$$

Because they must match, I can match the x terms on the left and right to form an equation.

$$2 = A + B$$

There are no coefficients on the left, so that is as if a 0 is in that position on the left. Here is a second equation for the coefficients.

$$0 = 3A - 2B$$

We now have this system of linear equations.

$$2 = A + B$$

$$0 = 3A - 2B$$

b) In the space below, solve this system for A and B.

c) You now know the values of A and B and can finish the decomposition.

$$\frac{A}{x-2} + \frac{B}{x+3} =$$

Try one.

d) Find the partial fraction decomposition of the following expression.

$$\frac{4x}{(x-1)(x+3)}$$

e) Before we look at our next idea, we need to do some work to help us better understand. In the space below, give the prime factorization for the number 12.

$$12 =$$

f) Solve the following equations:

$$\frac{x}{12} = \frac{1}{4} + \frac{2}{3}$$

g)
$$\frac{x}{12} = \frac{1}{2} + \frac{2}{3}$$

When decomposing a fraction, we run into trouble with repeated factors. Since 12 is made up of repeated factors of 2, you can create a fraction involving 12 with both 2 or $2^2 = 4$. So, if you are trying to decompose with repeated factors, you must include all the possibilities. Look at this problem:

$$\frac{2x+1}{(x-2)^2}$$

To decompose it, we must consider the repeated factors and we get:

$$\frac{2x+1}{(x-2)^2} = \frac{A}{(x-2)} + \frac{B}{(x-2)^2}$$

h) Use the process of decomposition to find A and B.

i) This next problem will involve three factors, and so we have three unknowns. (Notice it includes the repeated factor.)

$$\frac{3x^2 + 2x + 4}{x(x-1)^2} = \frac{A}{x} + \frac{B}{x-1} + \frac{C}{(x-1)^2}$$

With three variables, you should now generate three equations, an x^2 set of equations, an x set of equations, and a coefficient set of equations. Below, I'm giving you the three equations that you should generate. It would be easy to finish solving for the variables with these equations, but in the space below, try to generate the equations yourself.

$$3 = A + B$$
$$2 = -2A - B + C$$
$$4 = A$$

j) Try a final problem on your own. Find the partial fraction decomposition of the following expression.

$$\frac{2x^2 + x - 2}{x(x-2)^2}$$

Algebra II

Active Learning: 7.4b

Factor the following quadratics, if possible.

a) $x^2 + x + 1$

b) $x^2 + 1$

In this activity, we will continue with decomposition of fractions. Here, we will look at denominators which include quadratic factors that don't reduce. Look at the following:

$$\frac{3x^2 + 4x - 2}{x(x^2 + x + 1)}$$

We will break this down in a similar fashion as before. However, we have one additional step.

$$\frac{3x^2 + 4x - 2}{x(x^2 + x + 1)} = \frac{A}{x} + \frac{Bx + C}{x^2 + x + 1}$$

The numerator of the quadratic factor must be at least one degree lower. And, we can't be sure if it has both an x term and a coefficient term. However, since it is possible, we must search for them. Nothing else changes in the process, but the quadratic makes the work harder. I'll help you get started.

$$\frac{3x^2 + 4x - 2}{x(x^2 + x + 1)} = \frac{A(x^2 + x + 1)}{x(x^2 + x + 1)} + \frac{(Bx + C)x}{x(x^2 + x + 1)}$$

So, we generate the following:

$$3x^2 + 4x - 2 = Ax^2 + Ax + 1A + Bx^2 + Cx$$

And get the equations:

$$3 = A + B$$
$$4 = A + C$$
$$-2 = A$$

c) Finish solving for the variables and write the decomposed solution.

d) Try a problem on your own. Find the partial fraction decomposition of the following expression.

$$\frac{x^2 + 5x - 1}{x(x^2 + 1)}$$

Next, we will try a problem with a non-reducible quadratic which repeats. The set up is consistent with what you might expect.

$$\frac{2x^2 + 3x + 1}{(x^2 + 1)^2} = \frac{Ax + B}{(x^2 + 1)} + \frac{Cx + D}{(x^2 + 1)^2}$$

I'll get you started.

$$\frac{2x^2 + 3x + 1}{(x^2 + 1)^2} = \frac{(Ax + B)(x^2 + 1)}{(x^2 + 1)} + \frac{Cx + D}{(x^2 + 1)^2}$$

$$\frac{2x^2 + 3x + 1}{(x^2 + 1)^2} = \frac{Ax^3 + Ax + Bx^2 + B}{(x^2 + 1)} + \frac{Cx + D}{(x^2 + 1)^2}$$

e) Create your equations and solve for the variables. Then, show the final partial deconstruction. (Hint: There are no x^3 terms in the original problem and so you will instantly know that $A = 0$.)

f) Work one more. Find the partial fraction decomposition of the following expression.

$$\frac{5x^2 - 2x + 1}{(x^2 + 3)^2}$$

Algebra II

Active Learning: 7.5a

The second half of this chapter focuses on an area of mathematics involving matrices. A matrix is an array of numbers with many applications. Matrices can be any size and shape, but they are defined by *rows × columns*.

$$\begin{bmatrix} 3 & -2 & 7 \\ 1 & 7 & 10 \end{bmatrix}$$

This is a 2 × 3 matrix because it has 2 rows and 3 columns. Give the dimensions of the following matrices.

a) $\begin{bmatrix} -6 & 3 \\ 19 & -5 \end{bmatrix}$ b) $\begin{bmatrix} 4 & 21 \\ 7 & 1 \\ -2 & 9 \end{bmatrix}$ c) $[11 \;\; -56 \;\; -81]$ d) $\begin{bmatrix} 2 & 1 & 8 \\ 5 & -2 & -16 \\ 1 & 1 & 12 \end{bmatrix}$

Entries in a matrix will be defined by row, column.

$$A = \begin{bmatrix} a_{11} & a_{12} & a_{13} \\ a_{21} & a_{22} & a_{23} \\ a_{31} & a_{32} & a_{33} \end{bmatrix}$$

For instance, the entry a_{31} is the element in the third row and first column. Look at the following matrix.

$$\begin{bmatrix} 2 & 1 & 8 \\ 5 & -2 & -16 \\ 1 & 1 & 12 \end{bmatrix}$$

Give the value of the following entries:

e) a_{13}

f) a_{21}

g) a_{32}

Adding matrices is fairly easy. To do so, the matrices must have the same dimensions. But, if they do, simply add the matching entries.

$$\begin{bmatrix} 4 & 21 \\ 7 & 1 \\ -2 & 9 \end{bmatrix} + \begin{bmatrix} -5 & 19 \\ 6 & 12 \\ 5 & 7 \end{bmatrix} = \begin{bmatrix} 4+(-5) & 21+19 \\ 7+6 & 1+12 \\ -2+5 & 9+7 \end{bmatrix} = \begin{bmatrix} -1 & 40 \\ 13 & 13 \\ 3 & 16 \end{bmatrix}$$

Add the following matrices:

h) $\begin{bmatrix} -6 & 3 \\ 19 & -5 \end{bmatrix} + \begin{bmatrix} 5 & 7 \\ 8 & -14 \end{bmatrix}$

i) $[11 \ -56 \ -81] + [8 \ 2 \ 50]$

j) $\begin{bmatrix} 2 & 1 & 8 \\ 5 & -2 & -16 \\ 1 & 1 & 12 \end{bmatrix} + \begin{bmatrix} 11 & -1 & 18 \\ 15 & 12 & -16 \\ -8 & 3 & -4 \end{bmatrix}$

Subtracting matrices follows the same idea. Just keep in mind that it "distributes" the negative sign to each entry in the second matrix.

$$\begin{bmatrix} 4 & 21 \\ 7 & 1 \\ -2 & 9 \end{bmatrix} - \begin{bmatrix} -5 & 19 \\ 6 & 12 \\ 5 & 7 \end{bmatrix} = \begin{bmatrix} 4-(-5) & 21-19 \\ 7-6 & 1-12 \\ -2-5 & 9-7 \end{bmatrix} = \begin{bmatrix} 9 & 2 \\ 1 & -11 \\ -7 & 2 \end{bmatrix}$$

Subtract the following matrices:

k) $\begin{bmatrix} -6 & 3 \\ 19 & -5 \end{bmatrix} - \begin{bmatrix} 5 & 7 \\ 8 & -14 \end{bmatrix}$

l) $[11 \ -56 \ -81] - [8 \ 2 \ 50]$

m) $\begin{bmatrix} 2 & 1 & 8 \\ 5 & -2 & -16 \\ 1 & 1 & 12 \end{bmatrix} - \begin{bmatrix} 11 & -1 & 18 \\ 15 & 12 & -16 \\ -8 & 3 & -4 \end{bmatrix}$

We will see that multiplying matrices together is a bit involved. However, multiplying a matrix by a scalar (a single number) is not hard. To multiply by a scalar, simply multiply every entry by that scalar.

$$3\begin{bmatrix} 2 & 1 & 8 \\ 5 & -2 & -16 \\ 1 & 1 & 12 \end{bmatrix} = \begin{bmatrix} 3(2) & 3(1) & 3(8) \\ 3(5) & 3(-2) & 3(-16) \\ 3(1) & 3(1) & 3(12) \end{bmatrix} = \begin{bmatrix} 6 & 3 & 24 \\ 15 & -6 & -48 \\ 3 & 3 & 36 \end{bmatrix}$$

Multiply the following matrices by the scalar indicated:

n) $5\begin{bmatrix} -6 & 3 \\ 19 & -5 \end{bmatrix}$
o) $2\begin{bmatrix} 4 & 21 \\ 7 & 1 \\ -2 & 9 \end{bmatrix}$
p) $-3[11 \quad -56 \quad -81]$
q) $4\begin{bmatrix} 2 & 1 & 8 \\ 5 & -2 & -16 \\ 1 & 1 & 12 \end{bmatrix}$

Finally, it is possible to do multiple operations when working with matrices. In the following problems, you must first multiply by the scalar and then add your resulting matrices.

r) $2\begin{bmatrix} -6 & 3 \\ 19 & -5 \end{bmatrix} + 3\begin{bmatrix} 5 & 7 \\ 8 & -14 \end{bmatrix}$

s) $4\begin{bmatrix} 2 & 1 & 8 \\ 5 & -2 & -16 \\ 1 & 1 & 12 \end{bmatrix} - 2\begin{bmatrix} 11 & -1 & 18 \\ 15 & 12 & -16 \\ -8 & 3 & -4 \end{bmatrix}$

Algebra II

Active Learning: 7.5b

Multiplying two matrices involves a process that is a bit unexpected. We multiply the elements in a row from the first matrix times the elements in a column in the second. Add these to make the element in the new matrix. (Notice that the value goes into the new matrix where the two circles intersected.)

$$\begin{bmatrix} 3 & 2 \\ 6 & -1 \end{bmatrix} \begin{bmatrix} -2 & 1 \\ 5 & 7 \end{bmatrix}$$

$$\begin{bmatrix} 3(-2) + 2(5) & \end{bmatrix} = \begin{bmatrix} 4 & \end{bmatrix}$$

Then the row is multiplied by the next column.

$$\begin{bmatrix} 3 & 2 \\ 6 & -1 \end{bmatrix} \begin{bmatrix} -2 & 1 \\ 5 & 7 \end{bmatrix}$$

$$\begin{bmatrix} 4 & 3(1) + 2(7) \end{bmatrix} = \begin{bmatrix} 4 & 17 \end{bmatrix}$$

Next, we move down a row and repeat the idea.

$$\begin{bmatrix} 3 & 2 \\ 6 & -1 \end{bmatrix} \begin{bmatrix} -2 & 1 \\ 5 & 7 \end{bmatrix}$$

$$\begin{bmatrix} 4 & 17 \\ 6(-2) + 1(7) & 17 \end{bmatrix} = \begin{bmatrix} 4 & 17 \\ -5 & 17 \end{bmatrix}$$

$$\begin{bmatrix} 3 & 2 \\ 6 & -1 \end{bmatrix} \begin{bmatrix} -2 & 1 \\ 5 & 7 \end{bmatrix}$$

$$\begin{bmatrix} 4 & 17 \\ -5 & 6(1) + (-1)(7) \end{bmatrix} = \begin{bmatrix} 4 & 17 \\ -5 & -1 \end{bmatrix}$$

The reason for this process is easiest to understand as a series of transformation on vectors. But this is beyond us at this level.

Use matrix multiplication to multiply the following:

a)
$$\begin{bmatrix} 1 & 5 \\ -3 & 2 \end{bmatrix} \begin{bmatrix} 4 & -5 \\ 2 & 1 \end{bmatrix}$$

b)
$$\begin{bmatrix} 4 & 21 \\ 7 & 1 \\ -2 & 9 \end{bmatrix} \begin{bmatrix} 1 & 6 & -2 \\ 3 & -1 & 5 \end{bmatrix}$$

This next set of matrices can't be multiplied. Something happens when we try to match the elements.
c) What is it?

$$\begin{bmatrix} 3 & -2 & 7 \\ 1 & 7 & 10 \end{bmatrix} \begin{bmatrix} 5 & 4 & 1 \\ 2 & -4 & 8 \end{bmatrix}$$

In order to be certain that we can multiply matrices we need to have a proper match in their dimensions.

$$(3 \times 2)(2 \times 3)$$

If the number of columns of the first matrix matches the number of rows of the second, they can be multiplied. If not, the elements won't match up properly. The easiest way to check is to line up their dimensions and check the middle. These can be multiplied:

$$(5 \times 3)(3 \times 4)$$

$$(1 \times 6)(6 \times 3)$$

$$(3 \times 3)(3 \times 3)$$

These cannot.

$$(2 \times 3)(2 \times 4)$$

$$(3 \times 4)(2 \times 3)$$

$$(1 \times 5)(4 \times 1)$$

d) Circle any of the following matrices which can be multiplied. (You do not need to do the multiplication.)

$$\begin{bmatrix} 4 & 21 \\ 7 & 1 \\ -2 & 9 \end{bmatrix} \begin{bmatrix} -6 & 3 \\ 19 & -5 \end{bmatrix} \qquad \begin{bmatrix} 2 & 1 & 8 \\ 5 & -2 & -16 \\ 1 & 1 & 12 \end{bmatrix} \begin{bmatrix} 4 & 21 \\ 7 & 1 \\ -2 & 9 \end{bmatrix} \qquad \begin{bmatrix} 11 & -56 & -81 \end{bmatrix} \begin{bmatrix} 7 \\ 45 \end{bmatrix}$$

Commutative multiplication means we can reverse the order and we get the same answer. Multiply the following:

$$A = \begin{bmatrix} 1 & 3 \\ 4 & 5 \\ -2 & -1 \end{bmatrix} \qquad B = \begin{bmatrix} 4 & 2 & 1 \\ -1 & 3 & 5 \end{bmatrix}$$

e) $A \cdot B =$

f) $B \cdot A =$

g) Is matrix multiplication commutative? Yes or No.

Algebra II

Active Learning: 7_6a

There are many applications of matrices. However, the most relevant to this course is their use for solving systems of linear equations. Before we see how this is done, we first must learn about augmented matrices. Below are systems of linear equations and their augmented matrices. Look over the systems and their augmented matrix.

$$4x - 3y = 7$$
$$-2x + 6y = 14$$

$$\begin{bmatrix} 4 & 3 & | & 7 \\ -2 & 6 & | & 14 \end{bmatrix}$$

$$31x + 45y = 62$$
$$24x - 20y = 150$$

$$\begin{bmatrix} 31 & 45 & | & 62 \\ 24 & -20 & | & 150 \end{bmatrix}$$

a) Explain the connection between a system of linear equations and its augmented matrix.

Give the augmented matrix for the following systems of linear equations.

b) $-8x - 4y = 24$
$5x - 21y = 80$

c) $6x + 18y = 36$
$-15x + y = 42$

d) Now, go in reverse. Starting with the augmented matrix, write the system of linear equations.

$$\begin{bmatrix} -12 & 16 & | & 8 \\ 14 & 9 & | & -4 \end{bmatrix}$$

e) As we will see more of later, augmented matrices can extend beyond linear equations. Write the following system of equations as an augmented matrix.

$$4x + 6y - 8z = 20$$
$$2x - 12y - z = 15$$
$$-6x + 14y - 5z = 16$$

To solve a system of linear equations involving a matrix, we will use an approach called Gaussian Elimination. The goal in gaussian elimination is to have the equation portion of the matrix reduced down to a single diagonal of 1's. (Gaussian elimination may seem different, but we will basically be following the same idea we use when doing elimination to solve systems of linear equations.)

$$\begin{bmatrix} 1 & 0 & | & 3 \\ 0 & 1 & | & -5 \end{bmatrix}$$

When that occurs, this is what the system would look like outside of the matrix:

$$1x + 0y = 3$$
$$0x + 1y = -5$$

And this means that we know the values of x and y.

$$x = 3$$
$$y = -5$$

To reduce the matrix, we have three tools we can use:

1. We can switch any two rows.
2. We can multiply any value through a row to make a permanent change.
3. We can temporarily multiply a row and then add that changed row to a new row.

Here's an example of why you would choose each tool.

1. Switch any two rows.

$$\begin{bmatrix} 0 & 8 & | & 6 \\ 1 & 12 & | & -7 \end{bmatrix}$$

Here, switching R_1 with R_2 would put the 1 in the top left corner. (Getting that 1 in the top left should always be your first goal.)

$$R_1 \overset{\Delta}{\rightarrow} R_2 \quad \begin{bmatrix} 1 & 12 & | & -7 \\ 0 & 8 & | & 6 \end{bmatrix}$$

2. Multiply through to change a row.

$$\begin{bmatrix} 5 & 10 & | & -15 \\ 6 & 12 & | & 8 \end{bmatrix}$$

Our first goal is to make the 1 in the top left corner. Here, multiplying R_1 by $\frac{1}{5}$ would accomplish this.

$$\frac{1}{5} \cdot R_1 \quad \begin{bmatrix} 5 \cdot \frac{1}{5} & 10 \cdot \frac{1}{5} & | & -15 \cdot \frac{1}{5} \\ 6 & 12 & | & 8 \end{bmatrix}$$

$$\begin{bmatrix} 1 & 2 & | & -3 \\ 6 & 12 & | & 8 \end{bmatrix}$$

3. Temporarily multiply a row and add it to another.

$$\begin{bmatrix} 1 & 3 & | & 6 \\ 5 & 4 & | & -4 \end{bmatrix}$$

After you've gotten the 1 in the top left, you want a 0 below it. Using the 1 is the key. Multiply R_1 by -5 and adding it to R_2 would create the 0.

$$-5 \cdot R_1 + R_2 \quad \begin{bmatrix} 1 & 3 & | & 6 \\ 5-5 & 4-15 & | & -4-30 \end{bmatrix}$$

$$\begin{bmatrix} 1 & 3 & | & 6 \\ 0 & -11 & | & -34 \end{bmatrix}$$

Use tools 1 or 2 to first get a 1 in the top left corner of the following matrices:

f) $\begin{bmatrix} 8 & -16 & | & -32 \\ 5 & 9 & | & 18 \end{bmatrix}$

g) $\begin{bmatrix} 0 & 4 & | & 7 \\ 1 & 16 & | & 6 \end{bmatrix}$

h) $\begin{bmatrix} 2 & -6 & | & 22 \\ 1 & -4 & | & 16 \end{bmatrix}$

Use tool 3 to get a 0 in the bottom left corner of the following matrices:

i) $\begin{bmatrix} 1 & 2 & | & 5 \\ 4 & -6 & | & -1 \end{bmatrix}$

j) $\begin{bmatrix} 1 & -3 & | & 2 \\ -7 & 20 & | & -9 \end{bmatrix}$

Next, we will use these tools to finish an entire matrix.

$$\begin{bmatrix} 2 & -1 & | & -5 \\ 1 & 1 & | & -1 \end{bmatrix}$$

First, get a 1 in the top left. We could switch R_1 with R_2.

$$\begin{bmatrix} 1 & 1 & | & -1 \\ 2 & -1 & | & -5 \end{bmatrix}$$

Next, we want a zero in the bottom left. I've made the change.

$$\begin{bmatrix} 1 & 1 & | & -1 \\ 0 & -3 & | & -3 \end{bmatrix}$$

k) What did I do to make this change?

Once you have completed the first column, the goal is then to make a 1 in the bottom center. We can do so by using the second tool and multiplying through R_2 by $-\frac{1}{3}$.

$$\begin{bmatrix} 1 & 1 & | & -1 \\ 0 & 1 & | & 1 \end{bmatrix}$$

Finally, we need to make a 0 at the top center. Using tool #3, we can multiply R_2 by -1 and add it to R_1.

$$\begin{bmatrix} 1 & 0 & | & -2 \\ 0 & 1 & | & 1 \end{bmatrix}$$

(Because there is a 0 in the bottom left, multiplying R_2 and adding to R_1 doesn't change the value of 1 in the top left.)

l) The augmented matrix is now finished. Give the value of x and y.

$x = $

$y = $

Use Gaussian elimination to solve the following systems of linear equations:

m)
$$\begin{bmatrix} 1 & 1 & | & -1 \\ -1 & 1 & | & -5 \end{bmatrix}$$

n)
$$\begin{bmatrix} 2 & 4 & | & 4 \\ 1 & -2 & | & 6 \end{bmatrix}$$

(Unfortunately, Gaussian elimination often involves working with fraction. That is the case with the next problem.)

o)
$$\begin{bmatrix} 2 & 1 & | & 0 \\ 3 & 1 & | & 1 \end{bmatrix}$$

Previously, when we solved systems of linear equations, we could run into odd situations. Since this *is* solving systems, the same thing can happen here. Solve the following matrix by Gaussian elimination. What happens here is **No Solution**. (The textbook will refer to this as an inconsistent system.)

p)
$$\begin{bmatrix} 1 & -5 & | & 15 \\ -1 & 5 & | & 20 \end{bmatrix}$$

Solve again by Gaussian elimination, this will end with **Infinite Solutions**. (The textbook will refer to this as a dependent system.)

q)
$$\begin{bmatrix} 2 & 4 & | & 8 \\ 4 & 8 & | & 16 \end{bmatrix}$$

These situations occur for the same reason as it did when we first introduced systems of linear equations. What is happening here is that a variable is dropping out. If what remains is telling the truth, we have infinite solutions. For instance:

$$\begin{bmatrix} 1 & 6 & | & -5 \\ 0 & 0 & | & 0 \end{bmatrix}$$

The bottom row is saying that $0 = 0$, which is true and therefore "Infinite Solutions." If what remains is not telling the truth, we have no solution. For instance:

$$\begin{bmatrix} 1 & 7 & | & 12 \\ 0 & 0 & | & -6 \end{bmatrix}$$

Here, the bottom row is saying $0 = -6$, which is not true and therefore "no solution." (This typically happens in the bottom row, but if it happens in the top row, the same conclusions apply.)

Algebra II

Active Learning: 7.6b

In the last activity, we saw that the idea of an augmented matrix can be extended to three variables. In the same way, Gaussian elimination can also be extended.

$$\begin{bmatrix} 2 & -2 & 1 & | & 3 \\ 3 & 1 & -1 & | & 7 \\ 1 & -3 & 2 & | & 0 \end{bmatrix}$$

Here, we would try to create a matrix which looks like this:

$$\begin{bmatrix} 1 & 0 & 0 & | & Answer\ 1 \\ 0 & 1 & 0 & | & Answer\ 2 \\ 0 & 0 & 1 & | & Answer\ 3 \end{bmatrix}$$

And although the concept is identical to before, the process is time consuming. Instead, it is time to allow technology to help us.

First, we want to enter the matrix on our calculator.
- Select Matrix. It is above the Math button.
- Move over two to the right and select Edit.
- Tell the calculator the dimensions of your Matrix. Rows are first. Columns are second. So, we have a 3 x 4.
- After you have the dimensions, enter the values in your matrix. Hit enter after each entry. When everything has been entered, quit out of the matrix screen.

Second, we want to reduce the matrix.
- Select Matrix.
- Move one to the right and select Math.
- Go down until you find "rref." This stands for reduced row echelon form, which is the Gaussian elimination approach. Select "rref."
- "rref" should now be on the front screen.
- Once again, select Matrix. Select the Matrix you created, probably A.
- The front screen should now say rref([A].
- Hit Enter and your reduced matrix should appear.

a) Start with the matrix above, and in the space below, record the matrix which your calculator is displaying:

Hint: Sometimes your calculator will give a very small number in scientific notation. This is rounding error and the value is actually zero. That will happen here on some calculators.

b) There are now three solutions: x, y and z. Give the solutions in the space below:

$x =$

$y =$

$z =$

Use the same approach to solve this matrix.

$$\begin{bmatrix} 3 & 0 & 1 & | & 2 \\ 1 & -1 & -2 & | & 13 \\ 0 & 2 & -1 & | & 15 \end{bmatrix}$$

c) Record the reduced matrix below.

d) What is the solution to the system of equations?

$x =$

$y =$

$z =$

Often, working with three dimensional matrices will lead to decimal answers. That is the case with the following matrix.

$$\begin{bmatrix} 4 & -2 & 1 & | & 12 \\ 3 & -1 & 4 & | & 8 \\ 6 & 3 & 1 & | & 2 \end{bmatrix}$$

e) Write the reduced matrix and record the solutions in the space below.

f) Answers of "No Solution" or "Infinite Solutions" are also possible here. Reduce the following matrix, give the reduced matrix, and record whether it has "No Solution" or "Infinite Solutions."

$$\begin{bmatrix} 1 & 2 & -1 & | & 3 \\ 2 & 4 & -2 & | & 6 \\ 1 & 1 & 1 & | & 5 \end{bmatrix}$$

g) Finally, try another. The solution will be "No Solution" or "Infinite Solutions."

$$\begin{bmatrix} 1 & 3 & -4 & | & 8 \\ 2 & -1 & -9 & | & 4 \\ 2 & 6 & -8 & | & 21 \end{bmatrix}$$

Write the reduced matrix in the space below. Is the solution "No Solution" or "Infinite Solutions"?

Algebra II

Active Learning: 7.7a

When a matrix has nothing but a diagonal of ones and the rest zeros, it is called an identity matrix.

$$\begin{bmatrix} 1 & 0 \\ 0 & 1 \end{bmatrix}$$

$$\begin{bmatrix} 1 & 0 & 0 \\ 0 & 1 & 0 \\ 0 & 0 & 1 \end{bmatrix}$$

Use matrix multiplication to multiply the following:

a) $\begin{bmatrix} 3 & -2 \\ 1 & 4 \end{bmatrix} \cdot \begin{bmatrix} 1 & 0 \\ 0 & 1 \end{bmatrix}$

b) $\begin{bmatrix} 1 & 0 \\ 0 & 1 \end{bmatrix} \cdot \begin{bmatrix} 3 & -2 \\ 1 & 4 \end{bmatrix}$

c) What was true when you multiplied a matrix by the identity matrix?

d) Multiply the following matrices using matrix multiplication:

$$\begin{bmatrix} 4 & 3 \\ 3 & 2 \end{bmatrix} \cdot \begin{bmatrix} -2 & 3 \\ 3 & -4 \end{bmatrix}$$

e)
$$\begin{bmatrix} -2 & 3 \\ 3 & -4 \end{bmatrix} \cdot \begin{bmatrix} 4 & 3 \\ 3 & 2 \end{bmatrix}$$

When you can multiply two matrices and their answer is an identity matrix those two matrices are inverses of each other. (You must get the identity matrix when you multiply both $A \cdot B$ and $B \cdot A$.) Show that the following matrices are inverses.

f)
$$A = \begin{bmatrix} 1 & 2 \\ 4 & 7 \end{bmatrix} \text{ and } B = \begin{bmatrix} -7 & 2 \\ 4 & -1 \end{bmatrix}$$

An interesting thing occurs when you augment a matrix with the identity matrix. Use Gaussian elimination to turn the left side of the matrix into an identity matrix. Any changes you make to the left side must be carried through to the right.

g)
$$\begin{bmatrix} 1 & 2 & | & 1 & 0 \\ 4 & 7 & | & 0 & 1 \end{bmatrix}$$

h) What happened during this process?

i) Try another. (Hint: Something will go wrong.)

$$\begin{bmatrix} 1 & 2 & | & 1 & 0 \\ 4 & 8 & | & 0 & 1 \end{bmatrix}$$

On the left side, a row of zeros should have been created. This means the matrix doesn't have an inverse.

Although I will not attempt the proof, an inverse matrix can also be found by using a formula.

$$A = \begin{bmatrix} a & b \\ c & d \end{bmatrix}$$

$$A^{-1} = \frac{1}{ad - bc} \begin{bmatrix} d & -b \\ -c & a \end{bmatrix}$$

For instance:

$$A = \begin{bmatrix} 1 & 2 \\ 4 & 7 \end{bmatrix}$$

So:

$$A^{-1} = \frac{1}{1(7) - 2(4)} \begin{bmatrix} 7 & -2 \\ -4 & 1 \end{bmatrix}$$

$$A^{-1} = \frac{1}{-1} \begin{bmatrix} 7 & -2 \\ -4 & 1 \end{bmatrix} = \begin{bmatrix} -7 & 2 \\ 4 & -1 \end{bmatrix}$$

j) Use the formula to find the inverse of the following matrix.

$$A = \begin{bmatrix} 1 & 3 \\ 3 & 8 \end{bmatrix}$$

The easiest way to find an inverse matrix is, of course, your calculator. Enter your matrix into your calculator as we learned previously. Then, simply select the name of your matrix from the Matrix list. Once you've selected it, hit the x^{-1} key on your calculator. Finally, hit enter and you will be given the inverse matrix. (If an inverse matrix exists.)

k) Use your calculator to find the inverse matrix for the following 3×3 matrix.

$$\begin{bmatrix} 1 & 3 & 5 \\ 1 & -2 & 4 \\ 2 & -5 & 8 \end{bmatrix}$$

Algebra II

Active Learning: 7.7b

Instead of an augmented matrix, there is another method for writing systems of linear equations using matrices. We can create a coefficient matrix, a variable matrix, and a constant matrix.

$$3x + 2y = 4$$
$$-2x + y = 2$$

$$\begin{bmatrix} 3 & 3 \\ -2 & 1 \end{bmatrix} \cdot \begin{bmatrix} x \\ y \end{bmatrix} = \begin{bmatrix} 4 \\ 2 \end{bmatrix}$$

Write the following systems of linear equations as matrices using this approach:

a)
$$4x - 3y = 2$$
$$-3x + 2y = 1$$

b)
$$-x + 2y = 4$$
$$2x - 3y = 6$$

Find the inverse of the matrix below:

$$A = \begin{bmatrix} 3 & 3 \\ -2 & 1 \end{bmatrix}$$

c) $A^{-1} =$

d) What will happen if you multiply $A^{-1} \cdot A$? (You can do the work if you need to, however the result is something we've previously discussed.)

Next, multiply the following:

e) $A^{-1} \cdot \begin{bmatrix} 4 \\ 2 \end{bmatrix} =$

Look at our original system of linear equations written in matrix form.

$$\begin{bmatrix} 3 & 3 \\ -2 & 1 \end{bmatrix} \cdot \begin{bmatrix} x \\ y \end{bmatrix} = \begin{bmatrix} 4 \\ 2 \end{bmatrix}$$

If we multiply the left side of the equation by A^{-1}, we would be left with $\begin{bmatrix} x \\ y \end{bmatrix}$. But this is an equation, so if you do something to one side, you must do it to the other. (The order of matrix multiplication matters. When we do problems like this, we will multiply the inverse matrix on the left.)

$$A^{-1} \cdot \begin{bmatrix} 3 & 3 \\ -2 & 1 \end{bmatrix} \cdot \begin{bmatrix} x \\ y \end{bmatrix} = A^{-1} \cdot \begin{bmatrix} 4 \\ 2 \end{bmatrix}$$

$$\begin{bmatrix} 1 & 0 \\ 0 & 1 \end{bmatrix} \cdot \begin{bmatrix} x \\ y \end{bmatrix} = A^{-1} \cdot \begin{bmatrix} 4 \\ 2 \end{bmatrix}$$

However, the identity matrix times any other matrix, just gives the matrix.

$$\begin{bmatrix} x \\ y \end{bmatrix} = A^{-1} \cdot \begin{bmatrix} 4 \\ 2 \end{bmatrix}$$

f) The result on the left is the answer to the system of linear equations. Take your previous answer to give the values of x and y.

$x =$

$y =$

Let's try another.

$$4x - 3y = 2$$
$$-3x + 2y = 1$$

$$\begin{bmatrix} 4 & -3 \\ -3 & 2 \end{bmatrix} \cdot \begin{bmatrix} x \\ y \end{bmatrix} = \begin{bmatrix} 2 \\ 1 \end{bmatrix}$$

g) Find the inverse of $\begin{bmatrix} 4 & -3 \\ -3 & 2 \end{bmatrix}$.

Multiply the inverse on both sides of the equation.

$$A^{-1} \cdot \begin{bmatrix} 4 & -3 \\ -3 & 2 \end{bmatrix} \cdot \begin{bmatrix} x \\ y \end{bmatrix} = A^{-1} \cdot \begin{bmatrix} 2 \\ 1 \end{bmatrix}$$

h) Give the solution to the system of linear equations.

$x =$

$y =$

i) Try another. Solve the following system of linear equations using this approach.

$$-x + 2y = 4$$
$$2x - 3y = 6$$

The concept can be extended to three-dimensional systems.

j) Solve the following three-dimensional system using the same approach we used above. (Use your calculator to find the inverse.)

$x + 3y + 5z = 2$

$x - 2y + 4z = -1$

$2x - 5y + 8z = 3$

Algebra II

Active Learning: 7.8a

A scalar number can be calculated from a square matrix. This number is called a determinant. Matrices can be thought of as movement, and in that application, the determinant is a type of stretching factor. For us, the calculation of a determinant can be used to give another approach for solving systems of equations.

$$A = \begin{bmatrix} a & b \\ c & d \end{bmatrix}$$

For a matrix A, the determinant is the difference of the product of the diagonals.

$$\det(A) = \begin{bmatrix} a & b \\ c & d \end{bmatrix} = a \cdot d - c \cdot b$$

Here's the determinant of the following square matrix.

$$A = \begin{bmatrix} 3 & 3 \\ -2 & 1 \end{bmatrix}$$

$$\det(A) = 3 \cdot 1 - (-2) \cdot 3 = 3 + 6 = 9$$

Find the determinant of the following square matrices:

a) $\begin{bmatrix} 4 & -3 \\ -3 & 2 \end{bmatrix}$

b) $\begin{bmatrix} -1 & 2 \\ 2 & -3 \end{bmatrix}$

c) $\begin{bmatrix} 12 & 15 \\ -1 & -3 \end{bmatrix}$

For our purposes, the value of the determinant lies in Cramer's Rule. Cramer's is another approach for solving a system of equations. It only holds for square systems where there are the same number of equations as variables.

Cramer's Rule will find the solution to the system by finding special determinants and dividing them by the determinant for the original matrix. Look at this system:

$$3x + 2y = 4$$
$$-2x + y = 2$$

$$A = \begin{bmatrix} 3 & 2 \\ -2 & 1 \end{bmatrix}$$

There are two special determinants we need det (A_x) and det (A_y). To find det (A_x), we substitute the solution matrix in place of the x column and then find the determinant.

$$A_x = \begin{bmatrix} 4 & 2 \\ 2 & 1 \end{bmatrix}$$

d) Find det (A_x).

$$\det(A_x) =$$

To find det (A_y), we substitute the solution matrix in place of the y column and then find the determinant.

$$A_y = \begin{bmatrix} 3 & 4 \\ -2 & 2 \end{bmatrix}$$

e) Find det (A_y).

$$\det(A_y) =$$

Cramer's Rule finds the solutions by dividing the special determinants by the matrix's original determinant.

f) $x = \dfrac{\det(A_x)}{\det(A)} =$

g) $y = \dfrac{\det(A_y)}{\det(A)} =$

h) Solve this system by elimination to confirm that Cramer's Rule worked.

$$3x + 2y = 4$$
$$-2x + y = 2$$

Try another. Use Cramer's Rule to solve the system of linear equations.

$$4x - 3y = 2$$
$$-3x + 2y = 1$$

$$A = \begin{bmatrix} 4 & -3 \\ -3 & 2 \end{bmatrix}$$

i) $A_x =$

j) $A_y =$

k) $\det(A_x) =$

l) $\det(A_y) =$

m) $x = \dfrac{\det(A_x)}{\det(A)} =$

n) $y = \dfrac{\det(A_y)}{\det(A)} =$

The idea of a determinant can also be extended to 3 × 3 matrices.

$$A = \begin{bmatrix} 1 & 3 & 5 \\ 1 & -2 & 4 \\ 2 & -5 & 8 \end{bmatrix}$$

However, for the diagonals to work, the first two columns are repeated.

$$\begin{bmatrix} 1 & 3 & 5 | & 1 & 3 \\ 1 & -2 & 4 | & 1 & -2 \\ 2 & -5 & 8 | & 2 & -5 \end{bmatrix}$$

We then have three down diagonals and subtract three up diagonals.

$$\det(A) = 1 \cdot (-2) \cdot 8 + 3 \cdot 4 \cdot 2 + 5 \cdot 1 \cdot (-5) - 2 \cdot (-2) \cdot 5 - (-5) \cdot 4 \cdot 1 - 8 \cdot 1 \cdot 3 = -1$$

You should know how determinants are found, but unless I asked you to show your work, it would be okay to use your calculator. If you go under the Matrix menu, and move over to Math, the you will find a det() command.

Use your calculator to confirm the determinant of matrix A.

o) $\det(A) =$

Cramer's rule can also be extended to the third variable.

$$x + 3y + 5 = 2$$
$$x - 2y + 4 = -1$$
$$2x - 5y + 8 = 3$$

Extend the idea of Cramer's Rule to a third variable to solve this system of equations.

p) $A_x =$ s) $A_y =$ v) $A_z =$

q) $\det(A_x) =$ t) $\det(A_y) =$ w) $\det(A_z) =$

r) $x = \frac{\det(A_x)}{\det(A)} =$ u) $y = \frac{\det(A_y)}{\det(A)} =$ x) $z = \frac{\det(A_z)}{\det(A)} =$

y) Cramer's Rule requires that we divide by the determinant of the matrix. Sometimes that determinant can be zero. Confirm that the determinant for the following system is zero.

$$2x + 4y = 8$$
$$4x + 8y = 16$$

z) Since the determinant is zero, use elimination to determine if the system is **No Solution** or **Infinite Solutions**.

Algebra II

Active Learning: 7.8b

a) Find the determinant of the following matrix:

$$A = \begin{bmatrix} 1 & 3 & 5 \\ 0 & -2 & 4 \\ 0 & 0 & 8 \end{bmatrix}$$

$$\det(A) =$$

This matrix is in "upper-triangular form." This means it has all zeros in the bottom left of the matrix. Because of this, we come across our first property of determinants. **Matrices in Upper-Triangular Form have a determinant equal to the product of the main diagonal.** The reason is that a zero is involved in all of the other diagonals, making those products zero.

Find the determinant of the following matrices:

b)
$$A = \begin{bmatrix} 5 & 8 \\ -7 & 2 \end{bmatrix}$$

$$\det(A) =$$

c)
$$B = \begin{bmatrix} -7 & 2 \\ 5 & 8 \end{bmatrix}$$

$$\det(B) =$$

The difference in these two matrices was that their rows were interchanged. What occurred always holds true and is our second property of determinants. **When rows are interchanged, the value of the determinant changes signs.** This is true because we have reversed the diagonals. Now, we are subtracting the values which we previously had been adding.

Find the determinant of the following matrices:

d)
$$A = \begin{bmatrix} 1 & 3 & 5 \\ 1 & 3 & 5 \\ -2 & 6 & -1 \end{bmatrix}$$

$$\det(A) =$$

e)
$$B = \begin{bmatrix} 1 & 1 & 7 \\ 1 & 1 & -2 \\ 1 & 1 & -3 \end{bmatrix}$$

$$\det(B) =$$

In one matrix, two rows were identical. In the other matrix, two columns were identical. Whenever this occurs, we always get this result. **When two rows or two columns are identical in a matrix, we always get a determinant which is equal to zero.** This is true because the identical rows (or columns) make repeats of the diagonals which we add and subtract to find the determinant.

Find the determinant of the following matrices:

f)
$$A = \begin{bmatrix} -5 & -3 \\ 0 & 0 \end{bmatrix}$$

$$\det(A) =$$

g)
$$B = \begin{bmatrix} -1 & 0 & 9 \\ 4 & 0 & 6 \\ -3 & 0 & 7 \end{bmatrix}$$

$$\det(B) =$$

In the first matrix, we had a row of all zeros. In the second matrix, we had a column of all zeros. **A matrix with a row of zeros or a column of zeros will result in a determinant with a value of zero.**

Find the determinant of the following matrices:

h)
$$A = \begin{bmatrix} 1 & 2 \\ 3 & 4 \end{bmatrix}$$

$\det(A) =$

i)
$$A^{-1} = \begin{bmatrix} -2 & 1 \\ 1.5 & -.5 \end{bmatrix}$$

$\det(A^{-1}) =$

Your answers to the determinants should be inverses, and this is always true for inverse matrices. **The determinant of inverse matrices are always inverses of each other.**

Find the determinant of the following matrices:

j)
$$A = \begin{bmatrix} 1 & 3 & 5 \\ 1 & -2 & 4 \\ 2 & -5 & 8 \end{bmatrix}$$

$\det(A) =$

k)
$$B = \begin{bmatrix} 1 & 3 & 5 \\ 1 \cdot 3 & -2 \cdot 3 & 4 \cdot 3 \\ 2 & -5 & 8 \end{bmatrix}$$

$\det(B) =$

Matrix B is the same as matrix A except it has a row which has been multiplied by 3. As a result, the determinant of matrix B is the determinant of matrix A multiplied by 3. **If a matrix has a row or column multiplied by a constant, the determinant will be multiplied by that constant.** The reason for this is because each of the diagonals used to calculate the determinant will now be multiplied by that constant. You could then pull that constant out as a common factor.

$$3(\det(A))$$

Finally, we can see the value of these determinant properties while trying to solve the following system of linear equations using Cramer's rule.

$$x + 3y + 3 = 2$$
$$x - 2y - 2 = -1$$
$$2x - 5y - 5 = 3$$

l) What is true about the 2nd and 3rd columns of this matrix?

m) What does that tell us about the determinant of the matrix?

n) What would we need to do to determine if the solution to this system was **No Solution** or **Infinite Solutions**?

Algebra II

Active Learning: 8.1a

In this lab, we are going to examine the formula and graph of an ellipse. Use Geogebra to help you graph the following equation:

a)

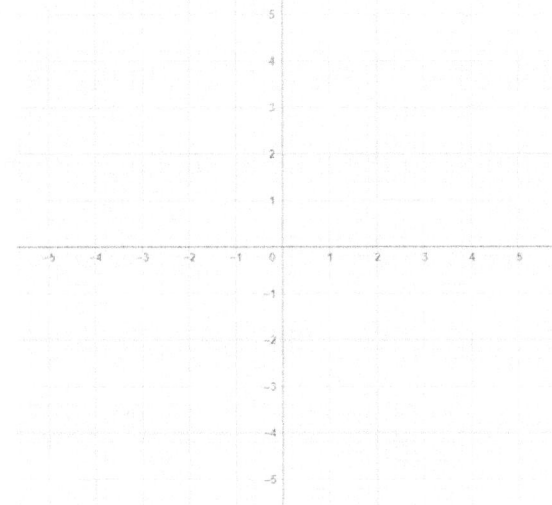

$$\frac{x^2}{25} + \frac{y^2}{9} = 1$$

b) An ellipse has a major axis and a minor axis. The major axis is the longer axis and the minor axis is the shorter. In this graph, which axis is the major axis: the x-axis or the y-axis?

Again, use Geogebra to help you graph the following equation:

c)

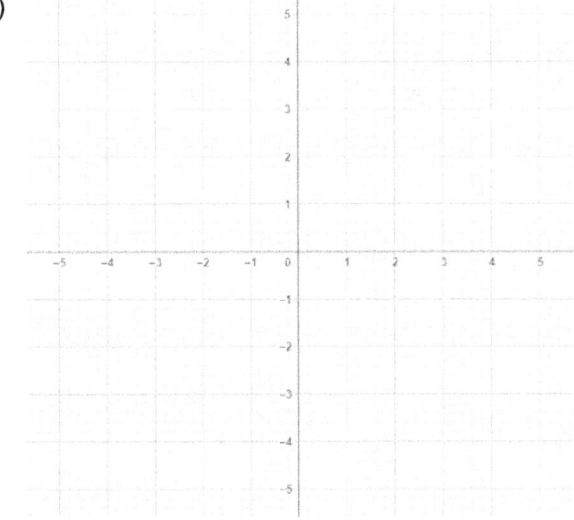

$$\frac{x^2}{9} + \frac{y^2}{25} = 1$$

d) In this graph, which axis is the major axis?

e) What is the connection between the major axis and the denominators in the equation?

The endpoints of each axis are called the vertices. Looking at your second graph, record the coordinates of the vertices for both the major and minor axis.

f) Major Axis: (__ , __) ; (__ , __)

g) Minor Axis: (__ , __) ; (__ , __)

Looking at your second graph, record the distance from the center of the ellipse to the endpoints along each axis.

h) Major Axis: Distance from the center to the endpoints _____.

i) Minor Axis: Distance from the center to the endpoints _____.

j) There is a connection between the equation for an ellipse and the distance from the center to the endpoints. What is that connection?

Without using Geogebra, graph the following equation:

k)

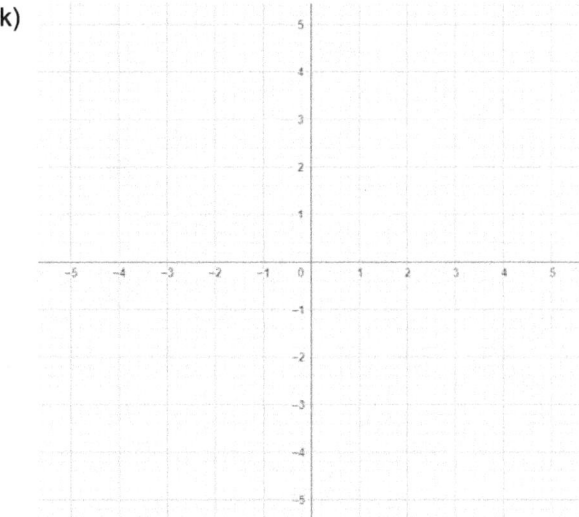

$$\frac{x^2}{9} + \frac{y^2}{4} = 1$$

The center of an ellipse isn't always at the origin. Use Geogebra to help you graph the following equations:

l)

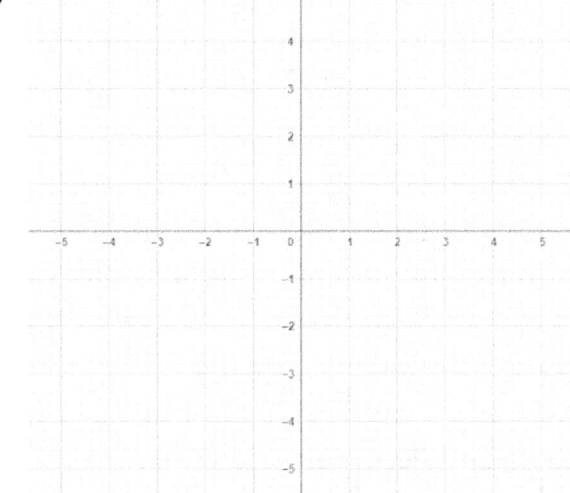

$$\frac{(x-2)^2}{9} + \frac{(y-1)^2}{4} = 1$$

m)

$$\frac{(x+2)^2}{9}+\frac{(y+1)^2}{4}=1$$

n)

$$\frac{(x+2)^2}{9}+\frac{(y-1)^2}{4}=1$$

The generic equation for an ellipse is recorded below.

$$\frac{(x-h)^2}{a^2}+\frac{(y-k)^2}{b^2}=1$$

In the blanks below, record the variables that represent the center of the ellipse.

o) (___, ___)

Graph the following ellipses without using Geogebra.

p) $$\frac{(x+1)^2}{36} + \frac{(y-2)^2}{16} = 1$$

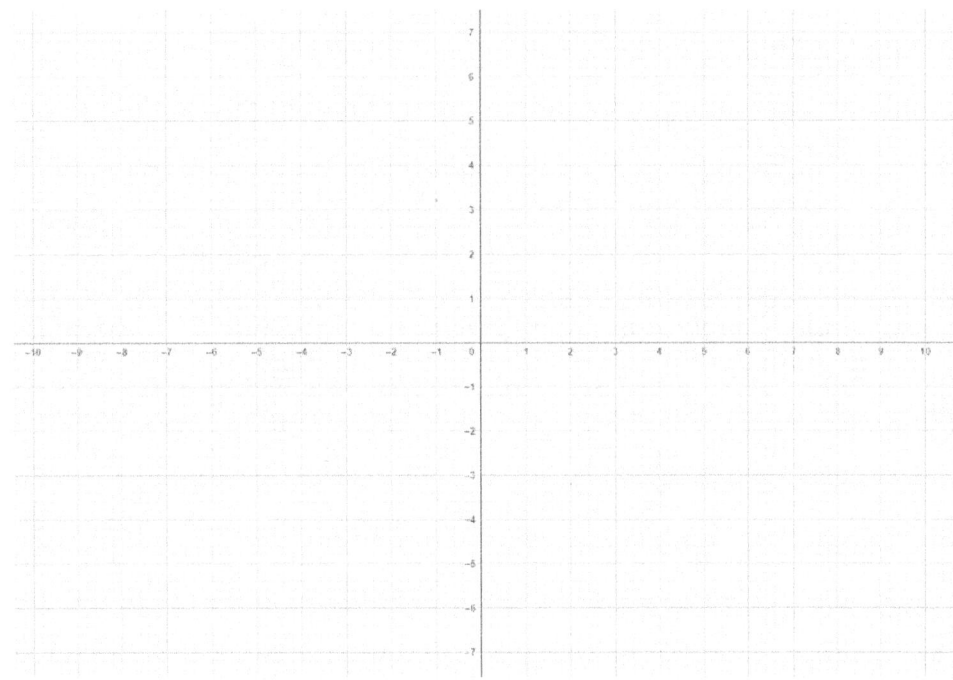

q) $$\frac{(x+5)^2}{9} + \frac{(y-1)^2}{25} = 1$$

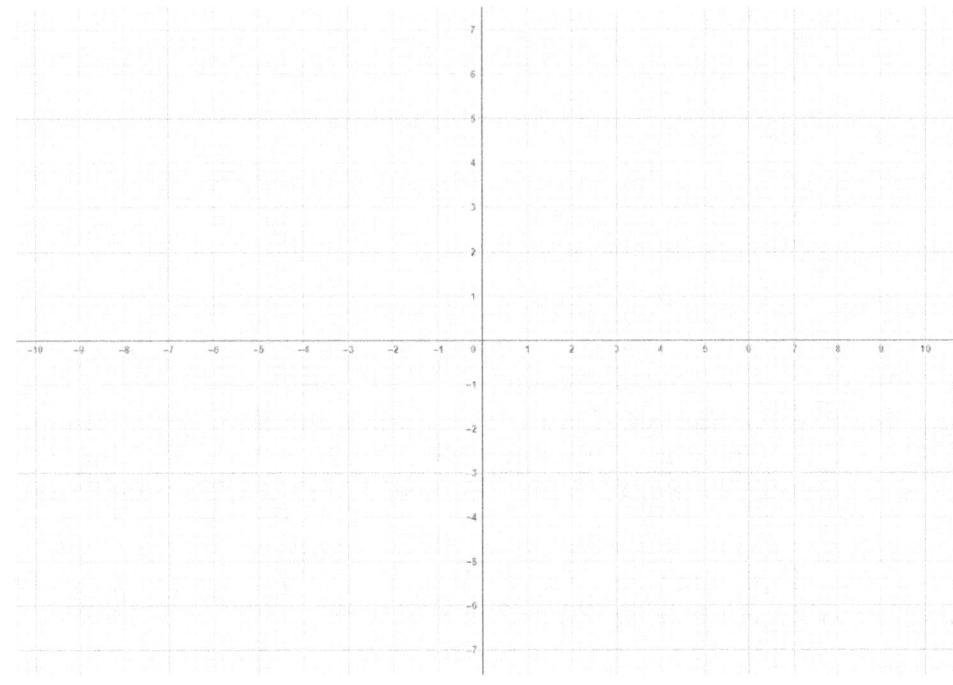

An ellipse has two key points called foci.

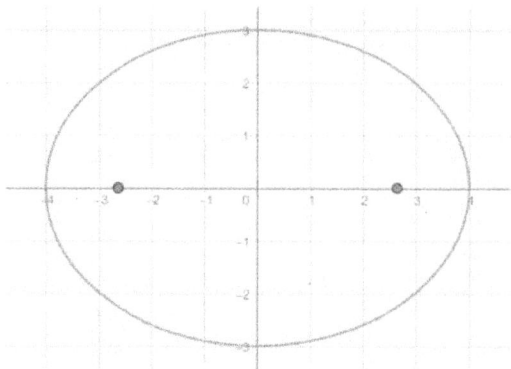

The foci are special because the distance from any point on the ellipse to the foci always adds to be the same.

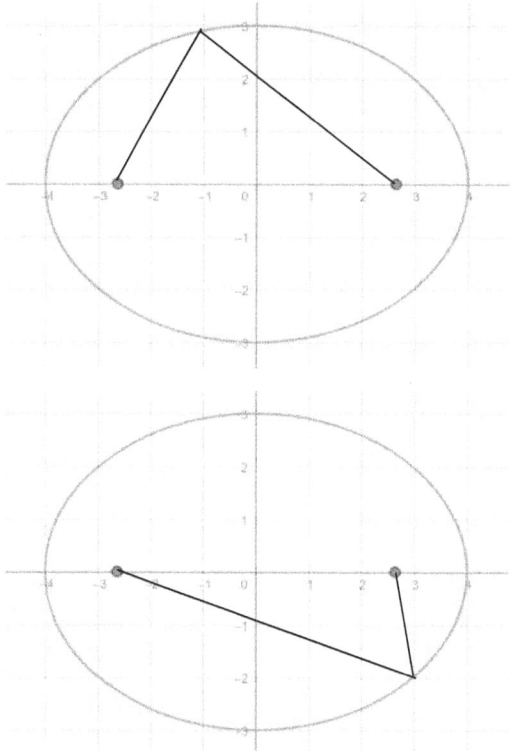

In both graphs, the sum of the line segments is the same. And this would be true for any line segments from the foci to any point on the ellipse. The foci are located along the major axis. If the x-axis is the major axis, they are located at the points $(c, 0), (-c, 0)$. If the y-axis is the major axis, they are located at the points $(0, \ c), (0, -c)$.

The relationship between a, b and c turns out to $c^2 = a^2 - b^2$. (It doesn't come from the Pythagorean Theorem but rather the distance from the center to a focus.)

If we know the location of the vertices and the foci, we can find the formula for the ellipse.

What is the standard form of an ellipse that has vertices at $(\pm 5, 0)$ and foci at $(\pm \sqrt{21}, 0)$?

The vertices and foci are along the x-axis. So, $a = 5$.

We also know that $c = \sqrt{21}$. And using the formula $c^2 = a^2 - b^2$ we can find b.

$$(\sqrt{21})^2 = (5)^2 - b^2$$
$$21 = 25 - b^2$$
$$b^2 = 25 - 21$$
$$b^2 = 4$$
$$b = 2$$

(b is a distance so we choose the positive value.)

Now that we know a and b, we can write the standard from.

$$\frac{x^2}{(5)^2} + \frac{y^2}{(2)^2} = 1$$
$$\frac{x^2}{25} + \frac{y^2}{4} = 1$$

Try one on your own.

r) What is the standard form of an ellipse that has vertices at $(\pm 7, 0)$ and foci at $(\pm \sqrt{13}, 0)$?

In a similar fashion, we can find the standard form of an ellipse centered somewhere other than the origin.

What is the standard form of an ellipse that has vertices at $(3, 8)$ and $(3, -4)$ and foci at $(3, 5)$ and $(3, -1)$?

First, the major axis is the y-axis. The center is at the midpoint of the vertices, which would be:

$$\left(\frac{3+3}{2}, \frac{8+(-4)}{2}\right) = (3, 2)$$

The value of a is the distance along the y-axis from the center to either vertex. So, $a = 6$.

Similarly, the value of c is the distance along the y-axis from the center to either focus. So, $c = 3$.

We can use our formula to find b.

$$(3)^2 = (6)^2 - b^2$$
$$b^2 = 36 - 9$$
$$b = \sqrt{27}$$

And the standard form of the ellipse would be:

$$\frac{(x-3)^2}{(\sqrt{27})^2} + \frac{(y-2)^2}{(6)^2} = 1$$

$$\frac{(x-3)^2}{27} + \frac{(y-2)^2}{36} = 1$$

Try one on your own.

s) What is the standard form of an ellipse that has vertices at $(3, 4)$ and $(-5, 4)$ and foci at $(1, 4)$ and $(-3, 4)$?

Algebra II

Warm Up: 8.1b

Most of this activity will be a review of ideas from the previous one. Let's begin by graphing an ellipse centered at the origin.

Graph the following ellipse. Label the center, vertices, co-vertices (along the minor axis), and foci.

$$\frac{x^2}{16} + \frac{y^2}{25} = 1$$

a)
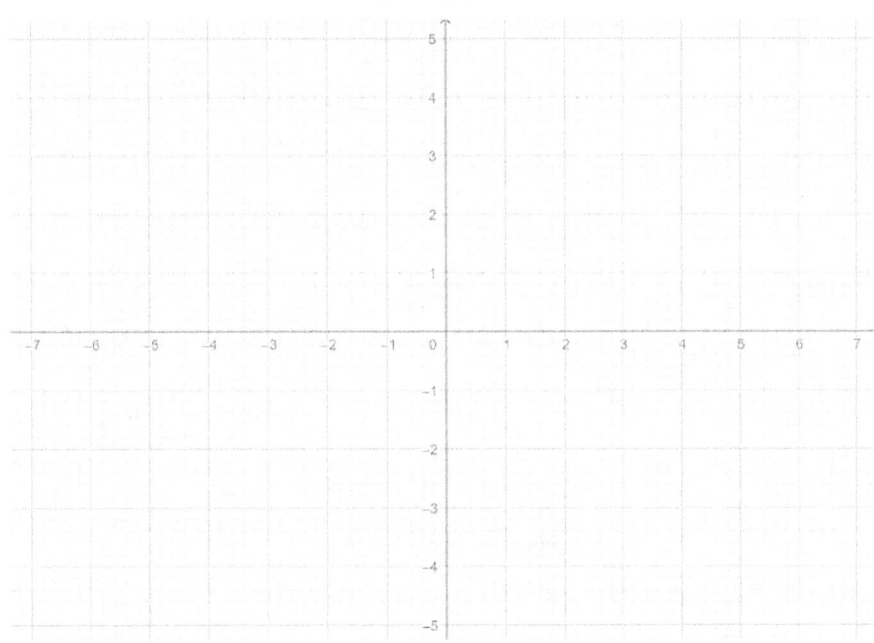

b)
- The larger value is beneath the y^2, so the major axis is the y-axis.
- The value of $a = 5$.
- The value of $b = 4$.
- To find the value of c in order to add the foci, use the formula $c^2 = a^2 - b^2$.

Try another on your own.

Graph the following ellipse. Label the center, vertices, co-vertices, and foci. (Your foci will involve a root. It is fine to approximate the locations on the graph.)

$$\frac{x^2}{36} + \frac{y^2}{9} = 1$$

c)

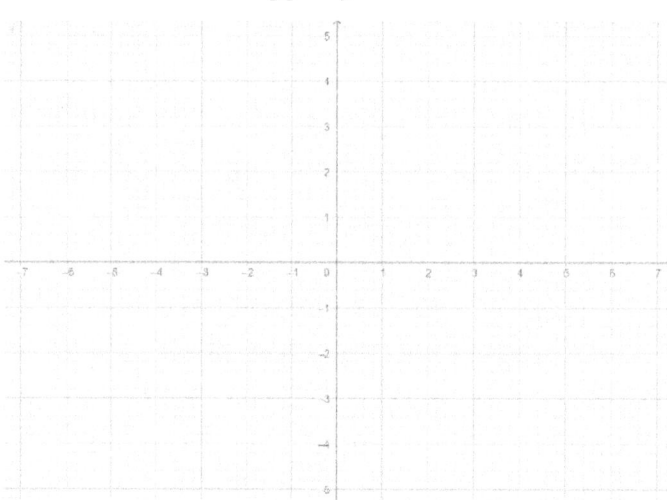

Sometimes, the equation for the ellipse will not be given in standard form.

$$16x^2 + 9y^2 = 144$$

Since we know that the equation of an ellipse must equal 1, we simply divide through by 144.

$$\frac{16x^2}{144} + \frac{9y^2}{144} = \frac{144}{144}$$

$$\frac{x^2}{9} + \frac{y^2}{16} = 1$$

d) From this point forward, the procedure would be the same. Convert the following equation into standard form. You do not need to graph it.

$$4x^2 + 36y^2 = 144$$

Next, we can graph an ellipse which is not centered at the origin.

Graph the following ellipse. Label the center, vertices, co-vertices, and foci.

$$\frac{(x-1)^2}{9} + \frac{(y+1)^2}{4} = 1$$

e)

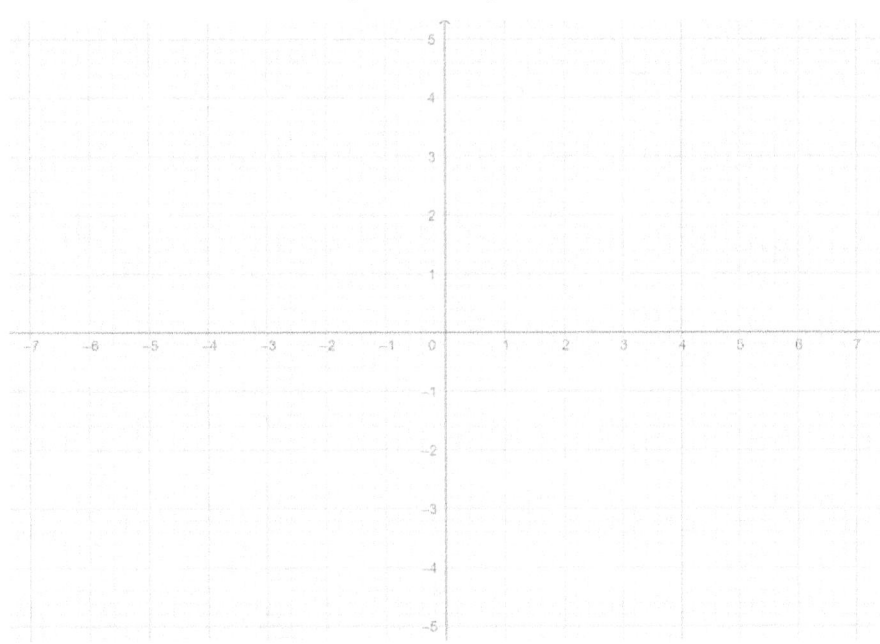

- The ellipse is centered at $(1, -1)$.
- The major axis is the x-axis.
- The value of $a = 3$.
- The value of $b = 2$.

f)
- To find the value of c in order to add the foci, use the formula $c^2 = a^2 - b^2$.

g)
- To find the vertices, add/subtract a to the x coordinate of the center.

h)
- To find the co-vertices, add/subtract b to the y coordinate of the center.

i)
- To find the foci, add/subtract c to the x coordinate of the center. (The value of c is a radical. Approximate it on the graph.)

Work another.

Graph the following ellipse. Label the center, vertices, co-vertices, and foci.

$$\frac{(x+2)^2}{9} + \frac{(y-1)^2}{16} = 1$$

j)
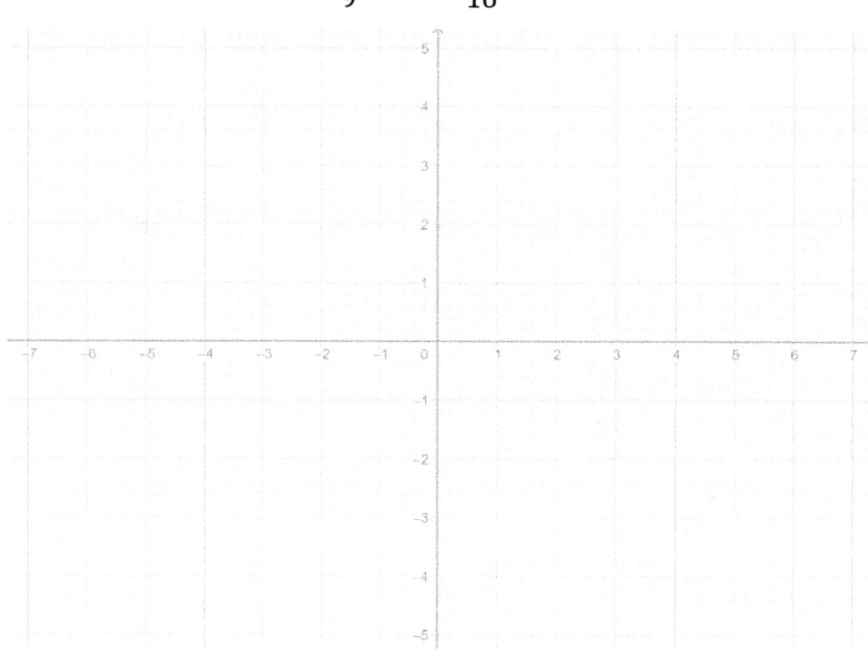

Algebra II

Active Learning: 8.1c

Let's begin with a skill we will need. Complete the square for each of the following polynomials.

a) $x^2 - 4x + \underline{}$

b) $y^2 - 2y + \underline{}$

In the last activity, we saw formulas for ellipses which were not in standard form. Those ellipses were centered at the origin. We will look at the same idea, however this time, the ellipses will be centered somewhere else.

$$3x^2 + 4y^2 - 12x - 8y + 4 = 0$$

Begin by organizing the x and y terms.

$$3x^2 - 12x + 4y^2 - 8y + 4 = 0$$

Next, send the last term to the other side.

$$3x^2 - 12x + 4y^2 - 8y = -4$$

pull out a GCF from the x terms and a GCF from the y terms.

$$3(x^2 - 4x) + 4(y^2 - 2y) = -4$$

Now, complete the squares. (Hint: we did this above.)

$$3(x^2 - 4x + 4) + 4(y^2 - 2y + 1) = -4$$

But this is out of balance. We can't add a number to one side without adding it to the other. However, we didn't just add a 4 and a 1. We actually added $3 \cdot 4$ and $4 \cdot 1$, because our numbers are in parenthesis. So, we add 12 and 4 to the left side.

$$3(x^2 - 4x + 4) + 4(y^2 - 2y + 1) = -4 + 12 + 4$$

$$3(x - 2)^2 + 4(y - 1)^2 = 12$$

c) Finally, divide through by 12 to get the formula in standard form.

Give the values of the center, vertices, co-vertices, and foci.

d) Center:

e) Vertices:

f) Co-vertices:

g) Foci:

Try one on your own. Write the ellipse in standard form and give the values of the center, vertices, co-vertices, and foci.

h)
$$2x^2 + 3y^2 + 12x - 12y + 24 = 0$$

Algebra II

Active Learning: 8.2a

In this lab, we are going to examine the formula and graph of the hyperbola. Use Geogebra to help you graph the following equation:

a)

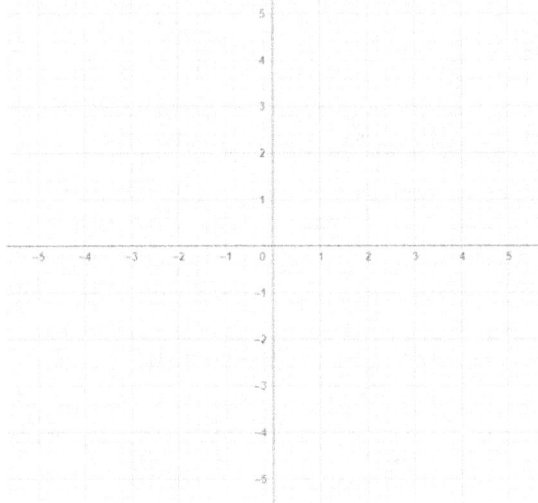

$$\frac{x^2}{16} - \frac{y^2}{25} = 1$$

Again, use Geogebra to help you graph the following equation:

b)

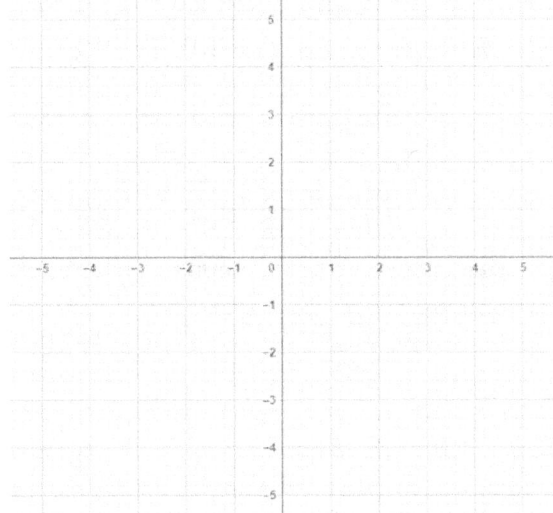

$$\frac{y^2}{16} - \frac{x^2}{25} = 1$$

For a hyperbola centered at the origin, there are two possible formulas. The formulas are listed below. One of the formulas creates a hyperbola which opens vertically and the other creates a hyperbola which opens horizontally. Indicate which formula is which.

c) $\dfrac{x^2}{a^2} - \dfrac{y^2}{b^2} = 1$

d) $\dfrac{y^2}{a^2} - \dfrac{x^2}{b^2} = 1$

A hyperbola has two vertices. Looking at your first graph, record the coordinates of the vertices.

e) Vertices: (___, ___) ; (___, ___)

f) The graph is centered at $(0, 0)$. What is the distance from the center to either of the vertices?

Looking at your second graph, record the coordinates of the vertices.

g) Vertices: (___, ___) ; (___, ___)

h) The graph is centered at $(0, 0)$. What is the distance from the center to either of the vertices?

i) Look back at the two possible equations for a hyperbola centered at the origin; what variable tells us the distance to the vertices?

Without using Geogebra, graph the following equation:

j)

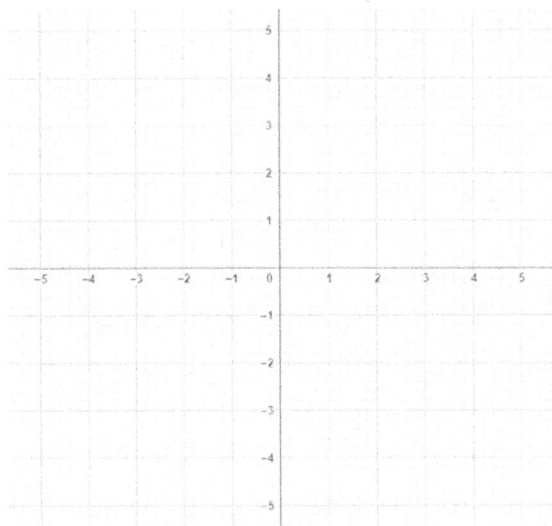

$$\frac{x^2}{9} - \frac{y^2}{4} = 1$$

The center of an ellipse isn't always at the origin. Use Geogebra to help you graph the following equations:

k)

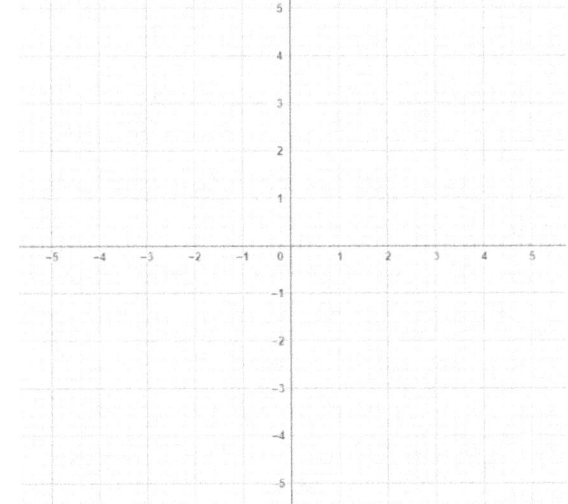

$$\frac{(x-1)^2}{9} - \frac{(y-2)^2}{4} = 1$$

l)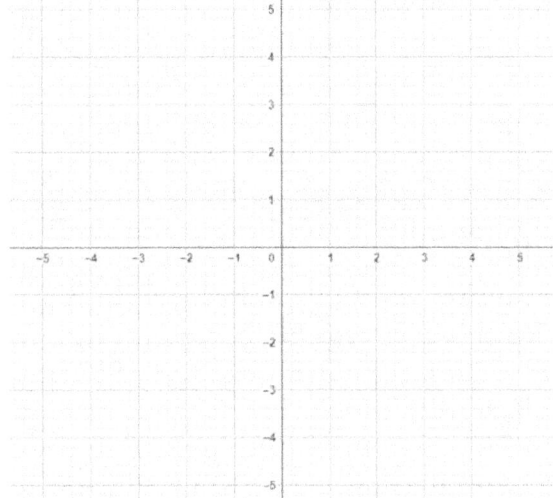

$$\frac{(x+2)^2}{9} - \frac{(y+2)^2}{4} = 1$$

m)

$$\frac{(x+1)^2}{9} - \frac{(y-2)^2}{4} = 1$$

The generic equations for a hyperbola are recorded below.

$$\frac{(x-h)^2}{a^2} - \frac{(y-k)^2}{b^2} = 1$$

$$\frac{(y-k)^2}{a^2} - \frac{(x-h)^2}{b^2} = 1$$

In the blanks below, record the variables that represent the center of the ellipse.

n) (___, ___)

Graph the following hyperbolas without using Geogebra.

$$\frac{(x+2)^2}{36} - \frac{(y-1)^2}{4} = 1$$

o)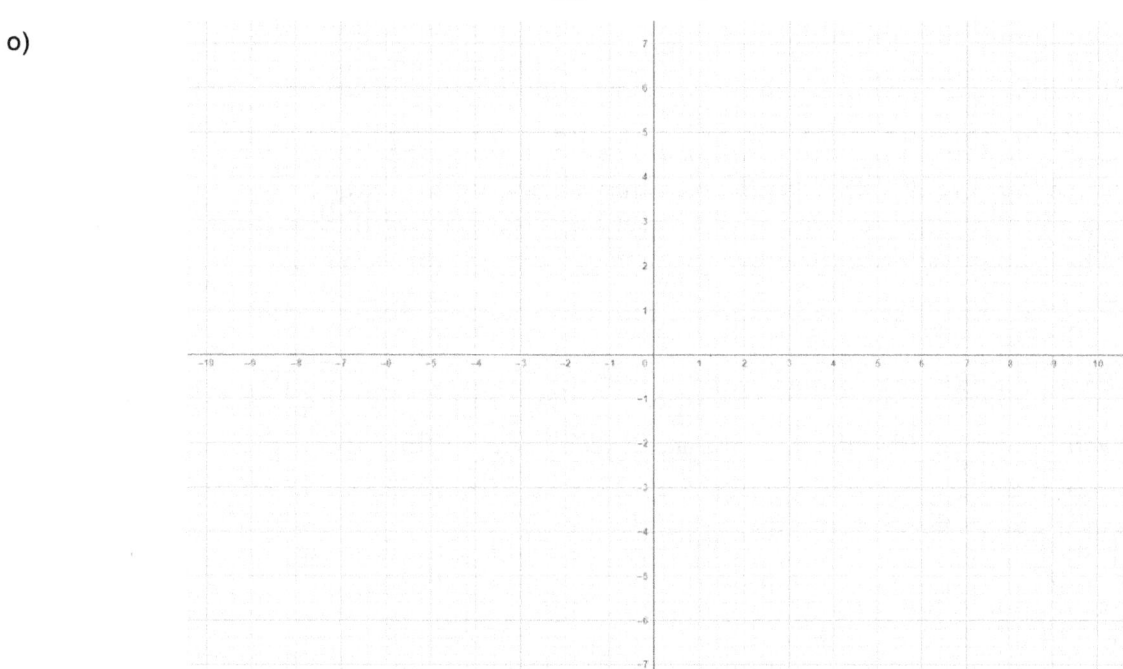

$$\frac{(y+1)^2}{9} - \frac{(x-3)^2}{25} = 1$$

p)

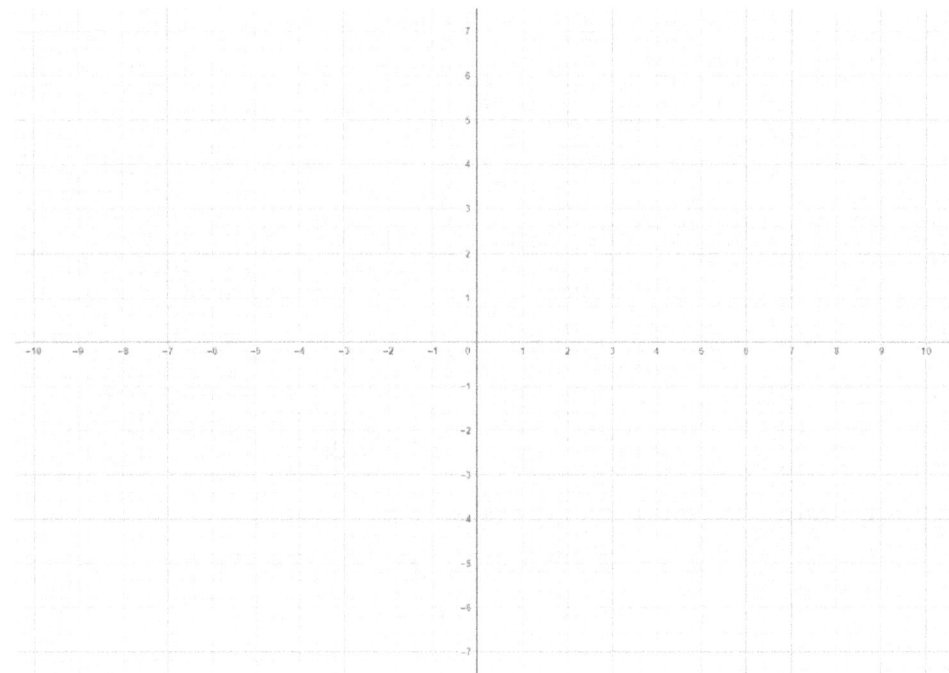

As with the ellipse, the hyperbola has foci. Here, the difference between the distance to any point on the graph and the two foci is constant. Foci in a hyperbola are past the vertices and inside the hyperbola's curve. Again, we can create the equation for a hyperbola in standard form given the vertex and the foci. To do so, we will need a new equation for the relationship between a, b and c.

$$b^2 = c^2 - a^2$$

What is the standard form of a hyperbola that has vertices $(\pm 5, 0)$ and foci $(\pm\sqrt{74}, 0)$?

- The vertices and foci are along the x-axis and so the $\frac{x^2}{a^2}$ term will be positive.
- We know $a = 5$ and $c = \sqrt{74}$.
- Use the formula $b^2 = c^2 - a^2$ to find b.

q)

r) In the space below, give the formula for the hyperbola.

Try another.

s) What is the standard form of a hyperbola that has vertices $(0, \pm 3)$ and foci $(0, \pm 5)$? (Hint: the vertices are giving the value of a and a is always part of the positive term in the standard form.)

Next, we will work a problem where the hyperbola is not centered at the origin.

What is the standard form of a hyperbola that has vertices $(2, 3)$ and $(2, -1)$ with foci $(2, 1 + \sqrt{13})$ and $(2, 1 - \sqrt{13})$?

t)
- The center of the hyperbola will be in the middle of the vertices. So $(2, 1)$.
- The vertices and foci are along the y-axis and so the $\frac{y^2}{a^2}$ term will be positive.
- We know $a = 2$ and $c = \sqrt{13}$.
- Use the formula $b^2 = c^2 - a^2$ to find b.

u) In the space below, give the formula for the hyperbola. (Remember to include the transformation for the center.)

Finally, try one more.

v) What is the standard form of a hyperbola that has vertices $(8, -4)$ and $(-2, -4)$ with foci $(3 + \sqrt{61}, -4)$ and $(3 - \sqrt{61}, -4)$?

Algebra II

Active Learning: 8.2b

As with the previous section, most of this second activity will be a review of ideas from the previous one. We will begin by graphing hyperbolas from their equation in standard form. One concept that was overlooked when we introduced hyperbolas was asymptotes. The curves of the hyperbolas are bounded by two diagonals. In the classroom, I don't typically require students to draw the asymptotes, but their equation follows the idea of a slope.

$$y = \pm \frac{\Delta y}{\Delta x} x$$

But here, we use either a or b in place of Δy or Δx. I'll illustrate in the first problem.

Graph the following ellipse. Label the center, vertices, co-vertices, foci, and give the equation for the asymptotes.

$$\frac{x^2}{25} - \frac{y^2}{16} = 1$$

a)

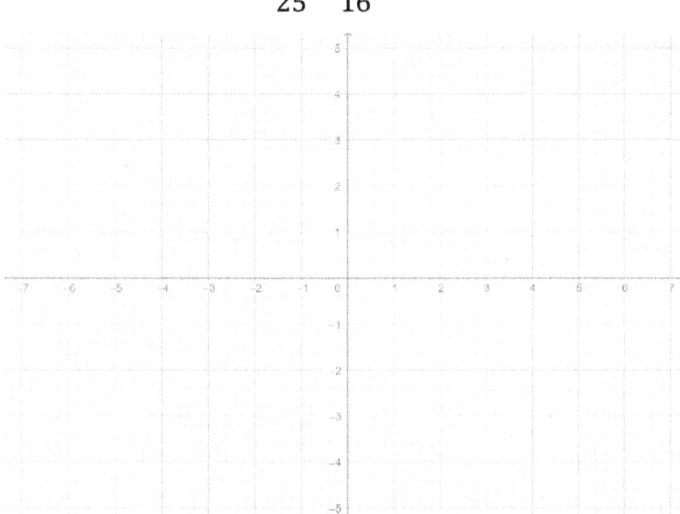

b)
- The positive value is beneath the x^2, so the major axis (called the transverse axis in this context) is the x-axis.
- The value of $a = 5$. The value of $b = 4$.
- To find the value of c in order to add the foci, use the formula $b^2 = c^2 - a^2$.

- The b is below the y. The a is below the x. So, following the slope formula, our asymptotes are:
$$y = \pm \frac{4}{5} x$$

Try another on your own.

Graph the following hyperbola. Label the center, vertices, co-vertices, foci, and give the equation for the asymptotes.

$$\frac{y^2}{36} - \frac{x^2}{25} = 1$$

The asymptotes would be $y = \pm \frac{6}{5}x$

c)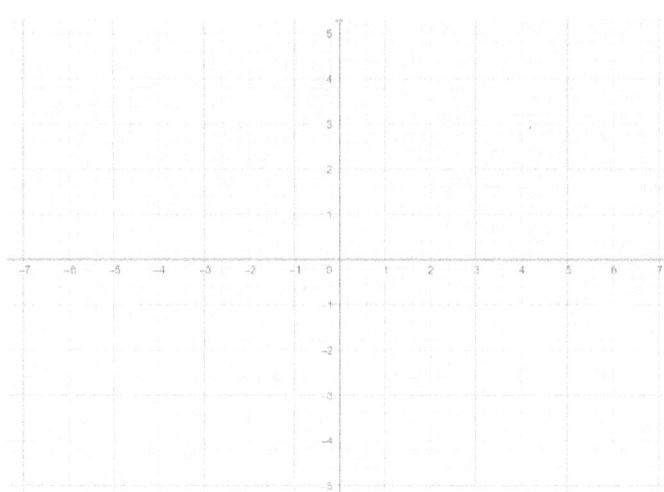

As we saw with the ellipse, we will not always be given the equation for the hyperbola in standard form.

$$16x^2 - 9y^2 = 144$$

Since we know that the equation of an hyperbolas must equal 1, we divide through by 144.

$$\frac{16x^2}{144} - \frac{9y^2}{144} = \frac{144}{144}$$

$$\frac{x^2}{9} - \frac{y^2}{16} = 1$$

From this point forward, the procedure would be the same.

d) Convert the following equation into standard form. You do not need to graph it.

$$4x^2 - 36y^2 = 144$$

Next, we can graph a hyperbola which is not centered at the origin.

Graph the following ellipse. Label the center, vertices, co-vertices, foci, and give the equation for the asymptotes.

$$\frac{(x-1)^2}{9} - \frac{(y+1)^2}{4} = 1$$

e)
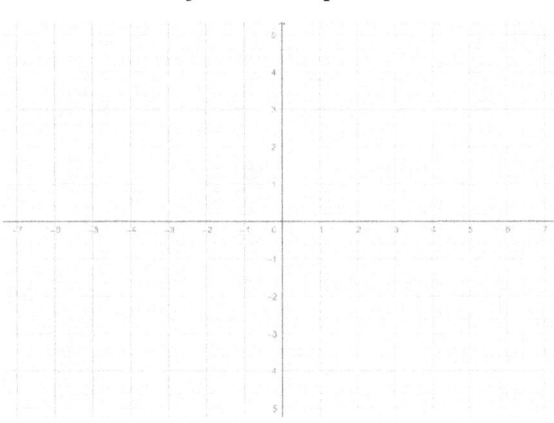

- The hyperbola is centered at $(1, -1)$.
- The transverse axis is the x-axis.
- The value of $a = 3$. The value of $b = 2$. They asymptotes would be at $y = \pm\frac{2}{3}(x-1) - 1$. The asymptotes have been shifted with the center.

f)
- To find the value of c in order to add the foci, use the formula $b^2 = c^2 - a^2$.

g)
- To find the vertices, add/subtract a to the x coordinate of the center.

h)
- To find the co-vertices, add/subtract b to the y coordinate of the center.

i)
- To find the foci, add/subtract c to the x coordinate of the center.

Work another.

Graph the following hyperbola. Label the center, vertices, co-vertices, foci, and give the equation for the asymptotes. (The y value is the larger value, however the x is positive making it the transverse axis.)

$$\frac{(x+2)^2}{9} - \frac{(y-1)^2}{16} = 1$$

j)
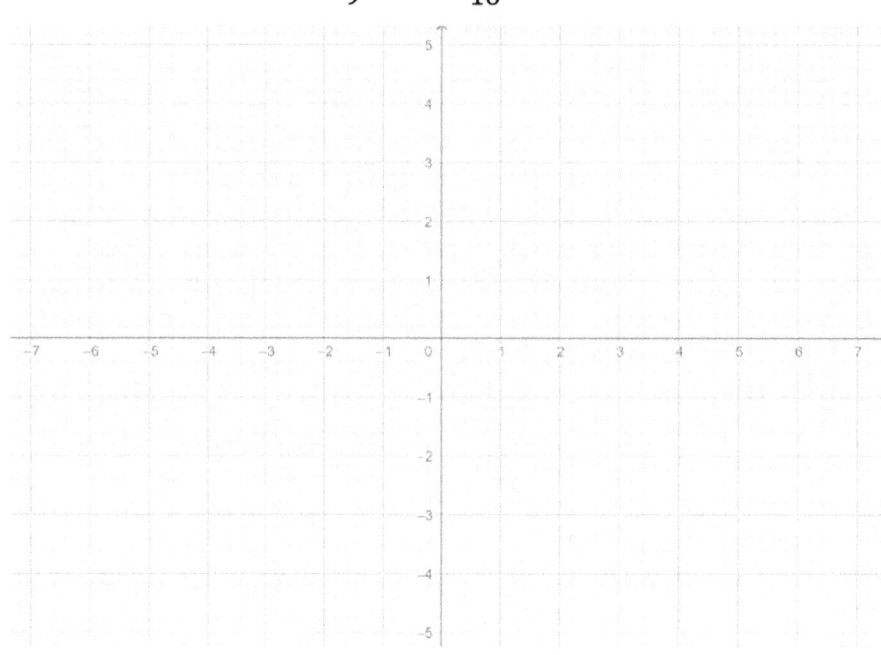

I'll give you the equation for the asymptotes: $y = \pm\frac{4}{3}(x+2) + 1$

Algebra II

Active Learning: 8.2c

Once again, complete the square for each of the following polynomials.

a) $x^2 - 4x + \underline{}$

b) $y^2 - 2y + \underline{}$

Here, we will look at the hyperbolas which are not centered at the origin.

$$3x^2 - 4y^2 - 12x + 8y - 4 = 0$$

Begin by organizing the x and y terms.

$$3x^2 - 12x - 4y^2 + 8y - 4 = 0$$

Next, send the last term to the other side.

$$3x^2 - 12x - 4y^2 + 8y = 4$$

Pull out a GCF from the x terms and a GCF from the y terms. (Notice here that it is a -4 and it will flip the sign of everything which follows.)

$$3(x^2 - 4x) - 4(y^2 + 2y) = 4$$

Now, complete the squares. (Hint: we did this above.)

$$3(x^2 - 4x + 4) - 4(y^2 - 2y + 1) = 4$$

But this is out of balance. We can't add a number to one side without adding it to the other. However, we didn't just add a 4 and a 1. We actually added $3 \cdot 4$ and $-4 \cdot 1$, because our numbers are in parenthesis. So, we add 12 and -4 to the left side.

$$3(x^2 - 4x + 4) - 4(y^2 - 2y + 1) = 4 + 12 - 4$$

$$3(x - 2)^2 - 4(y - 1)^2 = 12$$

c) Finally, divide through by 12 to get the formula in standard form.

Give the center, vertices, co-vertices, and foci.

d) Center:

e) Vertices:

f) Co-vertices:

g) Foci:

h) Try one on your own. Write the hyperbola in standard form and give the values of the center, vertices, co-vertices, and foci.

$$2x^2 - 3y^2 + 12x + 12y - 12 = 0$$

Algebra II

Active Learning: 8.3a

In this lesson, we are going to examine the formula and graph of the parabola. Use Geogebra to help you graph the following equation. Don't rearrange it, just type it in as it is written below:

a)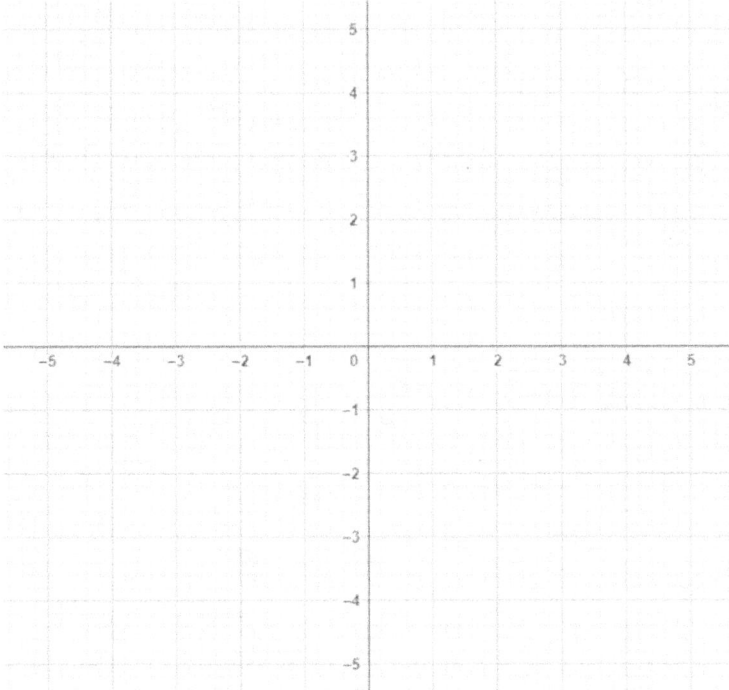

$$x^2 = 8y$$

b) Tell me how this parabola opens. Select the correct answer from the list below. (This is multiple choice.)
- Right
- Upward
- Left
- Downward

Again, use Geogebra to help you graph the following equation:

c)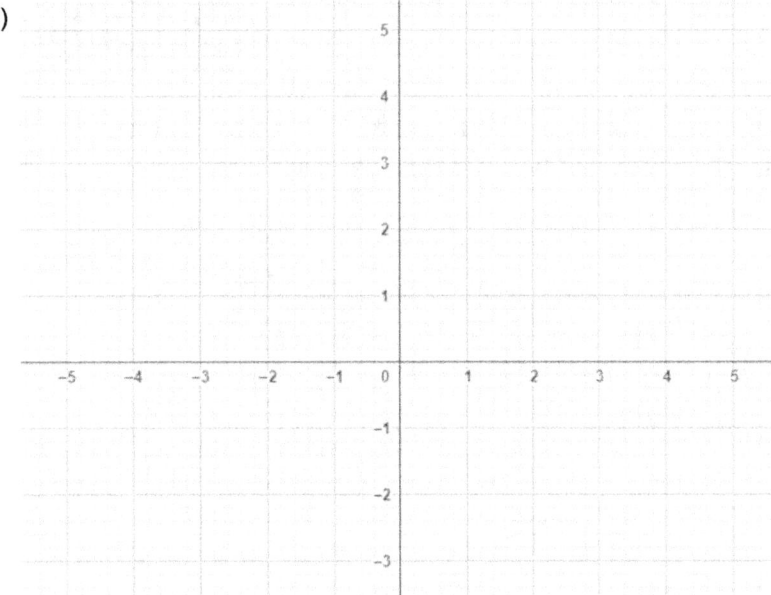

$$x^2 = -12y$$

d) Tell me how this parabola opens. Select the correct answer from the list below. (This is multiple choice.)

- Right
- Upward
- Left
- Downward

Next, graph the following equation:

e)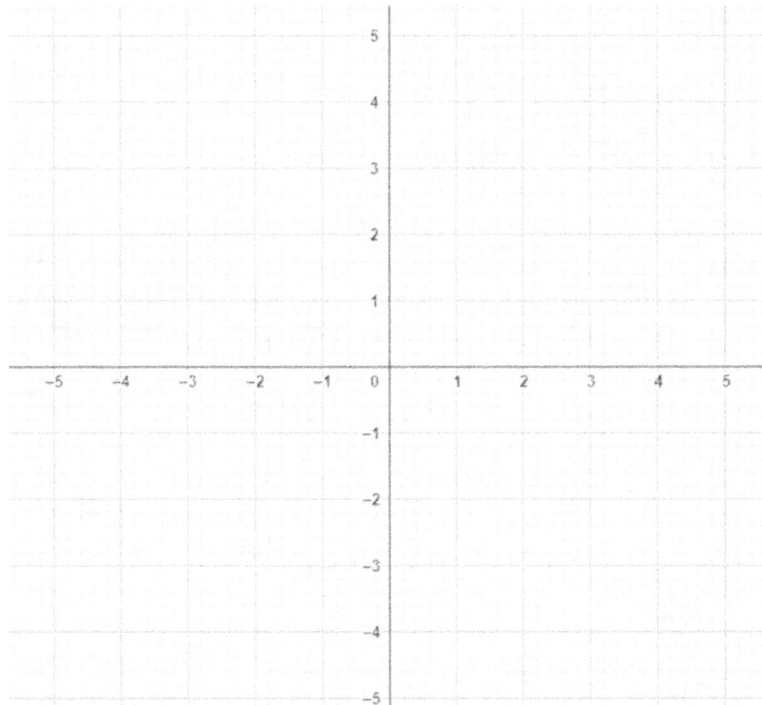

$$y^2 = -4x$$

f) Tell me how this parabola opens. Select the correct answer from the list below. (This is multiple choice.)

- Right
- Upward
- Left
- Downward

Next, graph the following equation:

g)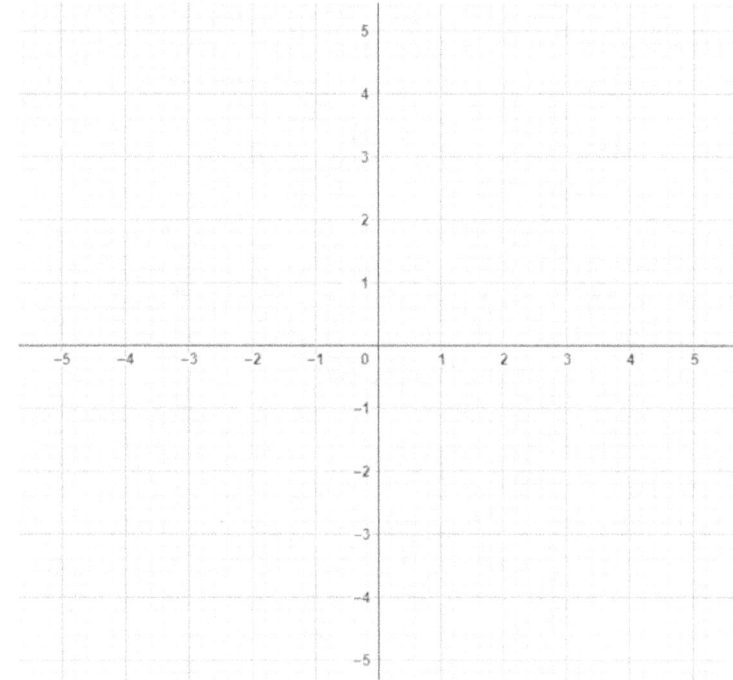

$$y^2 = 16x$$

h) Tell me how this parabola opens. Select the correct answer from the list below. (This is multiple choice.)

- Right
- Upward
- Left
- Downward

The definition of a parabola includes a fixed point called the focus. When the parabola is in this form, we can find the focus from the equation. Below I have listed the four equations we've used and their focus. Try to determine the relationship between the equation and the focus.

$$x^2 = 8y$$

Focus: $(0, 2)$

$$x^2 = -12y$$

Focus: $(0, -3)$

$$y^2 = -4x$$

Focus: $(-1, 0)$

$$y^2 = 16x$$

Focus: $(4, 0)$

i) What is the relationship between the equation and the focus?

(Continued on the next page.)

Next, look at the equations and indicate which way you believe the graph opens and where the vertex of the parabola would be located.

$$(x - 2)^2 = -16(y + 2)$$

j) Vertex:

k) Opens:
- Right
- Upward
- Left
- Downward

$$(y - 4)^2 = 4(x - 5)$$

l) Vertex:

m) Opens:
- Right
- Upward
- Left
- Downward

$$(x + 8)^2 = 24(y - 7)$$

n) Vertex:

o) Opens:
- Right
- Upward
- Left
- Downward

$$(y - 20)^2 = -8(x - 50)$$

p) Vertex:

q) Opens:
- Right
- Upward
- Left
- Downward

In the opposite direction of the focus, we have a vertical (or horizontal) line called the directrix. The directrix is important because all the points on the parabola are the same distance from both the directrix and the focus. Although I don't ask this in the classroom, you might find a question referring to the Latus Rectum. This is an imaginary line running from the focus directly up and down (or right and left) to the parabola. Finding the endpoints of the Latus rectum simply involves plugging the value of p into the equation in standard form. Substitute it in for x, if the parabola is horizontal. Substitute it in for y, if the parabola is vertical.

Given the formula in standard form, find the focus, directrix, and the endpoints of the latus rectum.

$$x^2 = -12y$$

- First, we have $4p = -12$. So, $p = -3$.
- Since we have x^2 and p is negative, the parabola will open downward. So, the focus is moving down the y-axis.
- The focus is at $(0, -3)$.
- The directrix will be a horizontal line going in the opposite direction of the focus: $y = 3$.
- To find the endpoints of the latus rectum, we would plug p into the equation.
$$x^2 = -12(-3) = 36$$
$$x = \pm 6$$

The endpoints would be $(6, -3)$ and $(-6, -3)$.

Try one on your own.

r) Given the formula in standard form, find the focus, directrix, and the endpoints of the latus rectum.

$$y^2 = 24x$$

We could also find the standard form if given the focus and directrix.

What is the equation of the parabola if the focus is $(-2, 0)$ and the directrix is $x = 2$?

- Because of the location of the focus, the parabola opens to the left. The basic form would be $y^2 = 4px$.
- We can see that $p = -2$.

s) In the space below, finish building the standard form equation for the parabola.

Algebra II

Active Learning: 8.3b

Graph the following parabola. Label the focus, directrix, and the endpoints of the latus rectum.

a) $$x^2 = -16y$$

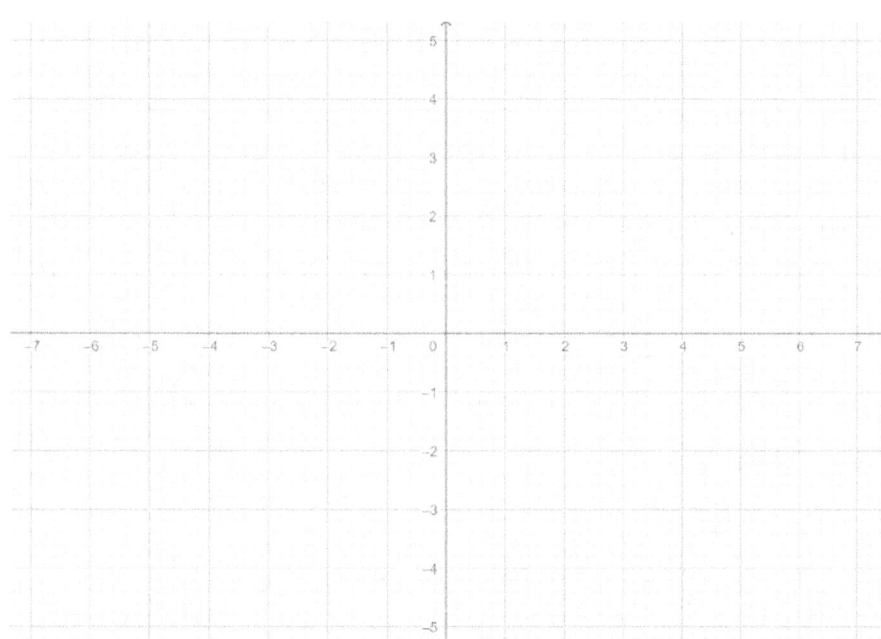

In the next graph, the center (vertex) will not be at the origin. This will create a new term called the axis of symmetry. The axis of symmetry is the horizontal (or vertical) line parallel to the axis which the focus has moved down.

Let's look at a problem.

Graph the following parabola. Label the focus and directrix.

$$(y-2)^2 = -12(x+1)$$

b)

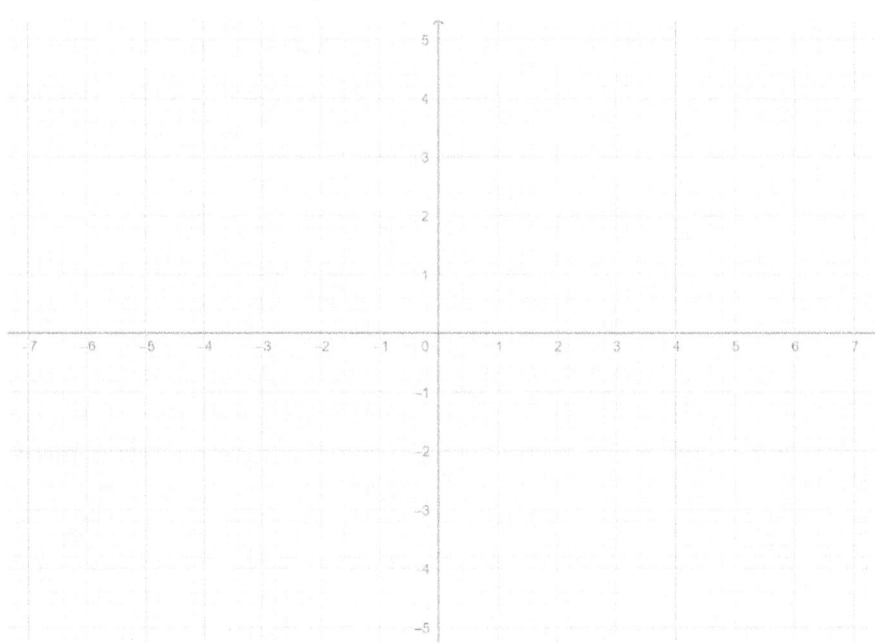

- The vertex is located at $(-1, 2)$.
- $p = -3$
- We have y^2 and negative p, so the parabola opens to the left.
- The focus is moving down the x-axis, so the focus would be $(-4, 2)$.
- The directrix would be in the other direction and would be the vertical line $x = 2$.
- The axis of symmetry would be parallel to the x-axis and through the focus. It is the line $y = 2$.

On the next page, try one on your own.

Graph the following parabola. Label the focus and directrix.

c) $(x+1)^2 = -16(y-1)$

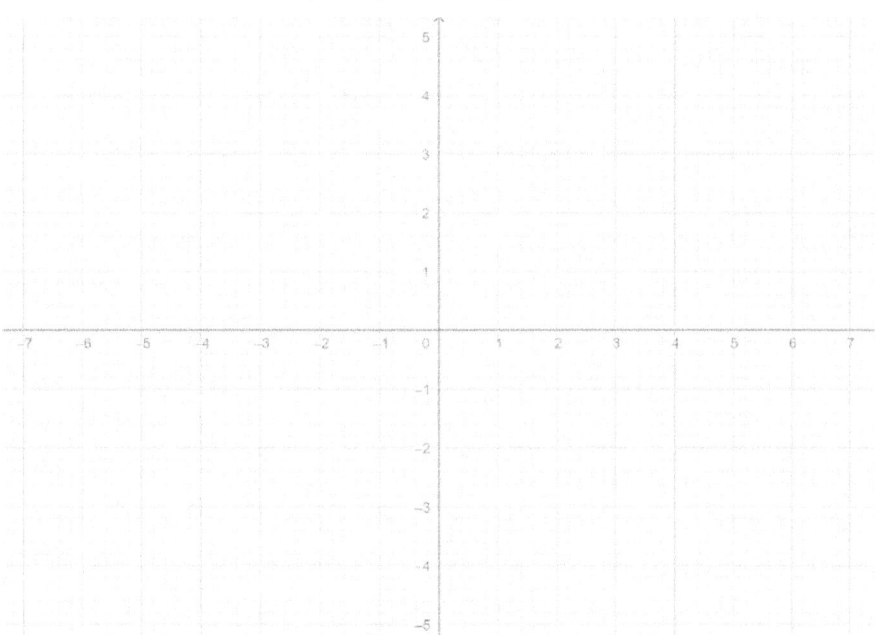

Algebra II

Active Learning: 8.3c

Here, we will convert the equation of a parabola from general form to standard form.

$$x^2 - 4x - 12y + 64 = 0$$

Organize the x-terms alone on the left.

$$x^2 - 4x = 12y - 64$$

a) Complete the square. (Remember, if you add to one side, you must add to the other.)

$$x^2 - 4x + \underline{} = 12y - 64 + \underline{}$$

b) Finish by making the left a perfect square binomial and pulling the GCF from the right. Give the final standard form below:

Identify the following:

c) Vertex:

d) Focus:

e) Directrix:

f) Axis of Symmetry:

On the next page, work one on your own.

g) Write the following parabola in standard form. Give the vertex, focus, directrix, and axis of symmetry.

$$y^2 + 6y - 8x + 25 = 0$$

Algebra II

Active Learning: 9.1a

Give the next number in each of the following:

a) $1, 3, 5, 7, \_\_$

b) $2, 4, 8, 16, 32, \_\_$

c) $\frac{1}{2}, \frac{1}{3}, \frac{1}{4}, \frac{1}{5}, \_\_$

d) $-5, 10, -15, 20, \_\_$

Above, the set of numbers followed a pattern. In this section, we want to look at a special type of number pattern called a sequence. Sequences are functions. Their domain is positive values of a variable n. They can be finite (coming to an end) or infinite (continuing forever).

$$a_1, a_2, a_3, a_4, \ldots a_n$$

a_1 is called the first term of the sequence, a_2 is the second, ect. a_n is referred to as the nth term.

Sometimes each term of a sequence can be defined by a formula (called an explicit formula). Here's an example:

$$a_n = -2n + 4$$

So, beginning at 1, we can find the terms of the sequence.

$a_1 = -2(1) + 4 = -2 + 4 = 2$

$a_2 = -2(2) + 4 = -4 + 4 = 0$

$a_3 = -2(3) + 4 = -2$

Find the next two terms of this sequence.

e) $a_4 =$

f) $a_5 =$

The first five terms make up a set of numbers. A set is contained within { }. Give the set of the first five terms of this sequence.

g)
$$\{2, 0, -2, \_\_, \_\_\}$$

Letter d) in our warm up problem involved terms which alternated signs. The trick to defining an explicit formula with alternating signs is the term -1^n.

In this next sequence, we have alternating signs and multiple n's in the explicit formula. Find the first five terms. (I have done the first two for you.)

$$a_n = \frac{(-1)^n n}{n+3}$$

$$a_1 = \frac{(-1)^1(1)}{(1)+3} = -\frac{1}{4}$$

$$a_2 = \frac{(-1)^2(2)}{(2)+3} = \frac{2}{5}$$

h) $a_3 =$

i) $a_4 =$

j) $a_5 =$

k) Give the first five terms as a set:

Explicit formulas can also be piecewise. The following sequence depends on different instructions depending on even or odd. Find the first five terms. (I will do the first two.)

$$a_n = \begin{cases} n^2 \text{ if even} \\ n-3 \text{ if odd} \end{cases}$$

$a_1 = (1) - 3 = -2$ (Since 1 is odd.)

$a_2 = (2)^2 = 4$ (Since 2 is even.)

l) $a_3 =$

m) $a_4 =$

n) $a_5 =$

o) Give the first five terms as a set:

Often, we will be given the terms for the sequence and be asked to find the explicit formula. This can be a bit tricky, but becomes easier with practice. I'll walk you through a problem.

Find an explicit formula for the nth term of the sequence.

$$\left\{-\frac{3}{5}, \frac{4}{7}, -\frac{5}{9}, \frac{6}{11}\right\}$$

To find the explicit formula, we need to go one step at a time. (Remember, we are starting with $n = 1$.)

- Because of the alternating signs, we will need to include $(-1)^n$.
- Next, we look at the numerators: 3, 4, 5, 6. $n + 2$ would make this.
- Finally, we look at the denominators: 5, 7, 9, 11. This is tricky, but this would work: $2n + 3$.

Our explicit formula for the nth term would be $a_n = \frac{(-1)^n(n+2)}{(2n+3)}$.

Try one on your own.

p) Find an explicit formula for the nth term of the sequence.

$$\left\{-\frac{2}{7}, \frac{3}{9}, -\frac{4}{11}, \frac{5}{13}\right\}$$

Try these.

q)
$$\{e^3, e^4, e^5, e^6 \ldots\}$$

On this problem, the numerator stays constant.

r)
$$\left\{\frac{3}{9}, \frac{3}{27}, \frac{3}{81}, \frac{3}{243} \ldots\right\}$$

This final problem is easier than it appears. The denominator will switch to the numerator when the exponent switches from a negative to a positive.

s)
$$\left\{\frac{1}{e^3},\frac{1}{e^2},\frac{1}{e},1,e,e^2\ldots\right\}$$

Algebra II

Active Learning: 9.1b

a) In our last activity, we learned about sequences with explicit formulas. Find the first five terms of the following sequence:

$$a_n = \frac{(-1)^n(n+3)}{(2n+1)}$$

b) Find an explicit formula for the nth term of the sequence.

$$\left\{-\frac{1}{2}, \frac{1}{4}, -\frac{1}{8}, \frac{1}{16} \ldots\right\}$$

In this activity, we will see a second type of formula for a sequence. Here, we investigate recursive formulas. Recursive formulas will use the previous term as the input into the next term. For instance:

$$a_1 = 5$$

$$a_n = 2a_{n-1} + 3 \text{ for } a \geq 2$$

So, we input a_1 to find a_2.

$a_2 = 2(5) + 3 = 13$

Then the answer for a_2 into a_3.

$a_3 = 2(13) + 3 = 26 + 3 = 29$

Follow this procedure to find a_4 and a_5.

c) $a_4 =$

d) $a_5 =$

e) Try one on your own. Find the first five terms of the following recursive sequence.

$$a_1 = 3$$

$$a_n = 5a_{n-1} - 2 \text{ for } a \geq 2$$

Sometimes a recursive sequence will depend upon more than just the previous term. Here, we need the previous two terms.

$$a_1 = 1$$

$$a_2 = 3$$

$$a_n = a_{n-1} - a_{n-2} \text{ for } a \geq 3$$

Therefore, a_3 will depend upon the term one before ($a_{n-1} = a_2$) and the term two before ($a_{n-2} = a_1$)

$$a_3 = 3 - 1 = 2$$

It can be easy to lose track of what numbers you need, but to find a_4, we need a_3 and a_2.

$$a_4 = 2 - 3 = -1$$

Find terms a_5 and a_6.

f) $a_5 =$

g) $a_6 =$

h) Try this problem. Find the first six terms of the sequence by using the recursive formula.

$$a_1 = 3$$

$$a_2 = 5$$

$$a_n = 3(a_{n-1}) + 2(a_{n-2}) \text{ for } a \geq 3$$

We will return to explicit formulas again, however before we do, we need to look at an important mathematical idea called a factorial. Multiply the following:

i) $3 \cdot 2 \cdot 1 =$

j) $5 \cdot 4 \cdot 3 \cdot 2 \cdot 1 =$

When we multiply all the way from a number down to 1, it is called a factorial. It is given an exclamation mark.

$3! = 3 \cdot 2 \cdot 1$

$5! = 5 \cdot 4 \cdot 3 \cdot 2 \cdot 1$

Find the first five terms of this sequence using the explicit formula. I will work the first two.

$$a_n = \frac{n+1}{(n+2)!}$$

$$a_1 = \frac{1+1}{(1+2)!} = \frac{2}{3!} = \frac{2}{3 \cdot 2 \cdot 1} = \frac{2}{6} = \frac{1}{3}$$

$$a_2 = \frac{2+1}{(2+2)!} = \frac{3}{4!} = \frac{3}{4 \cdot 3 \cdot 2 \cdot 1} = \frac{3}{24} = \frac{1}{8}$$

k) $a_3 =$

l) $a_4 =$

m) $a_5 =$

Finally, work one more on your own.

n) Find the first five terms of this sequence using the explicit formula. I will work the first two.

$$a_n = \frac{(n+1)!}{n}$$

Algebra II

Active Learning: 9.2a

Find the next number in the sequence.

a) $\{2, 6, 10, 14, 18 ....\}$

b) $\{-3, 6, 15, 24 ...\}$

Sequences where there is a common difference between terms are called arithmetic sequences.

$$\{a_1, a_1 + d, a_2 + d, ...\}$$

c) Write the first five terms of an arithmetic sequence if the $a_1 = 12$ and $d = -3$.

An arithmetic sequence can also be written exclusively in terms of a_1 and d.

$$\{a_1, a_1 + d, a_1 + 2d, a_1 + 3d ...\}$$

So, for instance, a_4 could be written as follows:

$$a_4 = a_1 + 3d$$

We can use this approach to find a missing term in a sequence.

Given $a_1 = 4$ and $a_4 = 25$ find a_5.

Since $a_4 = a_1 + 3d$ we can find the common difference.

$$25 = 4 + 3d$$

$$21 = 3d$$

$$d = 7$$

Now that we know the common difference we can find a_5.

$$a_5 = 25 + 7 = 32$$

Try these.

d) Given $a_1 = 6$ and $a_5 = 42$ find a_6.

e) Given $a_1 = 12$ and $a_4 = 0$ find a_5.

A variation would be to start with terms like this. We still need to find the common difference.

Given $a_3 = 15$ and $a_5 = 25$ find a_2.

Between a_3 and a_5 we would add d twice. So, we could write a_5 as follows:

$$a_5 = a_3 + 2d$$
$$25 = 15 + 2d$$
$$10 = 2d$$
$$d = 5$$

Now, that we know d we could go back from a_3 to find a_2.

$$a_3 - d = a_2$$
$$a_2 = 15 - 5 = 10$$

Work the following.

f) Given $a_3 = 27$ and $a_5 = 15$ find a_2.

g) Given $a_4 = 33$ and $a_6 = 55$ find a_1.

Algebra II

Active Learning: 9.2b

Find the value of the common difference in the following sequence.

$$\{3, 12, 21, 30, ...\}$$

a) $d =$

Previously, we learned about recursive formulas, where a_n is defined based on one (or more) previous terms. With arithmetic sequences making a recursive formula is easy.

$$a_n = a_{n-1} + d \; n \geq 2$$

Give the recursive formula for $\{3, 12, 21, 30, ...\}$. (No values are entered for a_n and a_{n-1}. They remain generic.)

b) $a_n =$

Try another.

c) Give the recursive formula for the following arithmetic sequence.

$$\{36, 29, 22, 15, ...\}$$

With arithmetic sequences, it is easy to find an explicit formula. Look again at the following sequence:

$$\{3, 12, 21, 30, ...\}$$

d) For a_2 how many times do we add d?

e) For a_3 how many times do we add d?

f) For a_4 how many times do we add d?

Here is the explicit formula for an arithmetic sequence.

$$a_n = a_1 + d(n-1)$$

For each term in the sequence, we add one less d. In other words, for term a_3, we add 2 d's. For term a_8, we would add 7 d's.

Below is a series which we have already looked at:

$$\{2, 6, 10, 14, 18 ....\}$$

g) We previously wrote a recursive formula. Now, write an explicit formula. (Just insert the value of d into the equation below.)

$$a_n = a_1 + d(n-1)$$

Try another.

h) Write the explicit formula for the following sequence. (Notice that d will be negative.)

$$\{36, 29, 22, 15, ...\}$$

i) Next, several terms are missing from the following sequence. Find the missing terms.

$$\{-21, -16, -11, ... \; 4, 9\}$$

j) How many terms are in this sequence?

To find the number of terms in a finite sequence, a sequence that ends, the manual approach would always work. However, if the sequence is large, a more mathematical approach might be better. The explicit formula for this sequence is:

$$a_n = a_1 + 5(n-1)$$

Substituting in the first term and the last term, we get:

$$9 = -21 + 5(n-1)$$

Now, solve for n and you have the number of terms in the sequence. Find n.

k) $n =$

Work this problem.

l) Find the number of terms in the finite arithmetic sequence.

$$\{-5, 3, 11, \ldots, 59, 67\}$$

Next, let's look at a simple application of arithmetic sequences.

m) Someone training for a race runs 3 miles on the first day of training. Each subsequent day they add 2 miles. Write an explicit formula for this arithmetic sequence.

n) How many miles would the racer run on the 8th day of training?

Algebra II

Active Learning: 9.3a

Find the next number in the sequences.

a) $16, 8, 4, 2, \underline{}$

b) $1, 3, 9, 27, \underline{}$

Each of the above sequences have something called a common ratio. To find the common ratio, r, divide a term by the term before:

$$r = \frac{a_{n+1}}{a_n}$$

Find the common ratio for each of these sequences:

c) $\{16, 8, 4, 2\}$

$r =$

d) $\{1, 3, 9, 27\}$

$r =$

A sequence with a common ratio is called a geometric sequence. To be a geometric sequence the ratio must remain the same between each term and the previous term. One of the following sequences is
e) geometric and one is not. Circle the geometric sequence and give its common ratio.

$$\left\{1, \frac{1}{5}, \frac{1}{25}, \frac{1}{125}\right\}$$

$$\left\{1, \frac{1}{3}, \frac{1}{10}, \frac{1}{30}\right\}$$

To find a geometric sequence, you check for the common ratio. But to find the terms of a geometric sequence, you repeatedly multiply by that ratio.

$$\{a_1, a_1r, a_1r^2, a_1r^3, ...\}$$

In the next group of problems, you will be given the first term and the common ratio. Find the first four terms of the geometric sequence.

f) $a_1 = 3, r = 5$

g) $a_1 = 100, r = \frac{1}{2}$

h) $a_1 = 7, r = -2$

Algebra II

Active Learning: 9.3b

Find the next term in the geometric sequence:

$$\{2, 2.4, 2.88, 3.456, \ldots\}$$

a) $r =$

b) $a_5 =$

Recall that recursive formulas are those in which the next term in the sequence is defined by one (or more) previous terms. For a geometric sequence, the recursive formula is straightforward.

$$a_n = a_{n-1} \cdot r \quad n \geq 2$$

Give the recursive formula for the following geometric sequences. (Remember, you only need to add the value of r. a_n and a_{n-1} are left generic.

$$\{2, 2.4, 2.88, 3.456, \ldots\}$$

c) $a_n =$

$$\{16, 8, 4, 2\}$$

d) $a_n =$

$$\left\{1, \frac{1}{5}, \frac{1}{25}, \frac{1}{125}\right\}$$

e) $a_n =$

The explicit formula is straightforward as well. Look again at this sequence:

$$\{2, 2.4, 2.88, 3.456\}$$

f) How many times has a_2 been multiplied by r?

g) How many times has a_3 been multiplied by r?

h) How many times has a_4 been multiplied by r?

For this reason, the nth term of geometric sequence is multiplied by $n - 1$ values of r.

$$a_n = a_1 r^{(n-1)}$$

The following problem will require us to find r before we can find the term that is requested.

Find a_2 if $a_1 = 2$ and $a_4 = 54$.

Use the explicit formula to find r.

$$a_4 = a_1 r^{(4-1)}$$
$$54 = 2r^3$$

Finish solving for r. (You will need to take a cube root.)

i) $r =$

Find a_2 by multiplying a_1 by r.

j) $a_2 =$

Work a similar problem on your own.

k) Find a_2 if $a_1 = 3$ and $a_5 = 768$.

Next, we will write explicit formulas for geometric sequences.

$$\{2, 2.4, 2.88, 3.456, \ldots\}$$

In this sequence, first find r.

$$r = \frac{a_{n+1}}{a_n} = \frac{2.4}{2} = 1.2$$

Now, we plug r and a_1 into the explicit formula.

$$a_n = a_1 r^{(n-1)}$$

$$a_n = 2 \cdot (1.2)^{(n-1)}$$

Find the explicit formula for each of the following:

$$\{16, 8, 4, 2\}$$

l) $a_n =$

$$\left\{1, \frac{1}{5}, \frac{1}{25}, \frac{1}{125}\right\}$$

m) $a_n =$

Let's conclude with an application.

In 2015, the population of deer in a park was 250. The population is growing by 5% per year.

    Find an explicit formula for this geometric sequence.

Because we are given a percentage, in order for the population to be growing, we need a value of r above 100%.

$r = 1.05$

n)   Since we know a_1 and r, substitute them into the explicit formula.

What will be the population of the deer in the year 2021?

To find the population, we will plug into the explicit formula. However, we need to know the value of n for the year 2021. If we subtract the first year, 2015, from the year 2021, we will know what term the year 2021 is.

$$n = 2021 - 2015 = 6$$

o) 2021 is the 6th term. Substitute $n = 6$ into the explicit formula you found in part a to find the population in 2021.

$a_6 =$

Algebra II

Active Learning: 9.4a

Add the following:

a) $5 + 6 + 7 + 8 + 9 =$

b) $4 + 8 + 16 + 32 =$

Whenever mathematicians see a pattern, they look for a way to simplify it using math language. Let's look at each of the above series.

- Here, a number continues to increase by 1 and each of the values is added. For this situation, mathematicians invented the summation sign: $\sum k$. It will repeatedly increase the value of k by 1 and add. However, we need to define starting and stopping points. In this example, we started at 5 and ended at 9. So:

$$\sum_{5}^{9} k = 5 + 6 + 7 + 8 + 9$$

- In this problem, we have the addition but a different kind of increase. This could be defined by k^2, where k starts at 2 and ends at 5.

$$\sum_{2}^{5} 2^k = 4 + 8 + 16 + 32$$

Write your own summation expression for each of the following:

c) $2 + 3 + 4 + 5 + 6$

d) $27 + 81 + 243$

There are endless variations on the summation expression. Evaluate each of the following. Remember, k increases by one for each term and all the terms are added.

e) $\sum_{1}^{4}(2k+1)$

f) $\sum_{3}^{7} k^2 - 1$

g) $\sum_{3}^{6}(3k+4)$

Add the following:

h) $1 + 3 + 5 + 7 + 9 + 11 =$

i) $6 + 6 + 6 + 6 + 6 + 6 =$

Find the average for each of the following pairs of numbers:

j)   1, 11   k)   3, 9   l)   5, 7

An interesting thing happens when adding a finite arithmetic sequence. You get the same thing as if you repeatedly added the average of the pair of numbers. Look at this sequence:

$3 + 6 + 9 + 12 + 15 + 18$

m) Average the first and last number.

n) There are six numbers in the finite sequence so add that average repeatedly six times. (Repeated addition is multiplication. So, multiply the average you found by the six terms in the sequence.)

The sum of a finite arithmetic sequence (called a series) always follows this pattern and can be defined by the formula:

$$S_n = n \frac{(a_1 + a_n)}{2} = \frac{n(a_1 + a_n)}{2}$$

Where n is the number of terms you are adding. (This is the average of the first and last term multiplied by how many terms in the series.) Find the sum of the following arithmetic series:

o) $4 + 9 + 14 + 19 + 24 + 29 + 34$

p) $12 + 6 + 0 - 6 - 12 - 18 - 24 - 30$

q) Here, you are given the series in summation notation. (Hint: you don't need to find the entire series, just the first and last terms. You have 10 terms in the series.)

$$\sum_{1}^{10} 4k - 1$$

Next, let's revisit an applied problem which we saw previously.

Someone training for a race runs 3 miles on the first day of training. Each subsequent day they add 2 miles. How many total miles had they run after day five?

r) The difference between terms for this arithmetic sequence is $d = 2$. An explicit formula for the sequence would be $a_n = a_1 + d(n-1)$. Find the first term of the sequence.

s) Find the fifth term of the sequence.

Use the first term and the fifth term in the formula for summing the first five terms of the series.

$$S_n = \frac{n(a_1 + a_n)}{2}$$

t) How many total miles did the runner run through the first five days of training?

We next want to look at the sum of the terms in a geometric series. The proof of the sum of the terms of a geometric series is not intuitive. Rather, it involves some math tricks. The tricks aren't hard, but we will leave the idea for Calculus. Here is the formula for summing the first n terms of a geometric series.

$$S_n = \frac{a_1(1 - r^n)}{1 - r}$$

Notice that you only need the first term and the common ration, r. Use the formula to find the sum of the following:

$$\{8, 4, 2, 1\}$$

u) First find r by dividing a_2 by a_1. Since $n = 4$ and $a_1 = 8$, you have everything you need to use the formula to some the first four terms.

v)   Find S_8 of the geometric sequence $2, 6, 18, 54 \dots$

We want to sum the first 8 terms. Find r by dividing a_2 by a_1.

$n = 8$ and $a_1 = 2$.

w)   Find the sum of the following.

$$\sum_{1}^{6} 2 \cdot 3^k$$

It is easier to understand this problem if you expand out the summation. However, we can also reason it out. We can find a_1 and r.

$$a_1 = 2 \cdot 3^1$$

To find r, we could find a_2 and find the ratio with a_1. However, from the summation formula, we can see that each iteration of k involves multiplying by another 3. So, $r = 3$.

You have a_1 and r. And, $n = 6$, so find the sum using the formula.

x)   Find the sum of the following.

$$\sum_{1}^{8} 5 \cdot 2^k$$

Finally, let's end with an application.

A side job pays you $25000 per year. You get a 6% raise each year, how much money would you have earned at the end of 5 years?

This is the summation of the first five terms of a geometric sequence.

$a_1 = 25000$ $n = 5$

y)   And, $r = 1.06$. Because it was given as a percentage and the money grows each year, we need to add the percentage to 1. (We have 106% of our money at the end of each year.) Use the formula to find S_5.

Algebra II

Active Learning: 9.4b

Let's look at this problem: You put $500 in the bank and receive 3% interest at the end of every year.

Here is the explicit formula for this problem as a geometric sequence.

$$a_n = 500(1.03)^{n-1}$$

And, here is an exponential function for the same situation.

$$f(x) = 500(1.03)^x$$

A geometric sequence is exactly the same as an exponential function. One of the following exponential functions will grow and one will decrease. Circle the formula which will decrease.

$$g(x) = 1750(1.00015)^x$$

$$h(x) = 500(.9985)^x$$

a) What causes the function you picked to decrease?

b) If geometric sequences and exponential functions are the same then which of the following will cause the next term in a geometric sequence to decrease?

$r = 1$

$r < 1$

$r > 1$

If the terms of a geometric sequence keep getting smaller then they will eventually grow to be so small as to be insignificant. If that occurs, even an infinite geometric sequence could be summed. But it is only true for geometric sequences which have terms that are decreasing. Although this isn't easy to wrap your head around, it is also true for values of r which are between 0 and -1. Find the following:

$$h(x) = 500(-.9985)^x$$

c) $h(101) =$

d) $h(102) =$

e) $h(103) =$

With a negative value of r, the values oscillate between positive and negative, however they are approaching zero.

Look at the following infinite geometric sequences. Circle any which could be summed because their terms are decreasing. (You want to find the value of r and determine if it meets the criteria of decreasing: $1 > r > -1$)

f) $\{8, 4, 2, 1, ...\}$

g) $\{4, 8, 16, 32, 64, ...\}$

h) $\{6, -2, \frac{2}{3}, -\frac{2}{9}, ...\}$

i) $\{9, 3, 1, \frac{1}{3}, \frac{1}{9}, ...\}$

If the infinite geometric series can be summed, it is said that the series is defined. The formula to sum a defined infinite geometric series is:

$$S = \frac{a_1}{1-r}$$

Notice that you need the first term, a_1, and the ratio, r. (Again, to be defined $1 > r > -1$.) The following infinite geometric series are defined. Find their sum using the formula.

j) $\{8, 2, \frac{1}{2}, \frac{1}{8}, \frac{1}{32}, ...\}$

k) $\{25, 15, 9, \frac{27}{5}, ...\}$

l) $\{20, -\frac{40}{3}, \frac{80}{9}, -\frac{160}{27}, ...\}$

Sometimes, you can be asked to sum an infinite geometric sequence that has been written with a summation.

$$\sum_{1}^{\infty} \left(\frac{1}{3}\right)^n$$

We first need a_1 and r.

First, a_1.

$$a_1 = \left(\frac{1}{3}\right)^1 = \frac{1}{3}$$

Second, we need r. Since each term in the series is repeatedly being multiplied by $\frac{1}{3}$, $r = \frac{1}{3}$. (If you weren't sure, you could have written out the first few terms of the sequence.)

m) Now, use the formula to find the sum of this defined sequence.

Look at this summation.

$$\sum_{1}^{\infty} (.55)^n + 2$$

This is not defined. It may appear as if $r = .55$, however the $+2$ is causing there to be repeated addition as well as repeated multiplication. Therefore, it is not geometric. If you are unsure, look at the first few terms.

$$a_1 = 2.55$$
$$a_2 = 2.3025$$
$$a_3 = 2.166$$

Checking the ratios between terms does not give a common r.

Give the sum of the following infinite sequence *if it is defined*.

n)
$$\sum_{1}^{\infty} \left(\frac{3}{4}\right)^n$$

o)
$$\sum_{1}^{\infty} \left(-\frac{1}{2}\right)^n$$

p)
$$\sum_{1}^{\infty} \left(\frac{3}{5}\right)^n + 3$$

You have worked with repeating decimals since you were very young.

$$.\overline{3} = .3333\ldots$$

Interestingly, this can be written as an infinite geometric sequence which can be summed.

$$.\overline{3} = .3 + .03 + .003 + .003$$

Where there is a common ratio.

$$r = .1$$

q) Use the summation formula to find the sum of this infinite geometric sequence.

Finally, our last concept is an idea from financial math called an annuity. An annuity is an investment where you repeatedly pay into a retirement account. Let's look at an example.

Suppose you put $200 each month into an annuity which pays 4% interest each year. The interest is compounded monthly. If your money is in the account for 5 years, how much is it worth at the end?

Every month your money is making interest. The money that has been in longer is making more interest. However, it isn't 4% each month. Compounded monthly means that the 4% is split up over each month.

$$\frac{.04}{12}$$

We started with a percentage, and your money is growing, so we must add 1. (Remember, without 1, the money would be decreasing.)

$$1 + \frac{.04}{12}$$

Since we are multiplying the money by this number each month, this is r.

$$r = 1.004$$

At the end of the five years, considering 12 payments a year, you would have a finite geometric sequence with $n = 5 \cdot 12 = 60$ terms. It would look like this:

$$\{200, 200.8, 201.6032, 202.4096 \ldots 215.7598\}$$

r) Use the formula for the sum we learned in the last lesson to sum up a finite geometric sequence.

$$S_n = \frac{a_1(1 - r^n)}{1 - r}$$

$a_1 = 200$

$r = 1.004$

$n = 60$

Try a similar problem on your own.

s) Suppose you put $500 each month into an annuity which pays 8% interest each year. The interest is compounded monthly. If your money is in the account for 4 years, how much is it worth at the end?

Algebra II

Active Learning: 9.5

We now begin looking at several key ideas of probability which are frequently used in general mathematics.

a) You look in your freezer and find three microwave dinners, four different kinds of pizzas, and two bagged vegetable meals. How many different options do you have for dinner?

When we have one decision to make (in one category), we can use a principal called the Addition Principal to simply add up our choices. However, the next situation is more complex. There are two (or more) categories at the same time.

b) Below, show all the possible outcomes of flipping a coin and rolling a standard six-sided die. I've shown the first couple for you. (So, we have two categories at the same time.)

-Head/1

-Head/2

c) A child has three coins. How many possible outcomes could occur from flipping all three coins? I've shown the first couple for you. (Here, we have three categories.)

-Heads/Heads/Heads

-Heads/Heads/Tails

It turns out that a principal known as the counting principal tells us that the total number of outcomes is the multiplication of your options. For instance, a coin flip has two options. A six-sided die has six options. The total number of combinations would be:

$$2 \cdot 6 = 12$$

The idea can extend to any number of choices. Use the counting principal to answer each of the following:

d) How many possible outcomes could occur from flipping all three coins? (Each coin has two options. There are three coins.)

e) Three 5th graders, four 6th graders, and two 7th graders have been selected as their class representatives. How many possible student body councils could be made if one student must be chosen from each grade?

f) You flip five coins. How many possible outcomes could occur?

The counting principal is the key to an idea called a permutation. Imagine that there are five students and we want to see how many ways we can have them seated in a line of chairs. We could have any of the five students sit in the first chair. But, when someone sits down, there are only four students left for the next chair. And, when someone sits there, we only have three for the next chair, and so on.

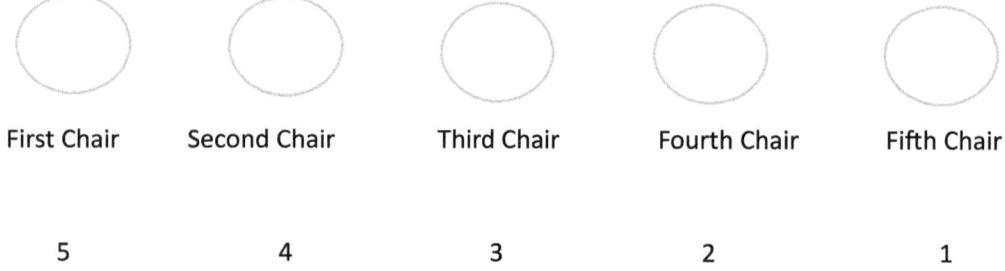

| First Chair | Second Chair | Third Chair | Fourth Chair | Fifth Chair |

5   4   3   2   1

The counting principal taught us if that we can multiply our choices for each category (the chairs). So:
$$5 \cdot 4 \cdot 3 \cdot 2 \cdot 1 = 120$$
And, we previously learned that another way of multiplying from five down to one is to use a factorial.
$$5! = 120$$

g) How many ways could we line up six friends in the first row of a movie theater?

In a permutation, the order in which students sit matters. If Becky is sitting in the first seat in the movie theater, it isn't the same as if she were sitting in the third seat. The way I teach students about a permutation is to look for a title. Is there a label in front of each spot in arrangement? At the theater, there was chair #1, chair #2, etc. Each of the following are permutations because they have a title/label.

h) How many ways can three students be arranged into the President, Vice President, and Secretary on a student council?

i) How many ways can five football players be arranged into an offensive line? (Each spot on the line has a unique title: Left Tackle, Left Guard, Center, Right Guard, and Right Tackle.)

Often with permutations, we have more people than spots. Suppose there were 8 students sitting in the line of five chairs in the classroom. There are eight people who can sit in the first chair, leaving seven for the next chair, six for the next, ect. But, after the last chair there are people remaining who didn't get a seat.

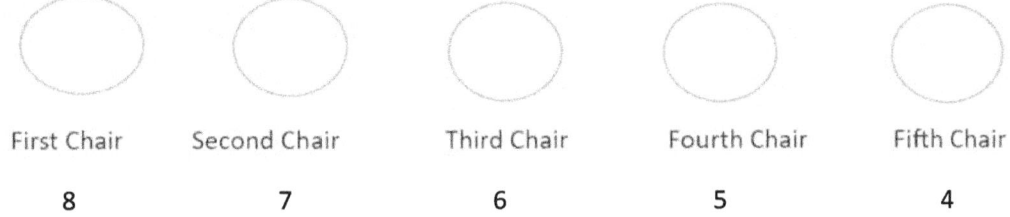

First Chair	Second Chair	Third Chair	Fourth Chair	Fifth Chair
8	7	6	5	4

The counting principal would give us the following possible outcomes.

$$8 \cdot 7 \cdot 6 \cdot 5 \cdot 4 = 6720$$

If we wanted to use factorials, we could find a permutation like this:

$$\frac{8!}{3!}$$

This may seem strange, but the idea is simple.

$$\frac{8 \cdot 7 \cdot 6 \cdot 5 \cdot 4 \cdot 3 \cdot 2 \cdot 1}{3 \cdot 2 \cdot 1}$$

Notice that the numbers in the denominator would simply cancel with some in the numerator and we would be left with exactly the same thing we had when we considered the counting principal.

$$\frac{8 \cdot 7 \cdot 6 \cdot 5 \cdot 4 \cdot \cancel{3} \cdot \cancel{2} \cdot \cancel{1}}{\cancel{3} \cdot \cancel{2} \cdot \cancel{1}} = 8 \cdot 7 \cdot 6 \cdot 5 \cdot 4 = 6720$$

Mathematicians love formulas and so they came up with the following:

$$P(n, r) = \frac{n!}{(n-r)!}$$

This may not look like an improvement, but it is really nothing more than the counting principal. The denominator is cancelling away the left-over people who didn't get a seat. The n stands for the number of available people and the r stands for the number of available chairs. Work the following permutations. You can use the formula or you can just reason it out with the counting principal.

j) How many ways can five students be selected to serve on the student council, which has a President, Vice-President, and Secretary?

k) How many ways can ten students be seated in the first row of the movie theater, if the first row has five seats?

l) How many ways can eight football players be arranged into an offensive line? (Remember, it has five positions.)

In a permutation, the order in which people are being arranged matters. However, that isn't always the case. Suppose we were selecting students for a committee, but there were no titles for the positions. If Becky, Steve, and Isaac are selected, the order doesn't matter. For instance, a committee of Becky, Steve, and Isaac is the same as a committee of Isaac, Becky, and Steve. When the order doesn't matter, we have something called a combination.

Suppose we have eight people for five spots on a field trip. The idea starts the same:

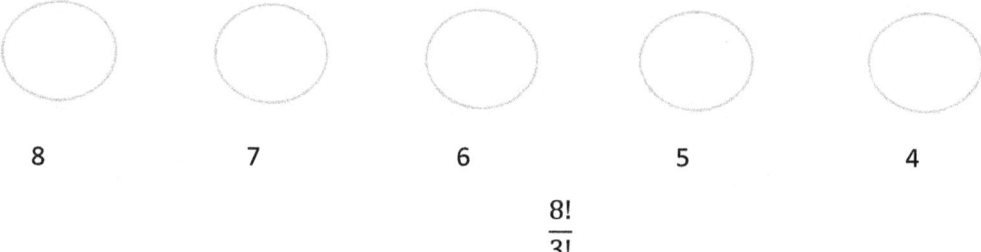

But, the order of the five people selected doesn't matter. And so, our current calculation has repeated arrangements. If you want to get rid of repeats, you divide away. There are 5! repeats because there are 5! meaningless ways in which you can arrange the five people for the trip. So, the math becomes:

$$\frac{8!}{5!\,3!}$$

The mathematician turns this into the following formula for a combination:

$$C(n,r) = \frac{n!}{r!\,(n-r)!}$$

Find the following combinations. Notice that order doesn't' matter in these problems. (There are no meaningful labels before each person.)

m) How many ways can you select six students to go on a field trip if there are ten students?

n) If there are twelve students to select from, how many ways can you make a dodgeball team that needs five students?

Before we conclude this section, let's look at two unique types of problems. The first is the subsets of a set. Suppose we had the following ice cream toppings: sprinkles, hot fudge, marshmallow, and chocolate chips. How many different ways are there to make your ice cream? So, the set has four toppings. We are making all the different subsets of the toppings.

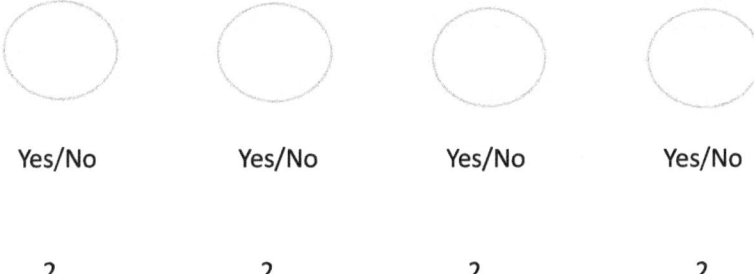

| Yes/No | Yes/No | Yes/No | Yes/No |

| 2 | 2 | 2 | 2 |

Basically, for each topping, you are asking if you will include it, Yes or No. This means two choices for each, and the counting principle would say:

$$2 \cdot 2 \cdot 2 \cdot 2 = 2^4$$

For subsets, we have the two choices for each item and so the problem always becomes:

$$2^{\#\ in\ the\ set}$$

o) Suppose an ice cream shop offers six different toppings for their ice cream. How many different ways are there to top an order of ice cream?

Finally, we want to find the number of permutations for non-distinct objects. For instance, how many ways could you rearrange the word SHIPS. The order matters here because you don't get the same word if letters are in a new arrangement. So, with five letters, the counting principle tells us:

$$5!$$

However, the letter S repeats. If we made all the arrangements, we would get two versions of HSSPI. One would have the first S in the second spot and the second S in the third spot. Another would have the second S in the second spot and the first S in the third spot. But, when you are arranging a word, the two arrangements are not unique. Because the S's are not distinct, we are getting repeats that we don't need. The solution is to divide away the unnecessary arrangements.

$$\frac{5!}{2!} = 5 \cdot 4 \cdot 3 = 60$$

If the word were CARRIER, we would have seven total letters with three non-distinct letters and we would calculate:

$$\frac{7!}{3!}$$

And if the word were SITTERS, we would have seven total letters with two non-distinct S's and two non-distinct T's.

$$\frac{7!}{2!\,2!}$$

Both S and T have unnecessary repeats which we would need to divide away.

Work the following.

p) Find the number of arrangements of the word COOKIE.

q) Find the number of arrangements of the word MISSISSIPPI.

Algebra II

Warm Up: 9.6

To understand this section, we need to know that there is an abbreviation for the combination formula which we learned in the last activity.

$$\binom{n}{r} = C(n,r) = \frac{n!}{r!\,(n-r)!}$$

a) Multiply out the following:

$(x+y)(x+y)(x+y)$

Find the following combinations.

b) $\binom{3}{0}$

c) $\binom{3}{1}$

d) $\binom{3}{2}$

e) $\binom{3}{3}$

f) Notice that there is a connection between the coefficients in your answer for $(x+y)(x+y)(x+y)$ and the combinations you just worked. What is the connection?

Multiply out the following:

g) $(x+y)(x+y)(x+y)(x+y)$

Find the following combinations.

h) $\binom{4}{0}$

i) $\binom{4}{1}$

j) $\binom{4}{2}$

k) $\binom{4}{3}$

l) $\binom{4}{4}$

Once again, you should find the same connection between the coefficients for your answer to $(x+y)(x+y)(x+y)(x+y)$ and the combinations you've worked here.

In each of the problems above, I was having you multiply binomials. It turns out that coefficients of the terms in your answer are always combinations. If we multiply three binomials, it is a combination with three people. The first coefficient is three people taken zero at a time. The second coefficient is three people taken one at a time. The third coefficient is three people taken two at a time. The fourth coefficient is three people taken three at a time.

There is also a connection with the rest of the terms, and it leads to something called the binomial theorem.

$$(x+y)^5 = \binom{5}{0}x^5 + \binom{5}{1}x^4y + \binom{5}{2}x^3y^2 + \binom{5}{3}x^2y^3 + \binom{5}{4}xy^4 + \binom{5}{5}y^5$$

Notice how the degrees of the x's decrease and the degrees of the y's increase. This will always be the case. Expand the following using the binomial theorem.

m) $(x+y)^6$ (Hint: You will start with 6 x's and work your way down to none. You will start with zero y's and work your way up to 6.)

Look at the following.

$(2x+3y)^3$

The terms in the original binomial don't have coefficients of 1. The binomial theorem can adjust for this.

$$(2x+3y)^3 = \binom{3}{0}(2x)^3 + \binom{3}{1}(2x)^2(3y) + \binom{3}{2}(2x)(3y)^2 + \binom{3}{3}(3y)^3$$

n) In the space below, finish multiplying out the terms to get the final answer.

Try one on your own. Expand the following using the binomial theorem.

o) $(3x-y)^3$ (Hint: you will be working with $(3x)(-y)$.)

Finally, you may be asked to find only a single term of the binomial. The following equation is nothing more than a shortcut to a single term from the binomial theorem. This formula will find the $(r+1)^{th}$ term. (We start with $r=0$. It is called the $(r+1)^{th}$ because otherwise the first term would be the zeroth.)

$$\binom{n}{r} x^{n-r} y^r$$

To find the third term of $(x+y)^5$, we could use the formula. The third term would require us to remember that $r=2$:

$$\binom{5}{2} x^{5-2} y^2$$

$$\binom{5}{2} x^3 y^2$$

p) Expand that out to find the term.

To find the fourth term of $(x+y)^7$, we would use $r=3$.

$$\binom{7}{3} x^{7-3} y^3$$

q) Expand out to find the term.

Finally, we could be asked to find the sixth term of $(2x-y)^{10}$. We would use $r=5$.

$$\binom{10}{5} (2x)^{10-5} (-y)^5$$

r) Expand out to find the term.

Algebra II

Active Learning: 9.7

In this final activity, we want to look at a variety of basic probability topics.

First, we want to write the probability that some event occurs. If we rolled a six-sided dice, what is the probability of rolling a 1. In math language, we would write the statement like this:

$$P(1)$$

Probability statements are fractions which indicated the number of successes (the outcomes we are looking for) over the total possibilities. For the six-sided dice we would have:

$$P(1) = \frac{success}{total} = \frac{1}{6}$$

a) There is one way to roll a 1, and there are six possible numbers that could be rolled. The chart below is called a probability distribution (or probability model). Probability distributions are a list of all the events which are possible and their probabilities. Complete the chart by adding the probability of rolling each of the numbers listed.

Event	P(Event)
1	
2	
3	
4	
5	
6	

Basic probability statements are pretty straightforward. Give the probabilities of the following events.

b) What is the probability of getting a Head when flipping a coin?

c) What is the probability of drawing a Queen from a standard deck of cards?

d) If there are 12 marbles in a bag and 3 of them are red, what is the probability of drawing a red marble?

e) What is the probability of rolling an even number when rolling a six-sided die?

f) What is the probability of rolling a 3 or greater when rolling a six-sided die?

Before looking at our next idea, we need a reminder of some mathematical notation. A Union (∪) means Or. For instance, $a \cup b$ means we are happy when either a or b occurs. An Intersection, (∩) means And. For instance, $a \cap b$ means we are only happy when both a and b occurs.

Suppose there were 30 students in a class and the teacher asked them to raise their hand if they liked country music. 12 students raised their hand. And then, the teacher asked them to raise their hand if they liked rock music. 23 students raised their hand. Give the following probabilities:

g) $P(country) =$

h) $P(rock) =$

What about the probability $P(country \cup rock)$? We would be happy with the students who like country *or* the students who like rock. But wait! No one said that a student didn't raise their hand twice. Perhaps they did. Suppose 4 students raised their hand for both country and rock.

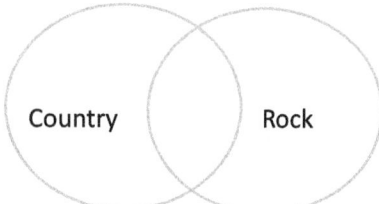

In the diagram above, there is overlap. These are the 8 students who raised their hand for both. In probability notation this would be $P(country \cap rock)$. I prefer the simpler notation $P(country \text{ and } rock)$.

What is the following probability?

i) $P(country \text{ and } rock) =$

So, if we want to find $P(country \text{ or } rock)$, we need to take into account that we have counted some people twice. The formula looks like something to memorize, but it isn't. It is simply the fact that we counted the overlap two times, so we need to subtract the overlap once.

$$P(country \text{ or } rock) = P(country) + P(rock) - P(country \text{ or } rock)$$

j) Find $P(country \text{ or } rock)$.

In a standard deck of cards, find the following probabilities.

k) $P(King) =$

l) $P(Spade) =$

m) $P(King\ or\ Spade) =$ (There is overlap here. You can have both a King and a Spade)

Returning to the illustration about the teacher asking students to raise their hand. Suppose the teacher asked students to raise their hand if they were born in the United States, and 25 students raised their hand. Then, the teacher asked students to raise their hand if they were born in Mexico, 3 students raised their hand. Find the following probabilities.

n) $P(United\ States) =$

o) $P(Mexico) =$

p) $P(U.S\ or\ Mexico) =$

In this situation, there is no overlap section. You can't be born in both the United States and Mexico at the same time. And so, this situation is referred to as Mutually Exclusive. The two events can't overlap. So, the formula for the union (or) of two mutually exclusive events drops the subtraction and just adds.

$$P(A \cup B) = P(A) + P(B)$$

q) In a standard deck of cards, what is the probability of drawing a Queen or a King? (This is mutually exclusive because you can't draw a Queen and King at the same time.)

Answer the following regarding rolling a standard six-sided die.

r) $P(Rolling\ Greater\ than\ 1) =$

s) $P(Not\ Rolling\ a\ 1) =$

t) $P(Rolling\ a\ 1) =$

u) $1 - P(Rolling\ a\ 1) =$

Because the probability of all the possible events must add to 100%, the following must be true:

$$P(Event) + P(Not\ That\ Event) = 1$$

The probability of "Not That Event" is called the Compliment. In probability terms, it is essentially the opposite. The compliment is abbreviated with a '.

$$P(Event') = P(Not\ That\ Event)$$

So:

$$P(Event) + P(Event') = 1$$

Solving this equation for $P(Event)$, we get:

$$P(Event') = 1 - P(Event)$$

If you want the compliment of an event just subtract the event from 1. Use the compliment to help you answer the following.

Suppose there were 30 students in a class and the teacher asked them to raise their hand if they liked country music. 12 students raised their hand. And then, the teacher asked them to raise their hand if they liked rock music. 23 students raised their hand.

v) $P(Not\ Liking\ Rock) = P(Rock') =$

w) $P(Country') =$

The compliment is helpful in many situations, but particularly in one like the following.

The $P(Rolling\ a\ 1\ on\ two\ Dice) = \frac{1}{36}$.

x) $P(Rolling\ 2\ or\ Greater) = $ (This is simply the compliment.)

We are going to conclude with some much more difficult problems. Read the following situation.

A bag contains 5 red marbles, 3 white marbles, and 4 yellow marbles. Find the probability that three marbles are drawn from the bag and they are all yellow.

You can do this problem using the counting principle or combinations. We will do both, but will start with the counting principle.

$P(yellow\ on\ the\ first\ draw) = \frac{4}{12}$ (There are 4 yellows and 12 marbles.)

$P(yellow\ on\ the\ second\ draw) = \frac{3}{11}$ (You already drew a yellow, so there are 3 left and 11 marbles)

$P(yellow\ on\ the\ 3rd\ draw) = \frac{2}{10}$ (You already drew 2 yellows, so there is 1 left and 10 marbles)

If you need all of these things to happen, you multiply them. Multiply the following and write it as a decimal:

$$\frac{4}{12} \cdot \frac{3}{11} \cdot \frac{2}{10} =$$

Or, you could work this problem using the definition of probability and combinations.

$$Probability = \frac{Success}{Total}$$

How many ways are there to draw three yellow marbles out of the four? That is a combination:

$$Success = \binom{4}{3}$$

How many total ways are there to draw a marble? There are 12 marbles and we are drawing 3.

$$Total = \binom{12}{3}$$

y)   Therefore, the probability definition is success divided by total. Calculate the following and write your answer as a decimal. (Remember, it is two combinations being divided.)

$$Probability = \frac{\binom{4}{3}}{\binom{12}{3}}$$

Use the same situation to answer the following. You may use whichever approach you prefer.

z) What is the probability of selecting three red marbles?

aa) What is the probability of selecting zero red marbles?

bb) What is the probability of selecting at least one red marble? (This is a compliment problem.)

Algebra II

Answer Key

1.1a

a) $\frac{121}{1}, \frac{12}{1000}, -\frac{6}{1}, \frac{8}{10}$

b) Rational: $4.\overline{125}, 6.25$

Irrational: 2.71828..., 7.1542...

c) Rational: $\sqrt{4}, \sqrt{225}$

Irrational: $\sqrt{48}, \sqrt{35}$

d) d, f

e) e, f

f) d, f

g) c, d, f

h) e, f

i) a, b, c, d, f

j) a, b, c, d, f

k) b, c, d, f

l) None

m) 49

n) 14

o) 34

p) 46

q) 4

r) 3

s) 6

1.1b

a) Both ae 14

b) Both are 21

c) Both are 15

d) Both are 80

e) $5x + 15$

f) 9

g) 7

h) 0

i) 1

j) Constant: 5   Variable: x

k) Constant: $\frac{1}{4}$ Variables: x, y

l) Constant: 7   Variable: m

m) Constant: 6   Variables: r, s

n) 19

o) 75

p) -14

q) 7

r) $\frac{1}{10}$

s) 20

t) $v = 150 \, in^3$

u) $8x + 2y$

v) $-7xy + 3x - 15y$

w) $24m + 14 - 31n$

x) $P = 2l + 2w$

y) $Surface\ Area = 6a^2$

1.2a

a) x^{13}

b) y^{28}

c) z^{10}

d) x^7

e) y^2

f) a^{-3}

g) x^{12}

h) y^{20}

i) z^{-4}

j) 1

k) 1

l) 1

m) $\frac{1}{x^7}$

n) $\frac{1}{y^6}$

o) $\frac{1}{a^4}$

p) $x^{12}y^{15}$

q) $a^{14}b^{21}$

r) $x^{-6}y^{10}$

s) $9y^6$

t) $\frac{x^{15}}{y^{20}}$

u) $\frac{8}{z^{12}}$

v) $\frac{m^4}{n^{24}}$

w) a^6

x) $\frac{4x^4}{y^6}$

y) $\frac{1}{x^3}$

z) $\frac{27y^{12}}{x^6}$

aa) $\frac{b^9}{a^6}$

1.2b

a) 100

b) 10000

c) 15400

d) 1540000

e) .01

f) .0001

g) 1.54

h) .0154

i) 564700

j) 93200000

k) .00176

l) .000002951

m) 3.156×10^7

n) 5.1067×10^9

o) 2.15×10^{-5}

p) 7.088×10^{-6}

q) 10^8

r) 10^9

s) 10^{-8}

t) 10^5

u) 8.32×10^{10}

v) 3.382×10^6

w) 3.4008×10^{-2}

x) 5.175×10^{-11}

y) 10^3

z) 10^{-10}

aa) 10^6

bb) 2.269×10^3

cc) 1.822×10^{-3}

dd) 1.727×10^{-1}

1.3a

a) 3

b) 13

c) 2

d) $4x$

e) $9x^2y$

f) $5m\sqrt{2}$

g) $4xy\sqrt{5x}$

h) 7

i) $2\sqrt{5}$

j) 24

k) $\frac{2}{11}$

l) $\frac{5\sqrt{2}}{9}$

m) $\frac{3x}{4y^2}$

n) $\frac{2}{3}$

o) $5x^2$

p) $\frac{x^4\sqrt{3}}{y}$

q) $21\sqrt{5}$

r) $4\sqrt{7}$

s) $-2\sqrt{11} - 2\sqrt{17}$

t) $11\sqrt{v}$

u) $6\sqrt{2}$

v) $9x\sqrt{x}$

w) $27x\sqrt{3}$

1.3b

a) $\frac{7\sqrt{5}}{15}$

b) $\frac{2\sqrt{3}}{3}$

c) $\frac{5\sqrt{2}}{4}$

d) $\frac{6+8\sqrt{2}}{-23}$

e) $\frac{5\sqrt{2}-2\sqrt{6}}{13}$

f) $\frac{7\sqrt{x}-2\sqrt{15x}}{-11}$

g) 2

h) x

i) 2x

j) $\frac{2x^2}{3}$

k) $y^{\frac{5}{3}}$

l) $y^{\frac{3}{5}}$

m) $x^{-\frac{3}{2}}$

n) 27

o) $20x^{\frac{7}{8}}$

p) $40y^{\frac{19}{12}}$

1.4a

a) 5

b) 7

c) $13x^3 + 16x^2 + 10x - 10$

d) $3x^4 + 5x^3 - 9x^2 + 10x - 28$

e) $2x^3 + 6x^2 + 4x + 6$

f) $x^3 + 4x^2 + 2x + 6$

g) $18x^3 - 24x^2 + 48x$

h) $30x^5 + 15x^3 - 50x^2$

i) $21x^3 + 54x^2 + 94x + 40$

j) $6x^3 - 28x^2 + 9x + 35$

k) $x^4 + 8x^3 - x^2 + 24x - 12$

1.4b

a) $x^2 + 18x + 81$

b) $y^2 - 10y + 25$

c) $9x^2 - 36x + 36$

d) $a^2 - 4ab + 4b^2$

e) $x^2 - 49$

f) $y^2 - 9$

g) $9x^2 - 49$

h) $16a^2 - 25b^2$

i) $12x^2 - 6xy + 38x - 10y + 30$

1.5a

a) $7xy(2xy - 4x + 3)$

b) $4a(3a^2 + 2a + 1)$

c) $(x + 10)(x + 2)$

d) $(x - 7)(x - 5)$

e) $(2x - 1)(3x - 4)$

f) $(4x - 1)(2x + 3)$

g) $(x + 5)(x + 5)$

h) $(x - 11)(x - 11)$

i) $(x - 11)(x + 11)$

j) $(3x - 2)(3x + 2)$

1.5b

a) $(x - 2)(x^2 + 2x + 4)$

b) $(y + 4)(y^2 - 4y + 16)$

c) $(x - y)(x^2 + xy + y^2)$

d) $x^{-\frac{1}{2}}(1 + x)$

e) $x^{-\frac{2}{3}}(1 + x)$

f) $(x + 5)^{-\frac{2}{3}}(9x + 35)$

g) $(x - 9)^{-\frac{1}{5}}(2 + 3x^2 - 27x)$

1.6a

a) $\frac{5}{z-2}$

b) $\frac{2x-3}{x(x-3)}$

c) $\frac{x}{6}$

d) $\frac{2x-3}{2x+3}$

e) $\frac{7}{x+4}$

1.6b

a) $\frac{5x+9}{2(x+4)(x+3)}$

b) $\frac{3x-34}{(x+5)(x-2)}$

c) $\frac{1}{2(x+4)}$

d) $\frac{1}{y(x-y)}$

2.1a

a)

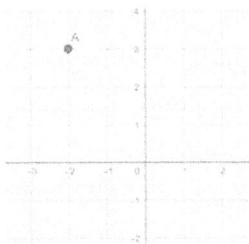

b)

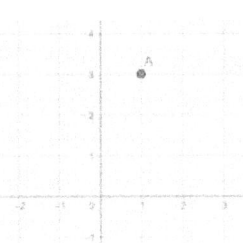

c)

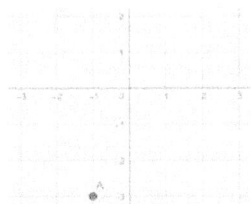

d) $(-1,-7); (2,5)$

e)

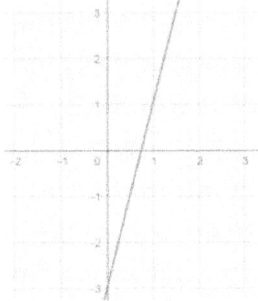

f) Your points may very: $(0,-4); (1,-1); (2,2)$

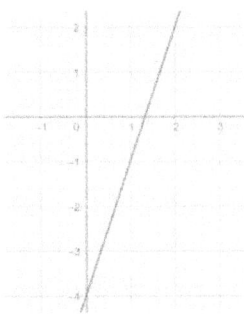

g) $(-1,0); (0,2)$

h) $(0,2); (-1,0)$

i) $(-6,0); (0,3)$

2.1b

a) 5

b) 5

c) 3

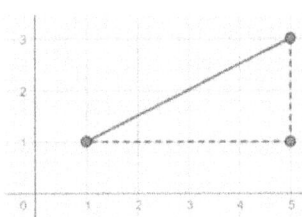

d) 2

e) $(8,9)$

f) $(2,-2)$

g) hypotenuse

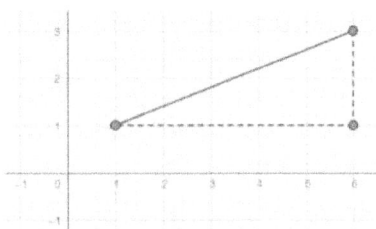

h) 5

i) 2

j) $\sqrt{29}$

k) $c = \sqrt{a^2 - b^2}$

l) 5

m) $\sqrt{52}$

2.2a

a) $x = 4$

b) $y = -\frac{25}{9}$

c) $x = -1$

d) $x = 51$

e) $x = -3$

2.2b

a) $A: (3, 3); B: (-1, -1)$

b) $m = 1$

c) $m = -3$

d) $y = \frac{1}{2}x - 4$

e) $y = -1x + 2$

f) $10y - 2x = -35$

g) $\frac{3-0}{2-2}$

h) The denominator will always be equal to zero since the x values are the same.

i) $\frac{3-3}{4+2}$

j) The numerator will always be equal to zero since the y values are the same.

2.2c

a) $y = -\frac{2}{3}x + \frac{7}{3}$

b) $y = -\frac{2}{3}x + \frac{1}{3}$

c) There slopes are the same.

d) $y = \frac{4}{3}x - 1$

e) $y = -\frac{3}{4}x + \frac{11}{4}$

f) The slopes are negative inverses of each other.

g) $m = \frac{4}{3}$

h) The slopes are the same.

i) $m = \frac{4}{3}$

j) $y = \frac{4}{3}x - 2$

k) $m = -\frac{2}{3}$

l)

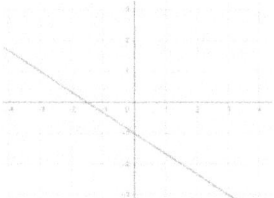

m) $m = \frac{1}{2}$

n) $y = \frac{1}{2}x$

o) $m = -\frac{2}{3}$

p) It is the negative inverse.

q) $m = \frac{3}{2}$

r)
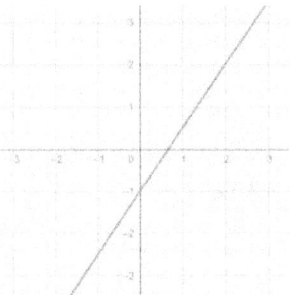

s) $m = \frac{1}{3}$

t)
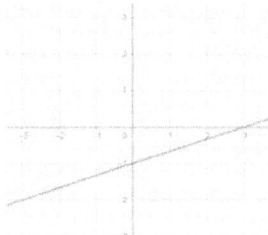

u) $m = -2$

v) $y = -2x + 5$

2.3a

a) $Cost = .15(Kwh) + 20$

b) $Cost = .25(Kwh) + 10$

c) Beta Electric

d) 100 kilowatt hours

e) $d = 2r;\ d = 2.5(r - 10)$

f) $r = 50$

g) 55 miles to the beach; 50 miles home

	rate	time	distance
To The Beach	r	5	d
Home	r-5	5.5	d

2.3b

a) $L = 40\ meters$

b) $L = 25\ feet$

c) $w = 35\ yards$

d) $w = 70\ meters$

e) $w = 10\ feet$

f) $w = 5\ meters$

g) $w = 10\ cm$

h) $l = 7.75\ inches$

2.4a

a) There are no two identical numbers which multiply to get a negative.

b) $11i$

c) $6i$

d) $2\sqrt{2}i$

e) $2\sqrt{6}i$

f) real: -5; complex: 17

g) real: 14; complex: -7

h) complex: 9

i)

j)

k)
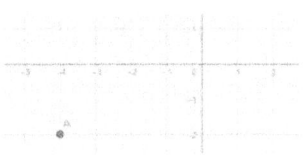

l) $12 + 12i$

m) $8 - 9i$

n) $1 - 12i$

o) $-6 - 9i$

p) $20 + 26i$

2.4b

a) $-9 - 9i$

b) $21 + 18i$

c) $6x + 15$

d) $-60y + 30$

e) distributive property

f) $-12 + 80i$

g) $20 - 30i$

h) $-18 + 12i$

i) $x^2 - x - 6$

j) $y^2 - 13y + 42$

k) F.O.I.L.

l) $\sqrt{-1} \cdot \sqrt{-1} = -1$

m) $-22 - 7i$

n) $34 - 8i$

o) 109

p) $\sqrt{-1} \cdot \sqrt{-1} \cdot \sqrt{-1} = -1 \cdot i = -i$

q) $\sqrt{-1} \cdot \sqrt{-1} \cdot \sqrt{-1} \cdot \sqrt{-1} = -1 \cdot -1 = 1$

r) $i^1 = i$

s) $i^2 = -1$

t) $i^3 = -i$

u) $i^0 = 1$

2.4c

a) $19 + 20i$

b) $9 - 38i$

c) $3x - 5$

d) $\frac{4}{3}x + 8$

e) $3 + 7i$

f) $3 + 4i$

g) $\frac{5}{3} + \frac{23}{3}i$

h) $121 + \sqrt{47}$

i) $19 - \sqrt{13}$

j) $x^2 - 9$

k) $4y^2 - 16$

l) $9x^2 - 25$

m) -1

n) $21 - 11i$

o) $-3 + 26i$

p) $10 - i$

q) $2i$

r) 25

s) 29

t) 58

u) $\frac{15-5\sqrt{2}}{7}$

v) $\frac{26-7i}{25}$

w) $\frac{7-22i}{41}$

2.5a

a)

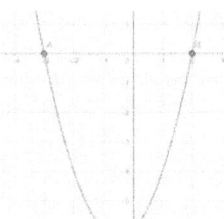

b)

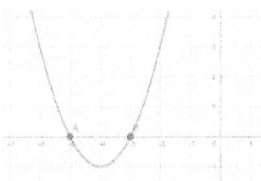

c) $(-5)^2 + 8(-5) + 15 = 25 - 40 + 15 = 0;\ (-3)^2 + 8(-3) + 15 = 9 - 24 + 15 = 0$

d) $x = 6; x = -1$

e) $x = 7; x = -3$

f) $x = -\frac{3}{4}; x = -3$

g) $x = -\frac{1}{4}; x = -\frac{2}{3}$

h) $x = 0; x = -4; x = 2$

i) ± 3

j) ± 11

k) ± 9

l) $x = \pm 4$

m) $x = \pm 2\sqrt{2}$

n) $x = \pm 5$

o) $x = 4 \pm \sqrt{5}$

p) $x = -2 \pm \sqrt{3}$

q) $x = 3 \pm \sqrt{22}$

2.5b

a) $x = \frac{-7 \pm \sqrt{41}}{2}$

b) $x = \frac{-1 \pm \sqrt{11}i}{2}$

c) $x = \frac{-3 \pm \sqrt{63}i}{18}$

d) 0

e) 100

f) 61

g) -80

h) 13 - two answers involving square roots.

i) -23 – two imaginary solutions

j) 0 – one rational solution

2.6a

a) x

b) y

c) $z^{\frac{1}{2}}$

d) $r^{\frac{1}{3}}$

e) $x^{\frac{1}{4}}$

f) 16

g) 32

h) $x = 8$

i) $y = 25$

j) $x = 27$

k) $x = 81$

l) $x = 127$

m) $x = 0; x = 1$

n) $x = 0; x = \frac{1}{16}$

o) $x = 0; x = 3; x = -3$

p) $x = 0; x = 3$

q) $x = 0$ does not work and should be discarded.

r) $x = \frac{5}{4}$

s) $x = 2$

2.6b

a) $x = 6; x = 2$

b) $x = 2; x = -3$

c) $x = 5; x = -13$

d) $x = 2$

e) No Solution

f) $x = \pm\sqrt{3}i; x = \pm\sqrt{2}$

g) $x = -5; x = 1$

h) $x = 51$

i) No solution. $x = -1$ but this value is undefined.

2.7a

a) $(-\infty, 3)$

b) $[12, \infty)$

c) $[-2, 7)$

d) $(-9, 11)$

e) $(-2, \infty)$

f) $[-13, \infty)$

g) $(-\infty, 96]$

h) cat, horse, dog

i) cat, horse, dog

j) mouse, cat, horse, dog, frog, cow, elephant, deer, pig, bird

k) $(-1, 6)$

l) $(-\infty, 2] \cup [10, \infty)$

m) No solution

n) $(-\infty, \infty)$

o) $(24, \infty)$

p) $(-\infty, -4] \cup (21, \infty)$

q) $(7, 15)$

r) $(8, 14]$

2.7b

a) $(-\infty, 1] \cup [6, \infty)$

b) $(-10, 8)$

c) $(-2, 6)$

d) $(-\infty, -2) \cup (1, \infty)$

e) $x = 6; x = -2$

f) $x = 1; x = -2$

g)

False ... *True* ... *False* (number line from -10 to 10)

h) $10 < 8$; False

i) $4 < 8$; True

j) $12 < 8$; False

k) $(-2, 6)$

l)

True ... *False* ... *True* (number line from -10 to 10)

m) $14 > 6$; True

n) $2 > 6$; False

o) $10 > 6$; True

p) $(-\infty, -2) \cup (1, \infty)$

q) $[-3, 5]$

r) $(-\infty, -2) \cup (3, \infty)$

s) c

t) $(-\infty, \infty)$

u) a

v) No solution. $\emptyset$

w) (2)

x) 7

y) $(-\infty, 7) \cup (7, \infty)$

z) $\emptyset$

aa) $(-\infty, 5) \cup (5, \infty)$

bb) (-7)

cc) $(-\infty, \infty)$

dd) $(-\infty, -\frac{5}{2}) \cup (\frac{11}{2}, \infty)$

ee) $|x - 3| \leq 5$

ff) $|x - 2| \leq 10$

3.1a

a) $y = -2x + 4$

b) $y = \pm\sqrt{16 - x^2}$

c)

x	y
-1	6
0	4
1	2
2	0
3	-2

d) At the end of each arrow is the y value from the table.

e)

x	y
-2	± 2.2
-1	± 2.8
0	± 3
1	± 2.8
2	± 2.2

f) At the end of each arrow are the two y values from the table, one positive and one negative.

g) The circle has two y's for each value of x.

h) A function has one value of y for each value of x.

i) Yes, price is a function of car.

j) Yes, cost is a function of item.

k) No, item is not a function of cost.

l) 13

m) 23

n) -12

o) 2

p) 18

q) 6

r) -2

s) 14

t) $a^2 + 3a + 1$

u) $a^2 - 2ah + h^2 + 3a - 3h + 1$

3.1b

a) function

b) not a function

c) not a function

d) function

e) They have the same value of y.

f) Not one-to-one

g) one-to-one

h) one-to-one

i) Not one-to-one

j) $f(1) = 1$

k) $f(1) = 1$

l) $f(-1) = 1$

m) $f(2) = 4$

n) $f(2) = 2$

o) $f(1) = 1, f(-1) = 1$

p) $f(-2) = 4; f(2) = 4$

3.2a

a) This has a mathematical problem. It is undefined.

b) This does not have a problem. You can have a negative under a cube root.

c) This has a mathematical problem. You can't have a negative under a square root.

d) This does not have a problem.

e) Values less than zero would "break" this function.

f) There are no values which would "break" this function.

g) A value of zero would "break" this function.

h) $(-\infty, \infty)$

i) $(5, \infty)$

j) $[5, \infty)$

k) $(-\infty, 5)$

l) $(-\infty, 5]$

m) Values less than 5 would cause a negative under the root.

n) The value -3 would cause the denominator to be undefined.

o) $(-\infty, 8) \cup (8, \infty)$

p) $(-\infty, 4) \cup (4, \infty)$

q) $(-\infty, 2) \cup (2, \infty)$

r) $[3, \infty)$

s) $(-\infty, 3]$

t) $(\frac{5}{2}, \infty)$

u) $[3, 4) \cup (4, \infty)$

v) $(-\infty, -1] \cup (1, \infty)$

w) $(-\infty, -1) \cup [1, \infty)$

x) $(-\infty, -1] \cup (1, \infty)$

y) $[-1, \infty)$

3.2b

a)

b)
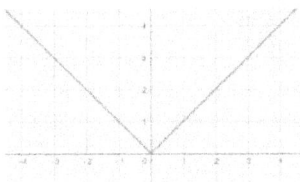

c)
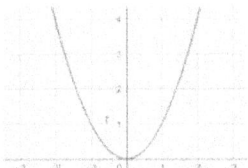

d)
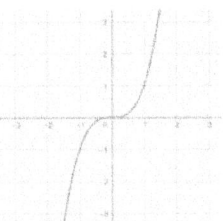

e)
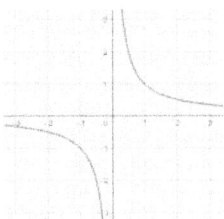

f)
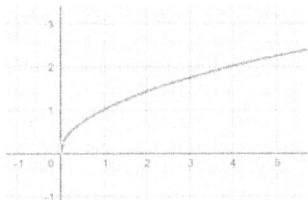

g)
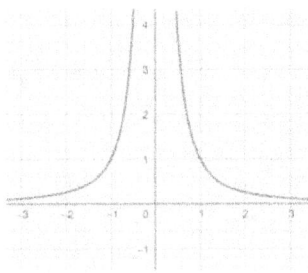

h)
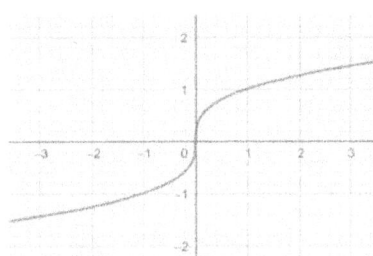

i)
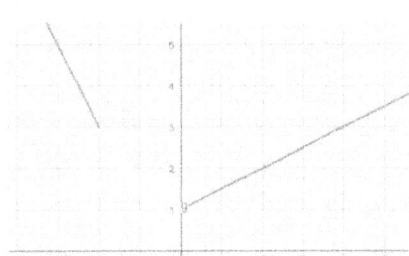

j)
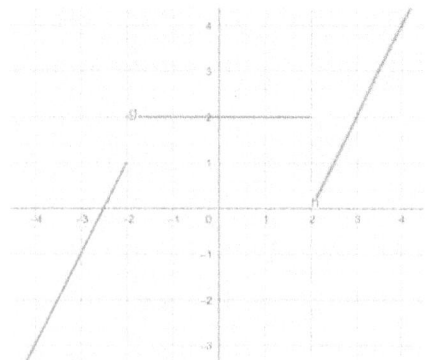

k) $f(5) = 13$

l) $f(0) = 5$

m) $f(10) = 28$

n) $f(-3) = 3$

o) $f(-2) = 2$

p) $f(1) = 5$

q) $f(2) = 4$

3.3a

a) $(2, 4)$

b) $(4, 1)$

c) $m = -\frac{3}{2}$

d) $b = 7$

e) $y = -\frac{3}{2}x + 7$

f) $(2, 4)$

g) $(4, 1)$

h) $avg\ rate\ of\ change = -\frac{3}{2}$

i) $(1, -2)$

j) $(4, 13)$

k) $avg\ rate\ of\ change = 5$

l) $avg\ rate\ of\ change = \frac{3}{4}$

m) $a + 3$

n) $a + 3$

3.3b

a) $(0, 0)$

b) $(-1.4, -4); (1.4, -4)$

c) $(-1.4, 0) \cup (1.4, \infty)$

d) $(-\infty, -1.4) \cup (0, 1.4)$

e) local maxima

f) absolute minima

g) $(-\infty, -1) \cup (0, 1)$

h) $(-1, 0) \cup (1, \infty)$

i) $(-1, 1); (1, 1)$

j) None

k) None

l) $(0, 0)$

m) $(0, 1)$

n) $(-2, 0)$

o) $(-2, 4)$

p) $(0, 0)$

q) $(1, 1)$

r) None

3.4a

a) $x - 5$

b) $5x$

c) $\frac{x}{5}$

d) $x^2 + x$

e) $x^2 - x$

f) x^3

g) x

h) $x^2 + x - 6$

i) $x^2 - x - 2$

j) $x^3 - 2x^2 - 4x + 8$

k) $x + 2$

l) $[-2, \infty)$

m) $(-\infty, \infty)$

n) $x\sqrt{x+2}$

o) $[-2, \infty)$

p) $\frac{\sqrt{x+2}}{x}$

q) 0

r) $[-2, 0) \cup (0, \infty)$

s) 3

t) 36

u) 0

v) -18

w) 2

x) -13

y) 2

z) -7

aa) 0

bb) -3

cc) 2

dd) 4

ee) 1

ff) 3

3.4b

a) 10

b) 34

c) 34

d) 27

e) 5

f) 5

g) $2x + 6$

h) $2x + 5$

i) $(-\infty, \infty)$

j) 2

k) $(-\infty, 2) \cup (2, \infty)$

l) $(-\infty, 3) \cup (3, 5) \cup (5, \infty)$

m) $[-3, \infty)$

n) $f(x) = \sqrt{x}; g(x) = \frac{1}{x}$

o) $f(x) = \sqrt{x}; g(x) = 3 + x$

p) $f(x) = \sqrt{x+1}; g(x) = \frac{x^2}{2-x}$

3.5a

a)

b)

c)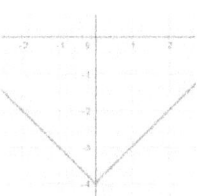

d)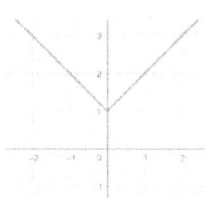

e) Adding a number to the end of the function will shift the function up. Subtracting a number at the end of the function will shift the function down.

f)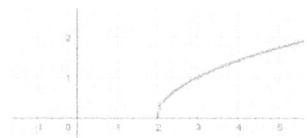

g)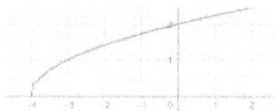

h)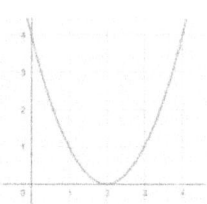

i)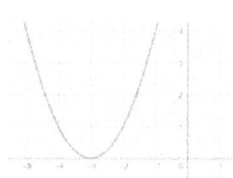

j) left

k) right

l) $f(x) = |x| - 3$

m) $f(x) = (x - 2)^2$

n) $f(x) = (x + 5)^3$

o) $f(x) = \sqrt{x} + 10$

p) Shift the function 5 to the left.

q) Shift the function up 5.

r) Shift the function three to the right and up two.

3.5b

a)

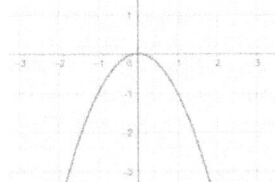

b)

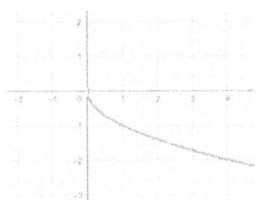

c) Flipped over the x-axis

d)

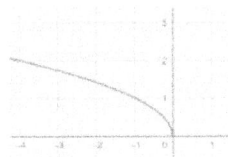

e) Flipped over the y-axis

f) $f(x) = -|x+3| - 2$

g) $f(x) = x^3 - 2$

h) $f(x) = \sqrt{-x} + 1$

i) $f(x) = \sqrt{-x} + 5$

j) $f(x) = -|x-2| + 1$

k) $f(x) = \frac{1}{(x+3)^2} + 6$

3.5c

a)

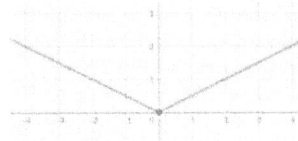

b)

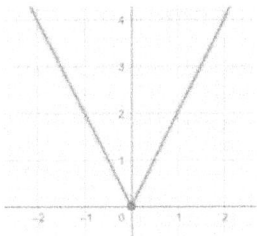

c) $g(x)$

d) $f(x)$

e)

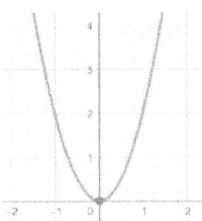

f)

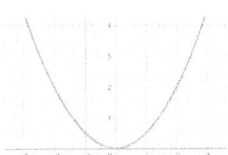

g) $h(x)$

h) $r(x)$

i)

j)

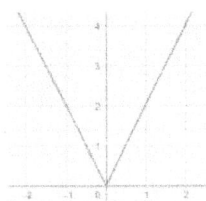

k) $f(x)$

l) $g(x)$

m)
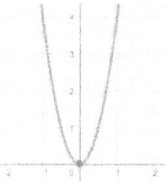

n)
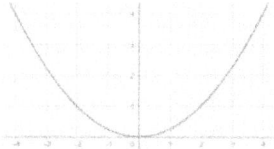

o) $r(x)$

p) $h(x)$

q)
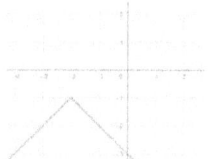

r)
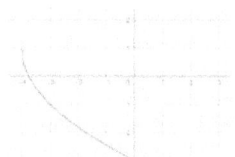

s) $f(x) = -(x-2)^3 + 3$

t) $f(x) = .25|x| - 4$

u) $f(x) = -3\sqrt[3]{x+1}$

v) $f(x) = x^2$

w) The same function.

x) $f(x) = x^4 + x^2$

y) The same function.

z) They are symmetrical

aa) It is -1 times the original function.

bb) $f(x) = -x^3 - x$

cc) It is -1 times the original function.

dd) They are symmetrical

ee) odd

ff) even

gg) even

hh) neither

<u>3.6</u>

a) $f(x) = -|x+4| + 7$

b) $f(x) = 2|x| - 5$

c) $f(x) = \left|\frac{1}{2}x\right|$

d)
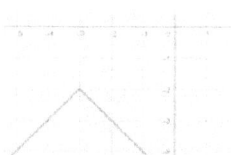

e)
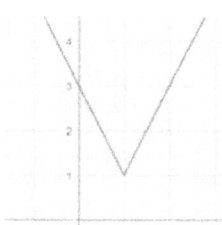

f) $f(x) = -|x+3| - 2$

g) $f(x) = 3|x-1| + 2$

h) Left 1; Flipped vertically; Down 3

i) $a = 2$

j) $x = 1$ and $x = -9$

k) $x = 1; x = 7$

l) $x = -2$ and $x = -10$

m) $x = 8$ and $x = -5$

n) $x = -5$ and $x = 4$

o) $|x + 6| = -6$; No solution

3.7a

a) x

b) x

c) x

d) x-2

e)

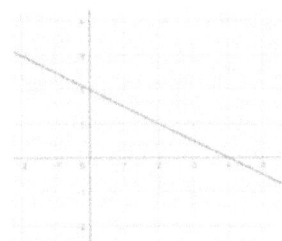

f) $y = x$

g) $(-\infty, 4)$

h) $(2, \infty)$

i) $f(1) = 2$

j) $f^{-1}(8) = 3$

k) $f(2) = -4$

l) $f^{-1}(-8) = 3$

m) $f^{-1}(-2) = 1$

n) $f(2) = 8$

o) $f(4) = 16$

p) $f^{-1}(4) = 1$

q) $f^{-1}(8) = 2$

3.7b

a) $y = 2x + 9$

b) $y = \frac{1}{x} + 3$

c) $y = (x - 5)^2 + 2$

d) $f^{-1}(x) = 2x + 9$

e) $f^{-1}(x) = \frac{1}{x} + 3$

f) $f^{-1}(x) = (x - 5)^2 + 2$

g) $f^{-1}(x) = x^3 - 3$

h) $f^{-1}(x) = \frac{2}{x-8} + 6$

i) $f^{-1}(x) = \frac{5x}{1+x}$

j) $f^{-1}(x) = \frac{-3x-2}{1-x}$

4.1a

a) Constant

b) Positive

c) Negative

d) Positive

e) Constant

f) Negative

g) $f(x) = 50x + 100$

h) $f(x) = 150 - 5x$

i) $f(x) = 25$

j) 200 per year

k) $f(x) = -\frac{3}{2}x + 2$

l) $f(x) = \frac{5}{2}x - 1$

m) $C(x) = 10x + 100$

n) $C(x) = 10 + 5x$

o) $340

p) $85

q) $S(x) = 110x + 2370$

r) $3140

4.1b

a)

x	f(x)
0	2
-2	5
2	-1

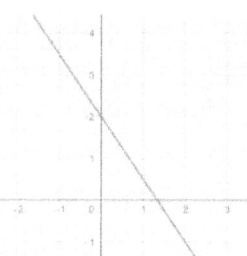

b)

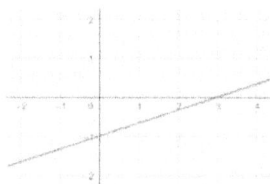

c) $(\frac{8}{3}, 0)$

d) $f(x) = 4$

e) $x = -2$

f)

g)

h)

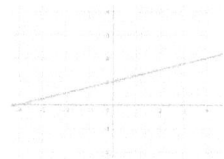

4.1c

a) Perpendicular

b) Neither

c) Parallel

d) $f(x) = -\frac{3}{2}x - 2$

e) $f(x) = -2x + 1$

f) $f(x) = -\frac{5}{2}x + 2$

4.2

a) $f(x) = \frac{35}{14}x + 25$

b) $45

c) $95

d) $Hefty: f(x) = .45x + 15; Speedy: f(x) = .35x + 20$

e) 50 miles

f) $127.50

g) $107.50

h) $d(t) = 13t$

i) 32.5 miles

4.3a

a) Linear

b) Positive

c) $r = .9$

d) $r = .9562$

e) Positive

f) Strong

g)

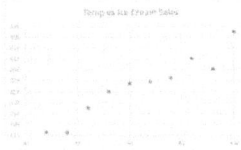

h) Answers could vary slightly. $m = 10$

4.3b

a) $\hat{y} = -55.29 + 3.96x$

b) 43.71

c) -47.37

d) A 2 month old child is predicted to know -47.37 words, which is impossible.

e) $\hat{y} = -567.17 + 9.079x$

f) $222.70

5.1a

a) $(2, 1)$

b) Vertex: $(3, 3)$

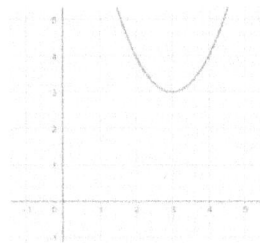

c) $(1, -2)$

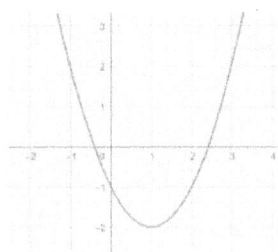

d) Vertex: $(-2, 1)$

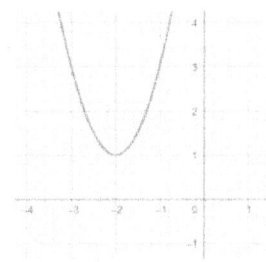

e) Vertex: $(-3, -1)$

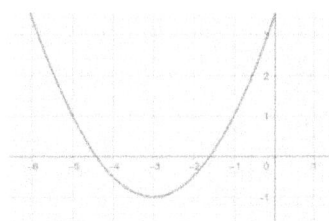

f) (h, k)

g) $h = -2$

h) $k = -3$

i) $(4, -46)$

j) Domain: $(-\infty, \infty)$ Range: $[-3, \infty)$

k) Domain: $(-\infty, \infty)$ Range: $(-\infty, 2]$

l) Domain: $(-\infty, \infty)$

m) Range: $(-\infty, 2]$

n) $(-1, -3)$

o) $a = 2$

p) $f(x) = 2(x + 1)^2 - 3$

q) $f(x) = -(x - 2)^2 + 5$

5.1b

a) $(0, -8)$

b) $\left(\frac{2}{3}, 0\right), (-4, 0)$

c) $\left(\frac{2}{5}, 0\right), (-1, 0)$

d) $h = 25$

e) $k = 1250$

f) $m = -4$

g) $Q = -4p + 900$

h) $Revenue = -4p^2 + 900p$

i) Max Price= 112.5; Max Revenue= 50625

5.1c

a) y-intercept: $(0, -2)$

b) $(-2 + \sqrt{6}, 0); (-2 - \sqrt{6}, 0)$

c) y-intercept: $(0, -3)$; x-intercepts: $(\frac{-3+\sqrt{33}}{4}, 0)$ and $(\frac{-3-\sqrt{33}}{4}, 0)$

d) $h = 1.2\ seconds$

e) $k = 44.96\ feet$

f) $t = -.427, t = 2.927$

g) $t = 2.927\ seconds$

5.2a

a) All are power functions except the last.

b) binomial

c) polynomial

d) monomial

e) trinomial

f) degree seven

g) degree three

h) degree three

i) degree two

j) $-12x^5$

k) -12

l) x^4

m) 1

5.2b

a) Positive first term; Even exponent

b) Negative first term; Even exponent

c) Positive first term; Odd exponent

d) Negative first term; Odd exponent

e) Diagonally downward

f) Opening downward

g) Opening upward

h) Diagonally upward

i) Diagonally upward

j) Opening upward

k) Diagonally downward

l) $(0,0); (2,0); (-3,0)$

m) $(0,0)$

n) x-intercepts: $(0,0); \left(\frac{3}{2},0\right); (-6,0)$ y-intercept: $(0,0)$

o) $(x-2)(x+2)(x-3)(x+3)$

p) x-intercepts: $(2,0); (-2,0); (3,0); (-3,0)$ y-intercept: $(0,36)$

q) 7

r) 5

s) 5

t) 3

u) 6

v) 4

w) 4

5.3a

a)

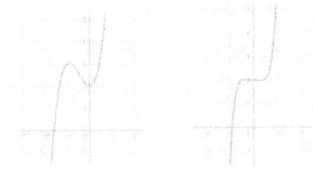

b) $x = 0; x = \pm\sqrt{3}; x = \pm\sqrt{2}$

c) $x = 0; x = \pm\sqrt{5}; x = \pm 2$

d) $x^2(x-4) - 4(x-4)$

e) $(x^2 - 4)(x - 4)$

f) $x = \pm 2, x = 4$

g) $(0,0), (3,0), (4,0)$

h) $(0)^3 - 7(0)^2 + 12(0) = 0; (3)^3 - 7(3)^2 + 12(3) = 0; (2)^3 - 7(2)^2 + 12(2) = 0$

i) $0 \times 2; -2 \times 3; 1 \times 1$

j) $0 \times 1; -12 \times 2; \frac{2}{3} \times 2$

k) $-1 \times 3; 0 \times 2; 1 \times 1$

l) $-1 \times 3; 0 \times 1; 1 \times 1$

m) $-1 \times 1; 0 \times 3; 1 \times 2$

5.3b

a) a

b) Diagonally downward

c) Opening downward

d) 4

e) 5

f) $0 \times 3; 4 \times 1; -5 \times 2$

g) $0 \times 2; -7 \times 1; 2 \times 2$

h) $0 \times 1; 1 \times 3; -3 \times 2$

i)

j)

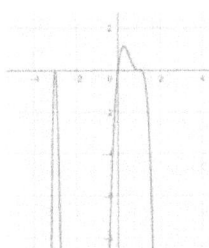

k) $a = -2$

l) $f(x) = -2(x-1)^2(x+2)$

5.4a

a) $21\ r7$

b) $x^3 + x^2 - x + 1 + \frac{1}{x}$

c) $x^4 - x^3 + x^2 + x - 1 + \frac{2}{x}$

d) $3x^3 + x^2 - 2x - \frac{2}{x-2}$

e) $4x^3 - 2x^2 - x - 3 + \frac{7}{2x+1}$

5.4b

a) $x = -3; x = 1$

b) $2x^2 - 4x + 4 + \frac{9}{x-1}$

c) $6x^2 + 15x + 22 + \frac{46}{x-2}$

d) $2x^3 - 8x^2 + 27x - 111 + \frac{453}{x+4}$

e) $8x^3 + 16x^2 + 32x + 64 + \frac{132}{x-2}$

5.5a

a) $x^2 - 4$

b) Remainder 0

c) $x^2 - 4$

d) $x^2 + 5$

e) Remainder 0

f) $x^2 + 5$

g) $-6x^4 - 2x^2 - 10x - 4 - \frac{6}{x-1}$

h) It is not a factor. The remainder was not zero.

i) $f(1) = -6$

j) They are the same.

k) 112

l) 112

m) They are the same.

n) 17

o) 0

p) Problem o involved a factor since the remainder was zero.

5.5b

a) -12

b) -6

c) 0

d) 0

e) 30

f) 0

g) 84

h) -12

i) 288

j) -96

k) 2100

l) -1260

m) 2, -2, -3

n) -3, -2, 1

o) 2, -1

p) 2, -1. -3

q) -5, -1, 1

r) $\frac{-6 \pm \sqrt{-108}}{8}$

s) $2, \pm i$

5.5c

a) $1 \pm \sqrt{3}i$

b) $f(x) = -2(x-2)(x+3i)(x-3i)$

c) $f(x) = 2(x-1)(x-2i)(x+2i)$

d) $+: 4, 2, 0\ -: 0$

e) $+: 2, 0\ -: 2, 0$

5.6a

a) $x = 7; x = -5$

b) $\dfrac{x+1}{(x-7)(x+5)}$

c) The vertical asymptotes are the values which make the denominator undefined.

d) $y = \dfrac{4}{5}$

e) $y = -\dfrac{2}{5}$

f) The coefficients in front of the leading terms in the numerator and denominator.

g) The degree of the polynomial in the denominator is higher.

h) $y = 2x + 7$

i) $2x + 7 + \dfrac{12}{x-3}$

j) .01204

k) -.01196

l) .0012

m) -.0012

n) .00012

o) -.00012

p) The value is becoming smaller and close to zero.

q) $x = 1; x = -1$

r) $\dfrac{1}{x+1}$

s) $x = 1$

t) The discontinuity is at $x = 2$

5.6b

a)

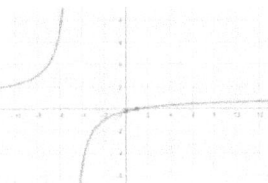

b)

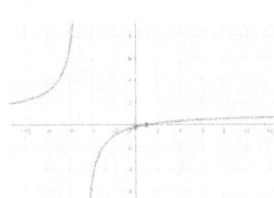

c)

d) $x = 3; x = -3$

e) $y = 0$

f) $(0, 0)$

g) $(0, 0)$

h) Yes, at $(0, 0)$

i) $a = 2$

j) $f(x) = 2\dfrac{(x-2)(x+3)}{(x+2)(x+5)}$

k) $f(x) = 3\dfrac{(x-2)(x+6)}{(x-3)(x+3)}$

5.7a

a) $g(f(x)) = \frac{1}{\frac{1}{x-5}} + 5 = x - 5 + 5 = x$

b) $g(f(x)) = 2\frac{(x-3)}{2} + 5 = x - 3 + 5 = x + 2$

c) $f^{-1}(x) = \sqrt[3]{4x - 2}$

d) $f^{-1}(x) = x^3 + 6$

5.7b

a) $(-\infty, \infty)$

b) $[0, \infty)$

c) They fail the horizontal line test.

d) $(-\infty, 0]$

e) $[0, \infty)$

f) $[0, \infty)$

g) $[0, \infty)$

h) They will reverse.

i) $[3, \infty)$

j) $[0, \infty)$

k) $[0, \infty)$

l) $[3, \infty)$

m) $[5, \infty)$

n) $[0, \infty)$

o) $[0, \infty)$

p) $[5, \infty)$

q) $(-\infty, -3]$

r) $[-2, \infty)$

s) $[-2, \infty)$

t) $(-\infty, -3]$

u) $[2, \infty)$

v) $[0, \infty)$

w) $f^{-1}(x) = \pm\sqrt{x} + 2$

x) $[0, \infty)$

y) $[2, \infty)$

z) $f^{-1}(x) = \sqrt{x} + 2$ for $x \geq 0$

aa) $(-\infty, -3]$

bb) $[-2, \infty)$

cc) $h^{-1}(x) = \pm\sqrt{x+2} - 3$

dd) $[-2, \infty)$

ee) $(-\infty, -3]$

ff) $h^{-1}(x) = -\sqrt{x+2} - 3$ for $x \geq -2$

gg) $[2, \infty)$

hh) $[0, \infty)$

ii) $f^{-1}(x) = x^2 + 2$ for $x \geq 0$

5.7c

a) $r = \sqrt[3]{\frac{3V}{4\pi}}$

b) $r = 6.9$

c) $n = \frac{100c - 30}{.5 - c}$

d) $n = 100$

e) $[-3, 2] \cup (4, \infty)$

f) $\sqrt{-1.55}$

g) $f(0) = \sqrt{1.5}$

h) $f(3) = \sqrt{-6}$

i) $f(5) = \sqrt{24}$

j) $[-3, 2] \cup (4, \infty)$

5.8

a) $k = 4.8; y = 33.6$

b) $k = .75; y = 27$

c) $k = 1.875; y = 120$

d) $k = 36; y = 9$

e) $k = 2.25; x = 3.79$

f) $k = 8; x = .5926$

6.1a

a)

Month	Money
0	100
1	200
2	400
3	800
4	1600
5	3200
6	6400

b) No Growth

c) Exponential Growth

d) Exponential Decay

e) 29.86

f) 331.776

g) $f(x) = 150(1.06)^x$

h) $b = 1.0696$

i) $b = 1.201; f(x) = 2(1.201)^x$

j) $b = 1.155; f(x) = 3(1.155)^x$

k) $b = 1.1447; f(x) = 4(1.1447)^x$

l) $f(x) = 4.579(1.1447)^x$

6.1b

a) We must include the 1 in order to have exponential growth.

b) $161.05

c) $b = 1.025$

d) $b = 1.0083$

e) $x = 20$

f) $x = 60$

g) $162.89

h) $163.86

i) $164.43

j) $7401.22

k) $10105.13

l) $30420.67

m) $3232.14

n) 1209.93; Since this is bacteria, I would round down to 1209.

o) 111.57 milligrams

6.2a

a) 3

b) 1

c) 2

d) $(-\infty, \infty)$

e) $(0, \infty)$

f) $(-\infty, \infty)$

g) $(1, \infty)$

h) $y = 1$

i) $(-\infty, \infty)$

j) $(-2, \infty)$

l) Vertical shift

m) Domain: $(-\infty, \infty)$; Range: $(4, \infty)$

n) Domain: $(-\infty, \infty)$; Range: $(-12, \infty)$

6.2b

a)
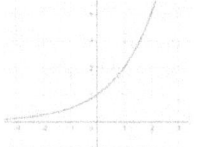

b)
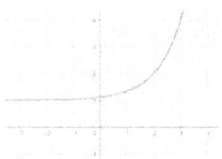

c)
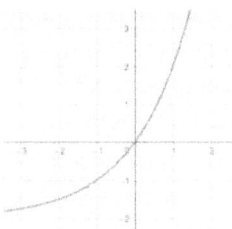

d) $(3, 1); (4, 0)$

e)
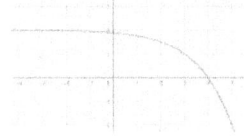

f)
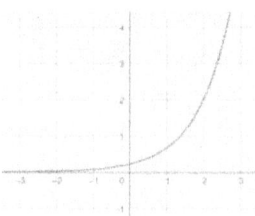

6.3

a) 3

b) 5

c) 4

d) 2

e) 3

f) 4

g) -3

h) -4

i) $5^3 = 125$

j) $10^4 = 10000$

k) $\log_3 81 = 4$

l) $\log_4 64 = 3$

m) 1.6532

n) -2.097

o) 2.708

p) 5.416

6.4a

a) Inverses

b)
$A: (0, 1); B: (1, 0); C: (2, 4); D: (4, 2); E: (8, 3); F: (3, 8)$

c) They reverse

d)

	Exponential		Logarithmic
X	Y	X	Y
-1	1/3	1/3	-1
0	1	1	0
1	3	3	1
2	9	9	2

e)

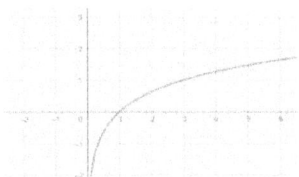

f)

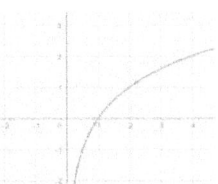

g)

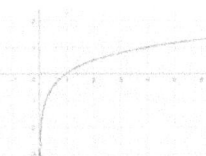

h) $(0, \infty)$

i) $(-\infty, \infty)$

j) $(9, \infty)$

k) $(4, \infty)$

l) $(-4, \infty)$

m) $(6, \infty)$

n) $x = 9$

o) $x = -4$

p) $x = 6$

q) $x = -4$

6.4b

a)

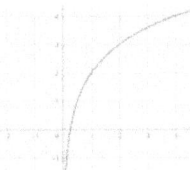

b)

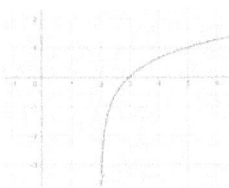

c)

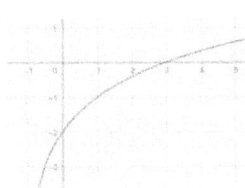

d)

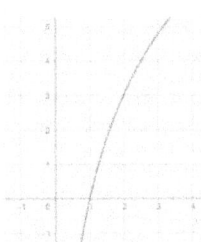

e)

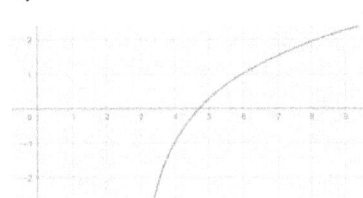

f) 3

g) 1

h) 2

6.5a

a) 5

b) 5

c) 4

d) 4

e) $\log_5 7 + \log_5 x$

f) $\log_3 2 + \log_3 x$

g) $\log_7 3 + \log_7 x + \log_7 y$

h) 4

i) 4

j) 3

k) 3

l) $\log_5 3 - \log_5 x$

m) $\log_4 r - \log_4 s$

n) 6

o) 6

p) 4

q) 4

r) $7 \log_3 y$

s) $3 \log_5 x$

t) $3 \log_3 2$

u) $-2 \log_2 y$

v) $\frac{1}{2} \log_3 x$

w) $\frac{2}{3} \log_8 x$

x) $\frac{1}{2} \log_3 x - 2 \log_3 y$

y) $\log_2 x + 5 \log_2 y - 3 \log_2 z$

z) $2 \log_5 x - \frac{1}{2} \log_5(y + 3)$

aa) $\frac{1}{3}\log_7(x+1) - \frac{1}{2}\log_7(y^2 + 5)$

6.5b

a) $\log_3(xy^2z^3)$

b) $\ln \frac{a^4 b}{c^9}$

c) $\log \frac{\sqrt{x-2}}{y^2}$

d) $\log_5 \sqrt[3]{xyz}$

e) $\log_7 \left(\frac{x^5}{y^3(z-1)^2}\right)$

f) 3

g) 3

h) 2

i) 2

j) $\frac{\log 51}{\log 3}$

k) $\frac{\log 121}{\log 7}$

l) $\frac{\log 37}{\log 2}$

6.6a

a) $x = 2$

b) $x = 8$

c) $x = 4$

d) $x = 1$

e) 2^{12}

f) 3^{10}

g) 5^{3x}

h) $x = -14$

i) $x = 3$

j) $x = 1$

k) $x = \frac{24}{5}$

l) $x = 16$

m) $x = \pm 2$

n) $x = \frac{\ln 6}{2}$

o) $x = \frac{\ln 5}{3}$

p) $x = \frac{\log 62600 + 1}{4}$

6.6b

a) $x = e^7$

b) $x = e^{10} - 4$

c) $x = \pm 3$

d) $x = 200$

e) $x = \ln 3$

f) No solution

7.1a

a) $(3, 1)$

b) $(1, -2)$

c) $(-5, -4)$

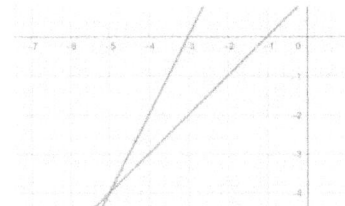

d)

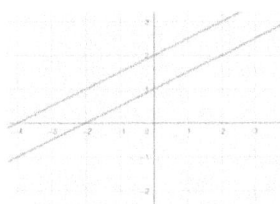

e) There is no solution because the lines don't intersect.

f) They are parallel.

g) $(-3, 1)$ is a solution to the system.

7.1b

a) $y = 15$

b) $x = 50$

c) $(-2, 1)$

d) $(-2, -4)$

e) $(4, -1)$

f) $20 = 15$; This is not true. There is no solution.

g) They are parallel lines.

h) $0 = 0$; This is true. Infinite solutions.

i) They are the same line.

7.1c

a) $7x$

b) $5x = 20$

c) $x = 4$

d) $y = -1$

e) $(3, -1)$

f) $(-5, -4)$

g) $(0, -3)$

h) $(1, 3)$

i) Infinite Solutions

j) $C(x) = .75x + 1250$

k) $R(x) = 2x$

l) $(1000, 2000)$

7.2

a) $y = -1; z = -3$

b) $x = 4$

c) $(-2, 6, 3)$

7.3a

a) $(2, 1)$

b) $(-1, 3); (2, 6)$

c) $x = 2; x = -1$

d) $y = 6; y = 3$

e) $(2, 6); (-1, 3)$

f) $(2, 3)$

g) $(-3, -4); (4, 3)$

h) $(\sqrt{13}, \sqrt{12}); (-\sqrt{13}, \sqrt{12}); (\sqrt{13}, -\sqrt{12}); (-\sqrt{13}, -\sqrt{?}$

i) $(2, 2\sqrt{2}); (2, -2\sqrt{2}); (-2, 2\sqrt{2}); (-2, -2\sqrt{2})$

7.3b

a)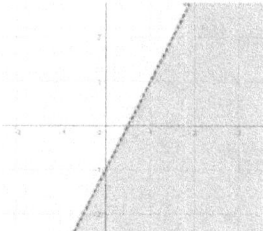

b)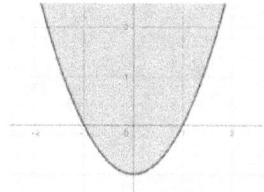

c)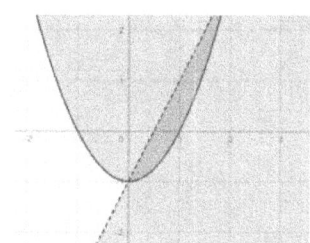

d) $(0, -1); (2, 3)$

e)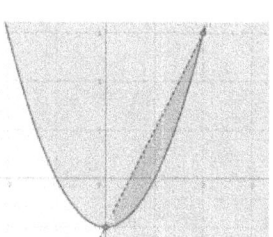

f)

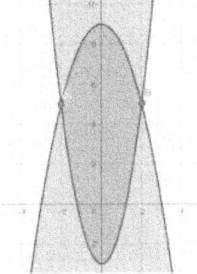

7.4a

a) $\frac{2x-24}{(x-2)(x+3)}$

b) $A = \frac{4}{5}; B = \frac{6}{5}$

c) $\frac{\frac{4}{5}}{(x-2)} + \frac{\frac{6}{5}}{(x+3)} = \frac{4}{5(x-2)} + \frac{6}{5(x+3)}$

d) $\frac{1}{(x-1)} + \frac{3}{(x+3)}$

e) $2 \cdot 2 \cdot 3$

f) $x = 11$

g) $x = 14$

h) $\frac{2}{(x-2)} + \frac{5}{(x-2)^2}$

i) $\frac{4}{x} - \frac{1}{(x-1)} + \frac{9}{(x-1)^2}$

j) $A = -\frac{1}{2}; B = \frac{5}{2}; C = 4; -\frac{1}{2(x)} + \frac{5}{2(x-2)} + \frac{4}{(x-2)^2}$

7.4b

a) Prime- Can't be factored.

b) Prime- Can't be factored.

c) $A = -2; B = 5; C = 6; \frac{2}{x} + \frac{5x+6}{x^2+x+1}$

d) $-\frac{1}{x} + \frac{2x+5}{x^2+1}$

e) $\frac{2}{(x^2+1)} + \frac{3x-1}{(x^2+1)^2}$

7.5a

a) 2×2

b) 3×2

c) 1×3

d) 3×3

e) 8

f) 5

g) 1

h) $\begin{bmatrix} -1 & 10 \\ 27 & -19 \end{bmatrix}$

i) $[19 \quad -54 \quad -31]$

j) $\begin{bmatrix} 13 & 0 & 26 \\ 20 & 10 & -32 \\ -7 & 4 & 8 \end{bmatrix}$

k) $\begin{bmatrix} -11 & -4 \\ 11 & 9 \end{bmatrix}$

l) $[3 \quad -58 \quad -131]$

m) $\begin{bmatrix} -9 & 2 & -10 \\ -10 & -14 & 0 \\ 9 & -2 & 16 \end{bmatrix}$

n) $\begin{bmatrix} -30 & 15 \\ 95 & -25 \end{bmatrix}$

o) $\begin{bmatrix} 8 & 42 \\ 14 & 2 \\ -4 & 18 \end{bmatrix}$

p) $[-33 \quad 168 \quad 243]$

q) $\begin{bmatrix} 8 & 4 & 32 \\ 20 & -8 & 64 \\ 4 & 4 & 48 \end{bmatrix}$

r) $\begin{bmatrix} 3 & 27 \\ 62 & -52 \end{bmatrix}$

s) $\begin{bmatrix} -14 & 6 & -4 \\ -10 & -32 & -32 \\ 20 & -2 & 56 \end{bmatrix}$

7.5b

a) $\begin{bmatrix} 14 & 0 \\ -8 & 17 \end{bmatrix}$

b) $\begin{bmatrix} 67 & 3 & 97 \\ 10 & 41 & -9 \\ 25 & -21 & 49 \end{bmatrix}$

c) Not enough rows in the second matrix to match elements.

d) Only the first two sets of matrices can be multiplied. The third cannot.

e) $\begin{bmatrix} 1 & 11 & 16 \\ 11 & 23 & 29 \\ -7 & -7 & -7 \end{bmatrix}$

f) $\begin{bmatrix} 10 & 21 \\ 1 & 7 \end{bmatrix}$

g) No

7.6a

a) The coefficients from the system make the elements of the matrix.

b) $\begin{bmatrix} -8 & -4 & 24 \\ 5 & -21 & 80 \end{bmatrix}$

c) $\begin{bmatrix} 6 & 18 & 36 \\ -15 & 1 & 42 \end{bmatrix}$

d) $-12x + 16y = 8;\ 14x + 9y = -4$

e) $\begin{bmatrix} 4 & 6 & -8 & 20 \\ 2 & -12 & -1 & 15 \\ -6 & 14 & -5 & 16 \end{bmatrix}$

f) Multiply through row 1 by 1/8. $\begin{bmatrix} 1 & -2 & -4 \\ 5 & 9 & 18 \end{bmatrix}$

g) Switch row 2 and row 1. $\begin{bmatrix} 1 & 16 & 6 \\ 0 & 4 & 7 \end{bmatrix}$

h) Switch row 2 and row 1. $\begin{bmatrix} 1 & -4 & 16 \\ 2 & -6 & 22 \end{bmatrix}$

i) Multiply row 1 by -4 and add to row 2. $\begin{bmatrix} 1 & 2 & 5 \\ 0 & -14 & -21 \end{bmatrix}$

j) Multiply row 1 by 7 and add to row 2. $\begin{bmatrix} 1 & -3 & 2 \\ 0 & -1 & 5 \end{bmatrix}$

k) Multiplied row 1 by -2 and added to row 2.

l) $x = -2;\ y = 1$

m) $x = 2;\ y = -3$

n) $x = 4;\ y = -1$

o) $x = 1;\ y = -2$

p) $\begin{bmatrix} 1 & -5 & 15 \\ 0 & 0 & 35 \end{bmatrix}$

q) $\begin{bmatrix} 1 & 2 & 4 \\ 0 & 0 & 0 \end{bmatrix}$

7.6b

a) $\begin{bmatrix} 1 & 0 & 0 & 2 \\ 0 & 1 & 0 & 0 \\ 0 & 0 & 1 & -1 \end{bmatrix}$

b) $x = 2;\ y = 0;\ z = -1$

c) $\begin{bmatrix} 1 & 0 & 0 & 3 \\ 0 & 1 & 0 & 4 \\ 0 & 0 & 1 & -7 \end{bmatrix}$

d) $x = 3;\ y = 4;\ z = -7$

e) $\begin{bmatrix} 1 & 0 & 0 & 1.645 \\ 0 & 1 & 0 & -2.658 \\ 0 & 0 & 1 & .101 \end{bmatrix}; x = 1.645, y = -2.658; z = .101$

f) $\begin{bmatrix} 1 & 0 & 3 & 7 \\ 0 & 1 & -2 & -2 \\ 0 & 0 & 0 & 0 \end{bmatrix}$ Infinite solutions.

g) $\begin{bmatrix} 1 & 0 & -4.4 & 0 \\ 0 & 1 & .14 & 0 \\ 0 & 0 & 0 & 1 \end{bmatrix}$ No Solution

7.7a

a) $\begin{bmatrix} 3 & -2 \\ 1 & 4 \end{bmatrix}$

b) $\begin{bmatrix} 3 & -2 \\ 1 & 4 \end{bmatrix}$

c) We get the same matrix.

d) We get the identity matrix.

e) We get the identity matrix.

f) $A \cdot B = B \cdot A = \begin{bmatrix} 1 & 0 \\ 0 & 1 \end{bmatrix}$

g) We get matrix B from above.

h) We are getting the inverse matrix.

i) $\begin{bmatrix} 1 & 2 & 1 & 0 \\ 0 & 0 & -4 & 0 \end{bmatrix}$

j) $A^{-1} = \begin{bmatrix} -8 & 3 \\ 3 & -1 \end{bmatrix}$

k) $A^{-1} = \begin{bmatrix} -4 & 49 & -22 \\ 0 & 2 & -1 \\ 1 & -11 & 5 \end{bmatrix}$

7.7b

a)
$\begin{bmatrix} 4 & -3 \\ -3 & 2 \end{bmatrix} \cdot \begin{bmatrix} x \\ y \end{bmatrix} = \begin{bmatrix} 2 \\ 1 \end{bmatrix}$

b)
$\begin{bmatrix} -1 & 2 \\ 2 & -3 \end{bmatrix} \cdot \begin{bmatrix} x \\ y \end{bmatrix} = \begin{bmatrix} 4 \\ 6 \end{bmatrix}$

c) $\begin{bmatrix} \frac{1}{9} & -\frac{1}{3} \\ \frac{2}{9} & \frac{1}{3} \end{bmatrix}$

d) You get the identity matrix.

e) $\begin{bmatrix} -\frac{1}{9} \\ \frac{14}{9} \end{bmatrix}$

f) $x = -\frac{1}{9}; y = \frac{14}{9}$

g) $\begin{bmatrix} -2 & -3 \\ -3 & -4 \end{bmatrix}$

h) $x = -7; y = -10$

i) $x = 24; y = 14$

j) $x = -123; y = -5; z = 28$

7.8a

a) $det = -1$

b) $det = -1$

c) $det = -21$

d) $det = 0$

e) $det = 14$

f) $x = 0$

g) $y = 2$

h) $x = 0; y = 2$

i) $\begin{bmatrix} 2 & -3 \\ 1 & 2 \end{bmatrix}$

j) $\begin{bmatrix} 4 & 2 \\ -3 & 1 \end{bmatrix}$

k) $det = 7$

l) $det = 10$

m) $x = -7$

n) $y = -10$

o) $det = -1$

p) $\begin{bmatrix} 2 & 3 & 5 \\ -1 & -2 & 4 \\ 3 & -5 & 8 \end{bmatrix}$

q) $\begin{bmatrix} 1 & 2 & 5 \\ 1 & -1 & 4 \\ 2 & 3 & 8 \end{bmatrix}$

r) $\begin{bmatrix} 1 & 3 & 2 \\ 1 & -2 & -1 \\ 2 & -5 & 3 \end{bmatrix}$

s) 123

t) 5

u) -28

v) $x = -123$

w) $y = -5$

x) $z = 28$

y) $16 - 16 = 0$

z) Infinite Solutions

7.8b

a) -16

b) 66

c) -66

d) 0

e) 0

f) 0

g) 0

h) -2

i) $-\frac{1}{2}$

j) 131

k) 393

l) They are the same.

m) The determinant will be zero.

n) Gaussian Elimination on the matrix or regular elimination.

8.1a

a)

b) x-axis

c)

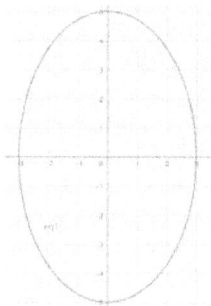

d) y-axis

e) The major axis corresponds to the larger denominator.

f) $(0, 5), (0, -5)$

g) $(3, 0), (-3, 0)$

h) 5

i) 3

j) It is the square root of the denominators.

k)

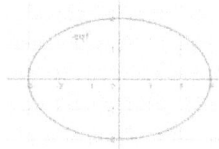

l)
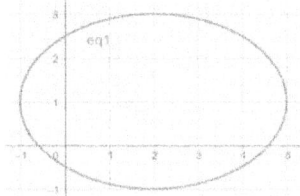

m)
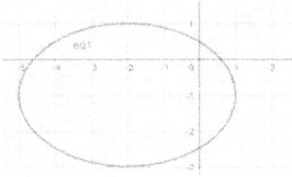

n)
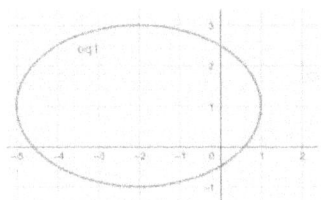

o) (h, k)

p)
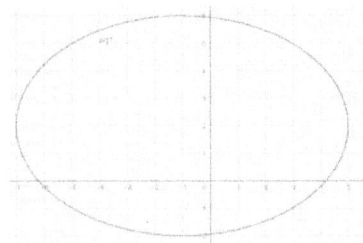

q)
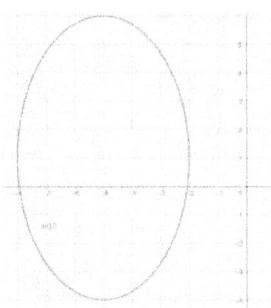

r) $\frac{x^2}{49} + \frac{y^2}{36} = 1$

s) $\frac{(x+1)^2}{16} + \frac{(y-4)^2}{12} = 1$

8.1b

a)
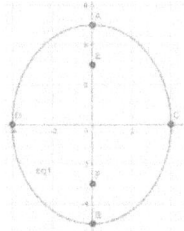

b) 3

c)
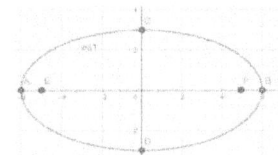

d) $\frac{x^2}{36} + \frac{y^2}{4} = 1$

e)
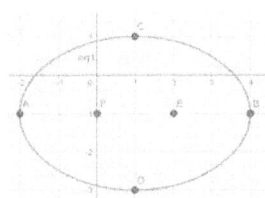

f) $\sqrt{5}$

g) $(4, -1), (-2, -1)$

h) $(1, 1), (1, -3)$

i) $(1 + \sqrt{5}, -1), (1 - \sqrt{5}, -1)$

j)

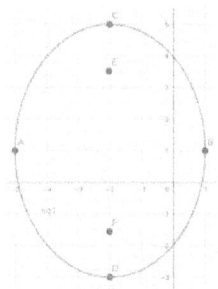

8.1c

a) 4

b) 1

c) $\frac{(x-2)^2}{4} + \frac{(y-1)^2}{3} = 1$

d) $(2, 1)$

e) $(0, 1), (4, 1)$

f) $(2, 1+\sqrt{3}), (2, 1-\sqrt{3})$

g) $(3, 1), (1, 1)$

h) $\frac{(x+3)^2}{3} + \frac{(y-2)^2}{2} = 1$

Center: $(-3, 2)$; Vertices: $(-3+\sqrt{3}, 2), (-3-\sqrt{3}, 2)$

Co-Vertices: $(-3, 2+\sqrt{2}), (-3, 2-\sqrt{2})$

Foci: $(-2, 2), (-4, 2)$

8.2a

a)

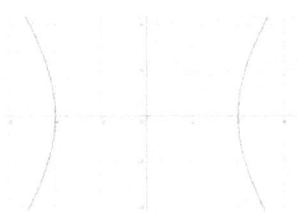

b)

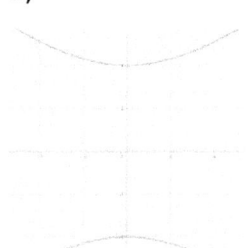

c) Horizontally

d) Vertically

e) $(4, 0); (-4, 0)$

f) 4

g) $(0, 4); (0, -4)$

h) 4

i) The square root of the positive denominator.

j)

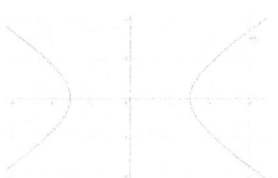

k)

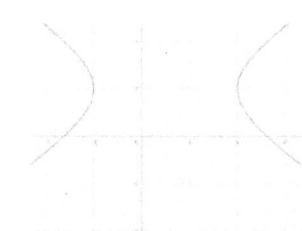

l)

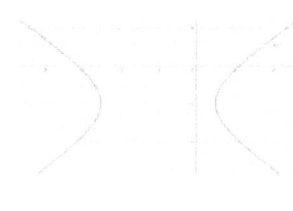

m)

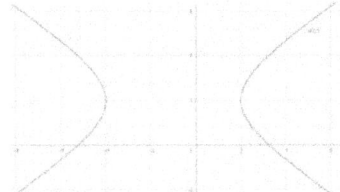

n) (h, k)

o)

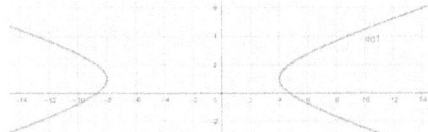

p)

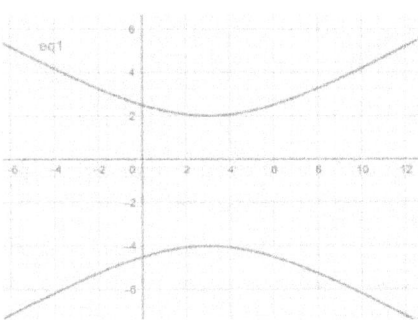

q) $b = 7$

r) $\frac{x^2}{25} - \frac{y^2}{49} = 1$

s) $\frac{y^2}{9} - \frac{x^2}{16} = 1$

t) $b = 3$

u) $\frac{(y-1)^2}{4} - \frac{(x-2)^2}{9} = 1$

v) $\frac{(x-3)^2}{25} - \frac{(y+4)^2}{36} = 1$

8.2b

a)

b) $c = \sqrt{41}$

c)

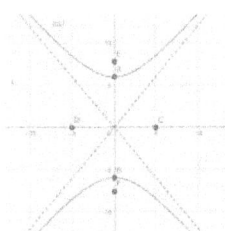

d) $\frac{x^2}{36} - \frac{y^2}{4} = 1$

e)

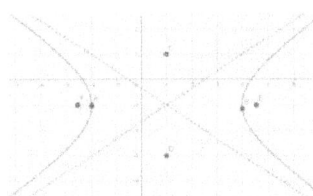

f) $c = \sqrt{13}$

g) $(4, -1), (-2, -1)$

h) $(1, 1), (1, -3)$

i) $(1 + \sqrt{13}, -1), (1 - \sqrt{13}, -1)$

j)

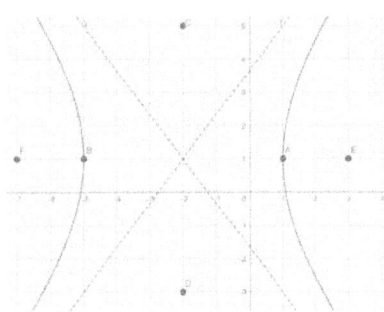

8.2c

a) 4

b) 1

c) $\frac{(x-2)^2}{4} - \frac{(y-1)^2}{3} = 1$

d) $(2, 1)$

e) $(4, 1), (0, 1)$

f) $(2, 1+\sqrt{3}), (2, 1-\sqrt{3})$

g) $(2+\sqrt{7}, 1), (2-\sqrt{7}, 1)$

h) $\frac{(x+3)^2}{9} - \frac{(y-2)^2}{6} = 1$, Center: $(-3, 2)$

Vertices: $(0, 2), (-6, 2)$ Co-Vert: $(-3, 2+\sqrt{6}), (-3, 2-\sqrt{6})$

Foci: $(-3+\sqrt{15}, 2), (-3-\sqrt{15}, 2)$

8.3a

a)

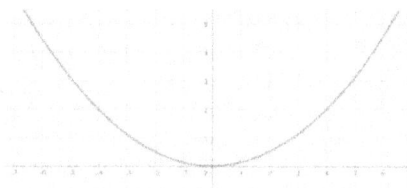

b) Upward

c)

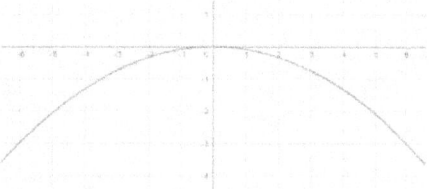

d) Downward

e)

f) Left

g)

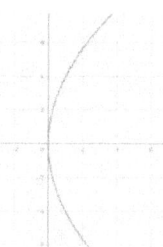

h) Right

i) Divide the value in front of the single variable by 4.

j) $(2, -2)$

k) Downward

l) $(5, 4)$

m) Right

n) $(-8, 7)$

o) Upward

p) $(50, 20)$

q) Left

r) $p = 6$; Focus: $(6, 0)$; Directrix: $x = -6$; Latus Rectum: $(6, 12), (6, -12)$

s) $y^2 = -8x$

8.3b

a)

b)

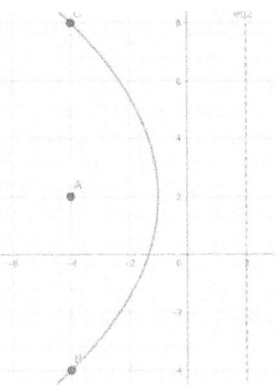

c)

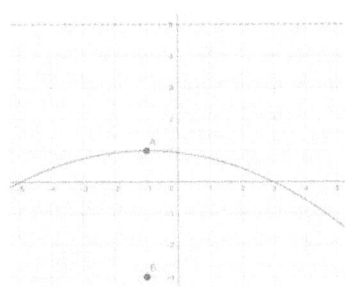

8.3c

a) $x^2 - 4x + 4 = 12y - 64 + 4$

b) $(x-2)^2 = 12(y-5)$

c) $(2, 5)$

d) $(2, 8)$

e) $y = 2$

f) $x = 2$

g) $(y+3)^2 = 8(y-2)$; $p = 2$; Vertex: $(2, -3)$; Focus: $(4, -3)$

Directrix: $x = 0$; Axis of Symmetry: $y = -3$

9.1a

a) 9

b) 64

c) $\frac{1}{6}$

d) -25

e) -4

f) -6

g) $\{2, 0, -2, -4, -6\}$

h) $-\frac{1}{2}$

i) $\frac{4}{7}$

j) $-\frac{5}{8}$

k) $\{-\frac{1}{4}, \frac{2}{5}, -\frac{1}{2}, \frac{4}{7}, -\frac{5}{8}\}$

l) 0

m) 16

n) 2

o) $\{-2, 4, 0, 16, 2\}$

p) $a_n = \frac{(-1)^n(n+1)}{(2n+5)}$

q) $a_n = e^{n+2}$

r) $a_n = \frac{3}{(3)^{n+1}}$

s) $a_n = e^{n-4}$

9.1b

a) $\{-\frac{4}{3}, 1, -\frac{6}{7}, \frac{7}{9}, -\frac{8}{11}\}$

b) $a_n = \frac{(-1)^n}{2^n}$

c) $a_4 = 61$

d) $a_5 = 125$

e) $\{3, 13, 63, 313, 1563\}$

f) -3

g) -2

h) $\{3, 5, 21, 73, 261\}$

i) 6

j) 120

k) $a_3 = \frac{1}{30}$

l) $a_4 = \frac{1}{144}$

m) $a_5 = \frac{3}{2520}$

n) $\{2, 3, 8, 30, 144\}$

9.2a

a) 22

b) 33

c) $\{12, 9, 6, 3, 0\}$

d) $d = 9;\ a_6 = 51$

e) $d = -4;\ a_5 = -4$

f) $d = -6;\ a_2 = 33$

g) $d = 11;\ a_1 = 0$

9.2b

a) 9

b) $a_n = a_{n-1} + 9, n \geq 2$

c) $a_n = a_{n-1} - 7$

d) 1

e) 2

f) 3

g) $a_n = a_1 + 4(n-1)$

h) $a_n = a_1 - 7(n-1)$

i) $d = 5;\ -6, -1$

j) 7

k) $n = 7$

l) $n = 10$

m) $a_n = 3 + 2(n-1)$

n) $a_8 = 17$

9.3a

a) 1

b) 81

c) $\frac{1}{2}$

d) 3

e) $\{1, \frac{1}{5}, \frac{1}{25}, \frac{1}{125}\}; r = \frac{1}{5}$

f) $\{3, 15, 75, 375, 1875\}$

g) $\{100, 50, 25, \frac{25}{2}, \frac{25}{4}\}$

h) $\{7, -14, 28, -56, 112\}$

9.3b

a) $r = 1.2$

b) $a_5 = 4.1472$

c) $a_n = a_{n-1}(1.2)$ for $x \geq 2$

d) $a_n = a_{n-1}(.5)$ for $x \geq 2$

e) $a_n = a_{n-1}(.2)$ for $x \geq 2$

f) 1

g) 2

h) 3

i) $r = 3$

j) $a_2 = 6$

k) $r = 4; a_2 = 12$

l) $a_n = 16\left(\frac{1}{2}\right)^{n-1}$

m) $a_n = 1\left(\frac{1}{5}\right)^{n-1}$

n) $a_n = 250(1.05)^{n-1}$

o) $a_6 = 319.07$

9.4a

a) 35

b) 60

c) $\sum_{2}^{6} k$

d) $\sum_{1}^{5} 3^k$

e) 24

f) 130

g) 70

h) 36

i) 36

j) 6

k) 6

l) 6

m) 10.5

n) 63

o) 133

p) -72

q) 210

r) 3

s) 11

t) 35

u) 15

v) 6560

w) 2184

x) 2550

y) $140927.33

9.4b

a) The rate is less than 1.

b) $r < 1$

c) -429.66

d) 429.02

e) -428.37

f) $r = \frac{1}{2}$ decreasing

g) $r = 2$ increasing

h) $r = -\frac{1}{3}$ decreasing

i) $r = \frac{1}{3}$ decreasing

j) 10.67

k) 62.5

l) 12

m) $\frac{1}{2}$

n) 3

o) $-\frac{1}{3}$

p) Not defined

q) $\frac{1}{3}$

r) 13532.04

s) 28034.78

9.5

a) 9

b) Head/1, Head/2, Head/3, Head/4, Head/5, Head/6, Tail/1, Tail/2, Tail/3, Tail/4, Tail/5, Tail/6

c) H/H/H, H/H/T, H/T/H, H/T/T, T/H/H, T/H/T, T/T/H, T/T/T

d) 8

e) 24

f) 32

g) 720

h) 6

i) 120

j) 60

k) 30240

l) 6720

m) 210

n) 792

o) 64

p) 360

q) 34650

9.6

a) $x^3 + 3x^2y + 3xy^2 + y^3$

b) 1

c) 3

d) 3

e) 1

f) The coefficients are the same as the combinations.

g) $x^4 + 4x^3y + 6x^2y^2 + 4xy^3 + y^4$

h) 1

i) 4

j) 6

k) 4

l) 1

m) $x^6 + 6x^5y + 15x^4y^2 + 20x^3y^3 + 15x^2y^4 + 6xy^5 + y^6$

n) $8x^3 + 36x^2y + 54xy^2 + 27y^3$

o) $27x^3 - 27x^2y + 9xy^2 - y^3$

p) $10x^3y^2$

q) $35x^4y^3$

r) $-8064x^5y^5$

9.7

a)

Event	P(Event)
1	$\frac{1}{6}$
2	$\frac{1}{6}$
3	$\frac{1}{6}$
4	$\frac{1}{6}$
5	$\frac{1}{6}$
6	$\frac{1}{6}$

b) $P(Head) = \frac{1}{2}$

c) $P(Queen) = \frac{1}{13}$

d) $P(Red) = \frac{1}{4}$

e) $P(Even) = \frac{1}{2}$

f) $P(3 \text{ or higher}) = \frac{2}{3}$

g) $P(Country) = \frac{6}{15}$

h) $P(Rock) = \frac{23}{30}$

i) $P(Country \text{ and } Rock) = \frac{4}{15}$

j) $P(Country \text{ or } Rock) = \frac{9}{10}$

k) $P(King) = \frac{1}{13}$

l) $P(Spade) = \frac{1}{4}$

m) $P(King \text{ or } Spade) = \frac{4}{13}$

n) $P(U.S) = \frac{5}{6}$

o) $P(Mexico) = \frac{1}{10}$

p) $P(U.S. \text{ or } Mexico) = \frac{28}{30}$

q) $P(King \text{ or } Queen) = \frac{2}{13}$

r) $P(Greater \text{ than } 1) = \frac{5}{6}$

s) $P(Not \text{ a } 1) = \frac{5}{6}$

t) $P(Rolling \text{ a } 1) = \frac{1}{6}$

u) $1 - P(Rolling \text{ a } 1) = \frac{5}{6}$

v) $P(not \text{ } rock) = \frac{7}{30}$

w) $P(not \text{ } country) = \frac{9}{15}$

x) $P(2 \text{ or greater}) = \frac{35}{36}$

y) $\frac{1}{55} = .01818$

z) $\frac{1}{22} = .04545$

aa) $\frac{7}{44} = .159$

bb) $\frac{37}{44} = .8409$

www.ingramcontent.com/pod-product-compliance
Lightning Source LLC
Chambersburg PA
CBHW081142290426
44108CB00018B/2415